Information Theory in Neuroscience

Information Theory in Neuroscience

Special Issue Editors

Stefano Panzeri
Eugenio Piasini

MDPI • Basel • Beijing • Wuhan • Barcelona • Belgrade

MDPI

Special Issue Editors

Stefano Panzeri
Istituto Italiano di Tecnologia
Italy

Eugenio Piasini
University of Pennsylvania
USA

Editorial Office
MDPI
St. Alban-Anlage 66
4052 Basel, Switzerland

This is a reprint of articles from the Special Issue published online in the open access journal *Entropy* (ISSN 1099-4300) from 2018 to 2019 (available at: https://www.mdpi.com/journal/entropy/special_issues/neuro)

For citation purposes, cite each article independently as indicated on the article page online and as indicated below:

LastName, A.A.; LastName, B.B.; LastName, C.C. Article Title. *Journal Name* **Year**, *Article Number*, Page Range.

ISBN 978-3-03897-664-6 (Pbk)
ISBN 978-3-03897-665-3 (PDF)

Cover image courtesy of Tommaso Fellin.

Contents

About the Special Issue Editors

Stefano Panzeri is a computational neuroscientist who works at the interface between theory and experiments, and investigates how circuits of neurons in the brain encode sensory information and generate behaviour. He graduated in Theoretical Physics at Turin University and then did a PhD in Computational Neuroscience at SISSA, Trieste, Italy. In previous years, he took Fellowships and/or Faculty jobs at the Universities of Oxford, Newcastle, Manchester and Glasgow in the UK, and Harvard Medical School in the US. He currently works as Senior Scientist with Tenure at the Istituto Italiano di Tecnologia in Rovereto, Italy.

Eugenio Piasini is interested in understanding how the brain performs inference and prediction in noisy, changing environments. In his work he aims to reconcile bottom-up statistical modelling of experimental data with a top-down perspective from normative theories of brain function. He holds a PhD from University College London, and he did postdoctoral work at the Italian Institute of Technology. He is now a Fellow (independent postdoc) of the Computational Neuroscience Initiative at the University of Pennsylvania.

Editorial

Information Theory in Neuroscience

Eugenio Piasini [1,*] and **Stefano Panzeri** [2,*]

1   Computational Neuroscience Initiative and Department of Physics and Astronomy, University of Pennsylvania, Philadelphia, PA 19104, USA
2   Neural Computation Laboratory, Center for Neuroscience and Cognitive Systems @UniTn, Istituto Italiano di Tecnologia, 38068 Rovereto (TN), Italy
*   Correspondence: epiasini@sas.upenn.edu (E.P.); stefano.panzeri@iit.it (S.P.)

Received: 26 December 2018; Accepted: 9 January 2019; Published: 14 January 2019

Abstract: This is the Editorial article summarizing the scope and contents of the Special Issue, Information Theory in Neuroscience.

Keywords: information theory; neuroscience

As the ultimate information processing device, the brain naturally lends itself to be studied with information theory. Because of this, information theory [1] has been applied to the study of the brain systematically for many decades and has been instrumental in many advances. It has spurred the development of principled theories of brain function [2–8]. It has led to advances in the study of consciousness [9]. It has also led to the development of many influential neural recording analysis techniques to crack the neural code, that is to unveil the language used by neurons to encode and process information [10–15].

The influence of information theory on the study of neural information processing continues today in many ways. In particular, concepts from information theory are beginning to be applied to the large-scale recordings of neural activity that can be obtained with techniques such as two-photon calcium imaging to understand the nature of the neural population code [16]. Advances in experimental techniques enabling precise recording and manipulation of neural activity on a large scale now enable for the first time the precise formulation and the quantitative test of hypotheses about how the brain encodes and transmits across areas the information used for specific functions, and information theory is a formalism that plays a useful role in the analysis and design of such experiments [17].

This Special Issue presents twelve original contributions on novel approaches in neuroscience using information theory, and on the development of new information theoretic results inspired by problems in neuroscience. The original contributions presented in this Special Issue span a wide range of topics.

Two papers use the concept of maximum entropy [18] to develop maximum entropy models to measure the existence of functional interactions between neurons and understand their potential role in neural information processing [19,20]. Kitazono et al. [21] and Bonmati et al. [22] develop concepts relating information theory to measures of complexity and integrated information. These techniques have potential for a wide range of applications, not least of which is the study of how consciousness emerges from the dynamics of the brain. Other work uses information theory as a tool to investigate different aspects of brain dynamics, from *latching* in neural networks [23], to the long-term development dynamics of the human brain studied using functional imaging data [24], to rapid information processing possibly mediated by the synfire chains [25] that have been reported in studies of simultaneously-recorded spike trains [26]. Other studies attempt to bridge between information theory and the theory of inference [27] and of categorical perception mediated by representation similarity in neural activity [28]. One paper [29] uses the recently-developed framework of partial information decomposition [30] to investigate the origins of synergy and redundancy in information representations, a topic of strong interest for the understanding of how neurons in the brain work together to represent information [31]. Finally, the two contributions of Samengo and colleagues examine applications of information theory to two specific problems of empirical importance in neuroscience: how to define how relevant specific response features are in a neural code [32], and what the code used by neurons in the temporal lobe to encode information is [33].

Author Contributions: E.P. and S.P. wrote the paper.

Acknowledgments: We are grateful to the contributing authors, to the anonymous referees, and to the Editorial Staff of Entropy for their excellent and tireless work, which made this Special Issue possible.

Conflicts of Interest: The authors declare no conflict of interest.

References

1. Shannon, C.E. A Mathematical Theory of Communication. *Bell Syst. Tech. J.* **1948**, *27*, 379–423. [CrossRef]
2. Srinivasan, M.V.; Laughlin, S.B.; Dubs, A.; Horridge, G.A. Predictive coding: a fresh view of inhibition in the retina. *Proc. R. Soc. Lond. Ser. B Biol. Sci.* **1982**, *216*, 427–459. [CrossRef]
3. Atick, J.J.; Redlich, A.N. Towards a Theory of Early Visual Processing. *Neural Comput.* **1990**, *2*, 308–320. [CrossRef]
4. Dong, D.W.; Atick, J. Temporal decorrelation: A theory of lagged and nonlagged responses in the lateral geniculate nucleus Network. *Netw. Comput. Neural Syst* **1995**, *6*, 159–178. [CrossRef]
5. Laughlin, S.B.; de Ruyter van Steveninck, R.R.; Anderson, J.C. The metabolic cost of neural information. *Nat. Neurosci.* **1998**, *1*, 36–41. [CrossRef] [PubMed]
6. Hermundstad, A.M.; Briguglio, J.J.; Conte, M.M.; Victor, J.D.; Balasubramanian, V.; Tkačik, G. Variance predicts salience in central sensory processing. *eLife* **2014**, *3*, e03722. [CrossRef] [PubMed]
7. Billings, G.; Piasini, E.; Lőrincz, A.; Nusser, Z.; Silver, R.A. Network Structure within the Cerebellar Input Layer Enables Lossless Sparse Encoding. *Neuron* **2014**, *83*, 960–974. [CrossRef]
8. Chalk, M.; Marre, O.; Tkačik, G. Toward a unified theory of efficient, predictive, and sparse coding. *Proc. Natl. Acad. Sci. USA* **2018**, *115*, 186–191. [CrossRef]
9. Tononi, G.; Sporns, O.; Edelman, G.M. A measure for brain complexity: Relating functional segregation and integration in the nervous system. *Proc. Natl. Acad. Sci. USA* **1994**, *91*, 5033–5037. [CrossRef]
10. Strong, S.P.; Koberle, R.; de Ruyter van Steveninck, R.R.; Bialek, W. Entropy and information in neural spike trains. *Phys. Rev. Lett.* **1998**, *80*, 197–200. [CrossRef]
11. Borst, A.; Theunissen, F.E. Information theory and neural coding. *Nat. Neurosci.* **1999**, *2*, 947–957. [CrossRef] [PubMed]
12. Schneidman, E.; Berry, M.J.; Segev, R.; Bialek, W. Weak pairwise correlations imply strongly correlated network states in a neural population. *Nature* **2006**, *440*, 1007–1012. [CrossRef] [PubMed]
13. Quian Quiroga, R.; Panzeri, S. Extracting information from neural populations: information theory and decoding approaches. *Nat. Rev. Neurosci.* **2009**, *10*, 173–185. [CrossRef] [PubMed]
14. Victor, J.D. Approaches to information-theoretic analysis of neural activity. *Biol. Theory* **2006**, *1*, 302–316. [CrossRef] [PubMed]
15. Tkačik, G.; Marre, O.; Amodei, D.; Bialek, W.; Berry, M.J. Searching for Collective Behavior in a Large Network of Sensory Neurons. *PLoS Comput. Biol.* **2014**, *10*, e1003408. [CrossRef] [PubMed]
16. Runyan, C.A.; Piasini, E.; Panzeri, S.; Harvey, C.D. Distinct timescales of population coding across cortex. *Nature* **2017**, *548*, 92–96. [CrossRef] [PubMed]
17. Panzeri, S.; Harvey, C.D.; Piasini, E.; Latham, P.E.; Fellin, T. Cracking the Neural Code for Sensory Perception by Combining Statistics, Intervention, and Behavior. *Neuron* **2017**, *93*, 491–507. [CrossRef]
18. Jaynes, E.T. Information theory and statistical mechanics. *Phys. Rev.* **1957**, *106*, 620–630. [CrossRef]
19. Cofré, R.; Maldonado, C. Information Entropy Production of Maximum Entropy Markov Chains from Spike Trains. *Entropy* **2018**, *20*, 34. [CrossRef]
20. Cayco-Gajic, N.A.; Zylberberg, J.; Shea-Brown, E. A Moment-Based Maximum Entropy Model for Fitting Higher-Order Interactions in Neural Data. *Entropy* **2018**, *20*, 489. [CrossRef]
21. Kitazono, J.; Kanai, R.; Oizumi, M. Efficient Algorithms for Searching the Minimum Information Partition in Integrated Information Theory. *Entropy* **2018**, *20*, 173. [CrossRef]

22. Bonmati, E.; Bardera, A.; Feixas, M.; Boada, I. Novel Brain Complexity Measures Based on Information Theory. *Entropy* **2018**, *20*, 491. [CrossRef]
23. Kang, C.J.; Naim, M.; Boboeva, V.; Treves, A. Life on the Edge: Latching Dynamics in a Potts Neural Network. *Entropy* **2017**, *19*, 468. [CrossRef]
24. Fan, Y.; Zeng, L.L.; Shen, H.; Qin, J.; Li, F.; Hu, D. Lifespan Development of the Human Brain Revealed by Large-Scale Network Eigen-Entropy. *Entropy* **2017**, *19*, 471. [CrossRef]
25. Abeles, M.; Bergman, H.; Margalit, E.; Vaadia, E. Spatiotemporal firing patterns in the frontal cortex of behaving monkeys. *J. Neurophysiol.* **1993**, *70*, 1629–1638. [CrossRef] [PubMed]
26. Xiao, Z.; Wang, B.; Sornborger, A.T.; Tao, L. Mutual Information and Information Gating in Synfire Chains. *Entropy* **2018**, *20*, 102. [CrossRef]
27. Isomura, T. A Measure of Information Available for Inference. *Entropy* **2018**, *20*, 512. [CrossRef]
28. Brasselet, R.; Arleo, A. Category Structure and Categorical Perception Jointly Explained by Similarity-Based Information Theory. *Entropy* **2018**, *20*, 527. [CrossRef]
29. Chicharro, D.; Pica, G.; Panzeri, S. The Identity of Information: How Deterministic Dependencies Constrain Information Synergy and Redundancy. *Entropy* **2018**, *20*, 169. [CrossRef]
30. Williams, P.L.; Beer, R.D. Nonnegative Decomposition of Multivariate Information. *arXiv* **2010**, arXiv:1004.2515.
31. Griffith, V.; Koch, C. Quantifying Synergistic Mutual Information. In *Guided Self-Organization: Inception*; Prokopenko, M., Ed.; Springer: Berlin, Germany, 2014; pp. 159–190.
32. Eyherabide, H.G.; Samengo, I. Assessing the Relevance of Specific Response Features in the Neural Code. *Entropy* **2018**, *20*, 879. [CrossRef]
33. Maidana Capitán, M.B.; Kropff, E.; Samengo, I. Information-Theoretical Analysis of the Neural Code in the Rodent Temporal Lobe. *Entropy* **2018**, *20*, 571. [CrossRef]

Article

Information Entropy Production of Maximum Entropy Markov Chains from Spike Trains

Rodrigo Cofré [1,*] and Cesar Maldonado [2]

[1] Centro de Investigación y Modelamiento de Fenómenos Aleatorios, Facultad de Ingeniería, Universidad de Valparaíso, Valparaíso 2340000, Chile
[2] IPICYT/División de Matemáticas Aplicadas, Instituto Potosino de Investigación Científica y Tecnológica, San Luis Potosí 78216, Mexico; cesar.maldonado@ipicyt.edu.mx
* Correspondence: rodrigo.cofre@uv.cl

Received: 7 November 2017; Accepted: 5 January 2018; Published: 9 January 2018

Abstract: The spiking activity of neuronal networks follows laws that are not time-reversal symmetric; the notion of pre-synaptic and post-synaptic neurons, stimulus correlations and noise correlations have a clear time order. Therefore, a biologically realistic statistical model for the spiking activity should be able to capture some degree of time irreversibility. We use the thermodynamic formalism to build a framework in the context maximum entropy models to quantify the degree of time irreversibility, providing an explicit formula for the information entropy production of the inferred maximum entropy Markov chain. We provide examples to illustrate our results and discuss the importance of time irreversibility for modeling the spike train statistics.

Keywords: information entropy production; discrete Markov chains; spike train statistics; Gibbs measures; maximum entropy principle

1. Introduction

In recent years, multi-electrode arrays and neuroimaging recording techniques have allowed researchers to record simultaneously from large populations of neurons [1]. Analysis carried on the recorded data has shown that the neuronal activity is highly variable (even when presented repeatedly the same stimulus). The observed variability is due to the fact that noise is ubiquitous in the nervous system at all scales, from ion channels through synapses up to the system level [2–4]. The nature of noise in the nervous system thus determines how information is encoded [5–7]. In spite of the different sources of noise, the spiking response is highly structured in statistical terms [8–10], for that reason many researchers have hypothesized that the population neural code is largely driven by correlations [10–15].

There are numerous sources of spike correlations that involve time delays, such as the activity of an upstream neuron projecting to a set of the observed neurons [16], top-down delayed firing rate modulation [17], among others. As discussed in [18], spike interactions in different times could have a non-negligible role in the spike train statistics. Indeed, there is strong evidence that interneuron temporal correlations play a major role in spike train statistics [19–22].

Since spikes are stereotyped events, the information about spikes is conveyed only by its times of occurrence. Considering small windows of time for each neuron, either a spike occurs in a given interval or not, producing in this way binary sequences of data easier to analyze statistically. However, traditional methods of statistical inference are useless to capture the collective activity under this scenario since the number of possible spike patterns that a neural network can take grows exponentially with the size of the population. Even long experimental recordings usually contain a very small subset of the entire state space, which makes the empirical frequencies poor estimators for the underlying probability distribution.

Since the spiking data are binary, it is natural to attempt to establish a link between neural activity and models of spins over lattices from statistical mechanics. Since the seminal work of Jaynes [23] a succession of research efforts have helped to develop a framework to characterize the statistics of spike trains using tools the maximum entropy principle (MEP). This approach is promising since the MEP provides a unique statistical model for the whole spiking neuronal network that is consistent with the average values of certain features of the data but makes no additional assumptions. In Schneidman et al. [10] and Pillow et al. [24], the authors used the maximum entropy principle focusing on firing rates and instantaneous pairwise interactions (Ising model) to describe the spike train statistics of the vertebrate retina responding to natural stimuli. Since then, the MEP approach has become a standard tool to build probability measures in this field [10,21,24,25]. Recently, several extensions of the Ising model have been proposed, for example, the triplet model, considering as an extra constraint, the correlation of three neurons firing at the same time [15], and the so-called *K*-pairwise model which consider *K* neurons firing at the same time bin [25]. These studies have raised interesting questions about important aspects of the neuronal code such as criticality, redundancy and metastability [25,26].

Although relatively successful in this field, this attempt of linking neural populations and statistical mechanics is based on assumptions that go against fundamental biological knowledge. In particular, most of these works have focused only on synchronous constraints and thus, modeling time-independent processes which are reversible in time. From a fundamental perspective, since a population of neurons is a living system it is natural to expect them not to be characterized by i.i.d. random variables. As such, the statistical description of spike trains of living neuronal networks should reflect irreversibility in time [27], and hence require a description based on out-of-equilibrium statistical mechanics. Thus, quantifying the degree of time irreversibility of spike trains becomes an important challenge, which, as we show here, can be approached using tools from the fruitful intersection between information theory and statistical mechanics. Given a stochastic system, the quantity that measures how far it is from its equilibrium state (in statistical terms) is called *information entropy production* (IEP) [28] (We distinguish the information entropy production with others forms of entropy production used in chemistry and physics).

The maximum entropy approach can be extended to include non-synchronous constraints within the framework of the thermodynamic formalism and Gibbs measures in the sense of Bowen [29] (the notion of the Gibbs measure extends also to processes with infinite memory [30], and have been used in the context of spike train statistics [31,32]). This opens the possibility to capture the irreversible character of the underlying biological process, and, thus, build more realistic statistical models. In this paper, we quantify the IEP of maximum entropy measures of populations of spiking neurons under arbitrary constraints and show that non-equilibrium steady states (NESS) emerge naturally in spike train statistics obtained from the MEP.

There is a vast body of theoretical work about the irreversibility of stochastic processes, for mathematical details we refer the reader to [28]. In particular, for discrete time Markov chains, Gaspard [33] deduced an explicit expression for the change in entropy as the sum of a quantity called entropy flow plus the entropy production rate. In this paper, we follow this expression adapted to Markov chains obtained from the MEP and we provide an explicit expression for the IEP of maximum entropy Markov chains (MEMC).

This paper is organized as follows: In Section 2, we introduce the setup of discrete homogeneous Markov chains and review the properties that we use further. In Section 3, we introduce the MEP within the framework of the thermodynamic formalism and Gibbs measures, discussing the role of the arbitrary constraints. We also provide the explicit formula to compute the IEP solely based on the spectral properties of the transfer matrix. In Section 4, we provide examples of relevance in the context of spike train statistics. We finish this paper with discussions pointing out directions for further research.

2. Generalities

To set a common ground for the analysis of the IEP of spike trains, which are time-series of action potentials (nerve impulses) emitted by neurons, these spikes are used to communicate with other neurons. Here, we introduce the notations and provide the basic definitions used throughout the paper.

2.1. Notation

We consider a finite network of $N \geq 2$ neurons. Let us assume that there is a natural time discretization such that at every time step, each neuron emits at most one spike (There is a minimal amount of time called "refractory period" in which no two spikes can occur. When binning, one could go beyond the refractory period and two spikes may occur in the same time bin. In those cases, the convention is to consider only one spike). We denote the *spiking-state* of each neuron $\sigma_k^n = 1$ whenever the k-th neuron emits a spike at time n, and $\sigma_k^n = 0$ otherwise. The spike-state of the entire network at time n is denoted by $\sigma^n := [\sigma_k^n]_{k=1}^N$, which we call a *spiking pattern*. For $n_1 \leq n_2$, we denote by σ^{n_1, n_2} to an ordered concatenation of spike patterns

$$\sigma^{n_1, n_2} = \sigma^{n_1} \sigma^{n_1+1} \ldots \sigma^{n_2-1} \sigma^{n_2},$$

that we call *spike block*. We call the sample of T spiking patterns a *spike train*, which is a spike block $\sigma^{0,T}$. We consider also infinite sequences of spike patterns that we denote $\tilde{\sigma}$. We denote the set of infinite binary sequences of N neurons Σ_N.

Let $L > 0$ be an integer, we write $\Sigma_N^L = \{0, 1\}^{N \times L}$ for the set of spike blocks of N neurons and length L. This is the set of $N \times L$ blocks whose entries are 0's and 1's. We introduce a symbolic representation to describe the spike blocks. Consider a fixed N, then to each spike block $\sigma^{0,L-1}$ we associate a unique number $\ell \in \mathbb{N}$, called *block index*:

$$\ell = \sum_{k=1}^{N} \sum_{n=0}^{L-1} 2^{n\,N+k-1} \sigma_k^n. \tag{1}$$

We adopt the following convention: neurons are arranged from bottom to top and time runs from left to right in the spike train. For fixed N and L, $\sigma^{(\ell)}$ is the unique spike block corresponding to the index ℓ.

2.2. Discrete-Time Markov Chains and Spike Train Statistics

Let Σ_N^L be the state space of a discrete time Markov chain, and let us for the moment use the following notation $\sigma_{(n)} := \sigma^{n,n+L-1}$, for the random blocks and analogously $\omega_{(n)} := \omega^{n,n+L-1}$ for the states. Consider the process $\{\sigma_{(n)} : n \geq 0\}$. If $\sigma_{(n)} = \omega_{(n)}$, we say that the process is in the state $\omega_{(n)}$ at time n. The transition probabilities are given as follows,

$$\mathbb{P}[\sigma_{(n)} = \omega_{(n)} \mid \sigma_{(n-1)} = \omega_{(n-1)}, \ldots, \sigma_{(0)} = \omega_{(0)}] = \mathbb{P}[\sigma_{(n)} = \omega_{(n)} \mid \sigma_{(n-1)} = \omega_{(n-1)}]. \tag{2}$$

We assume that this Markov chain is homogeneous, that is, (2) is independent of n. Consider two spike blocks $\sigma^{0,L-1}, \tilde{\sigma}^{1,L} \in \Sigma_N^L$ of length $L \geq 2$. Then, the transition $\sigma_{(0)} \to \tilde{\sigma}_{(1)}$ is *allowed* if they have the common sub-block $\sigma^{1,L-1} = \tilde{\sigma}^{1,L-1}$.

We consider Markov transition matrices $P : \Sigma_N^L \times \Sigma_N^L \to \mathbb{R}$, whose entries are given by:

$$P_{\sigma_{(0)}, \tilde{\sigma}_{(1)}} := \begin{cases} \mathbb{P}[\tilde{\sigma}_{(1)} \mid \sigma_{(0)}] > 0 & \text{if } \sigma_{(0)} \to \tilde{\sigma}_{(1)} \text{ is allowed} \\ 0, & \text{otherwise.} \end{cases} \tag{3}$$

Note that P has $2^{NL} \times 2^{NL}$ entries, but it is a sparse matrix since each row has, at most, 2^N non-zero entries. Observe that by construction, for any pair of states there is a path of maximum length L in the graph of transition probabilities going from one state to the other, therefore the Markov chain is irreducible.

2.3. Detailed Balance Equations

Consider a fix N and L. From the Markov property and the definition of the homogeneous transition matrix, one has for an initial measure ν, the following Markov measure $\mu(\nu, P)$

$$\mu[\sigma_{(0)} = \omega_{(0)}, \sigma_{(1)} = \omega_{(1)}, \ldots, \sigma_{(k)} = \omega_{(k)}] = \nu(\omega_{(0)}) P_{\omega_{(0)}, \omega_{(1)}} \cdots P_{\omega_{(k-1)}, \omega_{(k)}}, \tag{4}$$

for all $k > 0$. Here, again, we used the short-hand notation $\sigma_{(k)} := \sigma^{k, L+k-1}$ and $\omega_{(k)} := \omega^{k, L+k-1}$.

An *invariant probability measure* of a Markov transition matrix P is a row vector π such that

$$\pi P = \pi. \tag{5}$$

We recall that, for ergodic Markov chains (irreducible, aperiodic and positive recurrent), the invariant measure is unique.

Let us now consider a more general setting including non-stationary Markov chains. Let ν^n be the distribution of blocks $\sigma^{(\ell)} \in \Sigma_N^L$ at time n, then one has that the probability evolves in time as follows,

$$\nu^{n+1}(\sigma^{(\ell)}) = \sum_{\sigma^{(\ell')} \in \Sigma_N^L} \nu^n(\sigma^{(\ell')}) P_{\ell', \ell}.$$

For every $\sigma^{(\ell)} \in \Sigma_N^L$, one may write the following relation

$$\nu^{n+1}(\sigma^{(\ell)}) - \nu^n(\sigma^{(\ell)}) = \sum_{\sigma^{(\ell')} \in \Sigma_N^L} \left[\nu^n(\sigma^{(\ell')}) P_{\ell', \ell} - \nu^n(\sigma^{(\ell)}) P_{\ell, \ell'} \right]. \tag{6}$$

This last equation is related to the conditions of reversibility of a Markov chain. When stationarity and ergodicity are assumed, the unique stationary measure of the Markov chain π is said to satisfy detailed balance if:

$$\pi_\ell P_{\ell, \ell'} = \pi_{\ell'} P_{\ell', \ell} \quad \forall \sigma^{(\ell)}, \sigma^{(\ell')} \in \Sigma_N^L. \tag{7}$$

If the detailed balance equations are satisfied, then the quantity inside the parenthesis in the right-hand side of Equation (6) is zero.

2.4. Information Entropy Rate and Information Entropy Production

A well established measure of the amount of uncertainty of a probability measure ν is the *information entropy rate*, which we denote by $\mathcal{S}(\nu)$. In the case of independent sequences of spike patterns ($L = 1$), the entropy rate is given by:

$$\mathcal{S}(\nu) = - \sum_{\sigma^{(\ell)} \in \Sigma_N^1} \nu\left[\sigma^{(\ell)}\right] \log \nu\left[\sigma^{(\ell)}\right]. \tag{8}$$

In the setting of ergodic stationary Markov chains taking values in the state space $\Sigma_N^L; L \geq 2$ with transition matrix P and unique invariant measure π, the information entropy rate associated to the Markov measure $\mu(\pi, P)$ is given by:

$$\mathcal{S}(\mu) = - \sum_{\sigma^{(\ell)}, \sigma^{(\ell')} \in \Sigma_N^L} \pi_\ell P_{\ell, \ell'} \log P_{\ell, \ell'}, \quad L \geq 2, \tag{9}$$

which corresponds to the *Kolmogorov–Sinai entropy* (KSE) [34].

Here, we introduce the information entropy production as in [33]. For expository reasons, let us consider again the non-stationary situation. The information entropy of a probability measure ν in the state space Σ_N^L at time n be given by

$$\mathcal{S}_n(\nu) = - \sum_{\sigma^{(\ell)} \in \Sigma_N^L} \nu^n(\sigma^{(\ell)}) \log \nu^n(\sigma^{(\ell)}).$$

The *change of entropy rate* over one time-step is defined as follows:

$$\Delta\mathcal{S}_n := \mathcal{S}_{n+1}(\nu) - \mathcal{S}_n(\nu) = - \sum_{\sigma^{(\ell)} \in \Sigma_N^L} \nu^{n+1}(\sigma^{(\ell)}) \log \nu^{n+1}(\sigma^{(\ell)}) + \sum_{\sigma^{(\ell)} \in \Sigma_N^L} \nu^n(\sigma^{(\ell)}) \log \nu^n(\sigma^{(\ell)}).$$

Rearranging the terms, one has that the previous equation can be written as:

$$\Delta\mathcal{S}_n = - \sum_{\sigma^{(\ell)},\sigma^{(\ell')} \in \Sigma_N^L} \nu^n(\sigma^{(\ell')}) P_{\ell',\ell} \log \frac{\nu^{n+1}(\sigma^{(\ell')}) P_{\ell',\ell}}{\nu^n(\sigma^{(\ell)}) P_{\ell,\ell'}} +$$
$$\frac{1}{2} \sum_{\sigma^{(\ell)},\sigma^{(\ell')} \in \Sigma_N^L} [\nu^n(\sigma^{(\ell')}) P_{\ell',\ell} - \nu^n(\sigma^{(\ell)}) P_{\ell,\ell'}] \log \frac{\nu^n(\sigma^{(\ell')}) P_{\ell',\ell}}{\nu^n(\sigma^{(\ell)}) P_{\ell,\ell'}}, \tag{10}$$

where the first part on the r.h.s of this equation is called *information entropy flow* and the second *information entropy production* [33].

Observe that, in the stationary state, one has that $\nu^n = \nu^{n+1} = \pi$, thus the change of entropy rate is zero, meaning that information entropy flow equal information entropy production, therefore is possible to attain a steady state of fixed maximum entropy, but having positive IEP. In this case, we refer to NESS [35].

Here, since we are interested in the Markov chains that arise from the maximum entropy principle, we focus on the stationary case. In this case, the IEP of a Markov measure $\mu(\pi, P)$ is explicitly given by:

$$IEP(P, \pi) = \frac{1}{2} \sum_{\sigma^{(\ell)},\sigma^{(\ell')} \in \Sigma_N^L} [\pi_{\ell'} P_{\ell',\ell} - \pi_\ell P_{\ell,\ell'}] \log \frac{\pi_{\ell'} P_{\ell',\ell}}{\pi_\ell P_{\ell,\ell'}} \geq 0, \tag{11}$$

nevertheless, we stress the fact that one can obtain the information entropy production rate also in the non-stationary case.

3. Maximum Entropy Markov Chains

Usually, one only have access to a limited amount of experimental spiking data, which is a sampling of a very small subset of the entire state space. This makes that often the empirical frequencies are bad estimations of the elements of the Markov transition matrix. Here, we present how to use a variational principle from the thermodynamic formalism [36] to obtain the unique irreversible ergodic Markov transition matrix and its invariant measure having maximum entropy among those consistent with the constraints provided by data. This approach solves the problem of the bad estimations mentioned above and enables us to compute the IEP of the inferred Markov process, which is our main goal.

3.1. Inference of the Maximum Entropy Markov Process

The problem of estimating the Markov chain of maximum entropy constrained by the data is of general interest in information theory. It consists in solving a constrained maximization problem, from which one builds a Markov chain. The first step is choosing (arbitrarily) a set of indicator functions (also called monomials) and determine from the data the empirical average of these functions. This fixes the constraints of the maximization problem. After that, one maximizes the information entropy rate, which is a concave functional in the space of Lagrange multipliers associated to the constraints, obtaining the unique Markov measure that better approximates the statistics among all probability measures that match exactly the constraints [23]. To to our knowledge, previous approaches ignore how to deal with the inference of irreversible Markov processes in the maximum entropy context [37,38].

3.2. Observables and Potentials

Let us consider the space of infinite binary sequences Σ_N. An *observable* is a function $f : \Sigma_N \to \mathbb{R}$. We say that an observable f has *range R* if it depends only on R consecutive spike patterns, e.g., $f(\sigma) = f(\sigma^{0,R-1})$. We consider here that observables do not depend explicitly on time (*time-translation invariant observables*), i.e., for any time-step n, $f(\sigma^{0,R-1}) = f(\sigma^{n,n+R-1})$ whenever $\sigma^{0,R-1} = \sigma^{n,n+R-1}$. Examples of observables are products of the form:

$$f(\sigma^{0,T}) = \prod_{u=1}^{r} \sigma_{k_u}^{n_u}, \tag{12}$$

where $k_u = 1,\ldots,N$ (neuron index) and $n_u = 0,\ldots,T$ (time index). These observables are called *monomials* and take values in $\{0,1\}$. Typical choices of monomials are $\sigma_{k_1}^{n_1}$ which is 1 if neuron k_1 fires at time n_1 and 0 otherwise; $\sigma_{k_1}^{n_1}\sigma_{k_2}^{n_2}$ which is 1 if neuron k_1 fires at time n_1 and neuron k_2 fires at time n_2 and 0 otherwise. For N neurons and time range R there are 2^{NR} possible monomials. To alleviate notations, instead of labeling monomials by a list of pairs, as in (12), we label them by an integer index, l (the index is defined in the same way as the block index (1)), i.e., a monomial reads m_l.

A *potential* is an observable that can be written as a linear combination of monomials (the range of the potential is the maximum over the ranges of the m_l monomials considered). A potential of range R is written as follows:

$$\mathcal{H}(\sigma^{(\ell)}) := \sum_{l=1}^{2^{NR}} h_l m_l(\sigma^{(\ell)}) \quad \sigma^{(\ell)} \in \Sigma_N^R, \tag{13}$$

where the coefficients h_l real numbers. Some coefficients in this series may be zero. We assume throughout this paper that $h_\ell < \infty$ (here, we do not consider hard core potentials with forbidden configurations). One example of potential is the one considering as monomials the firing rates σ_i and the synchronous pairwise correlations $\sigma_i\sigma_j$.

$$\mathcal{H}(\sigma^{(\ell)}) = \sum_{i=1}^{N} h_i \sigma_i + \frac{1}{2}\sum_{i,j=1}^{N} J_{ij}\sigma_i\sigma_j \quad \sigma^{(\ell)} \in \Sigma_N^1$$

Additive Observables of Spike Trains

Let ϕ be the shift map $\phi : \Sigma_N \to \Sigma_N$, defined by $\phi(\sigma)_{(i)} = \sigma_{(i+1)}$. Let f be an arbitrary observable. We may consider the sequence $\{f \circ \phi^i(\sigma)\}$ as a random variable whose statistical properties depend on those of the process producing the samples of σ and the regularity of the observable f.

Given a spike train, one would like to empirically quantify properties of empirical averages and their fluctuation properties as a function of the sampling size. Consider a spike train σ, and let n be the sample length. The average of the observable f of range $R \geq 1$ in σ is given by,

$$A_n(f) = \frac{1}{n-R+1}\sum_{i=0}^{n-R} f \circ \phi^i(\bar{\sigma}),$$

in particular, for observables of range 1, one has

$$A_n(f) = \frac{1}{n}\sum_{i=0}^{n-1} f(\sigma^i). \tag{14}$$

3.3. Variational Principle

Let $A_n(f_k) = C_k$ be the average value of K observables for $k \in \{1, \dots, K\}$. As the empirical average of monomials is not enough to uniquely determine the spike train statistics (there are infinitely many probability measures sharing the same averages of monomials), we use the maximum entropy method to obtain the Markov measure μ that maximizes the KSE among all measures ν that match the expected values of all observables, i.e., $\nu[f_k] = C_k$, for all $k \in \{1, \dots, K\}$. This is equivalent to solve the following variational problem under constraints:

$$S[\mu] = \max \left\{ S[\nu] : \nu[f_k] = C_k \quad \forall k \in \{1, \dots, K\} \right\}. \tag{15}$$

Since the function $\nu \to S[\nu]$ is strictly concave, there is a unique maximizing Markov measure $\mu(\pi, P)$ given the set of values C_k. To solve this problem, we introduce the set of Lagrange multipliers $h_k \in \mathbb{R}$ in the potential $\mathcal{H} = \sum_{k=1}^{K} h_k f_k$, which is a linear combination of the chosen observables. Next, we study the following unconstrained problem, which is a particular case of the so-called *variational principle* of the thermodynamic formalism [36]:

$$\mathcal{P}[\mathcal{H}] = \sup_{\nu \in \mathcal{M}_{inv}} \left\{ S[\nu] + \nu[\mathcal{H}] \right\} = S[\mu] + \mu[\mathcal{H}], \tag{16}$$

where $\mathcal{P}[\mathcal{H}]$ is called the *free energy or topological pressure*, $\mathcal{M}_{inv}$ is the set of invariant measures with respect to the shift ϕ and $\nu[\mathcal{H}] = \sum_{k=1}^{K} h_k \nu[f_k]$ is the average value of $\mathcal{H}$ with respect to ν.

In this paper, we only consider potentials $\mathcal{H}$ of finite range, for which there is a unique measure μ attaining the supremum [39] and is a *Gibbs measure in the sense of Bowen*.

Gibbs measures in the sense of Bowen: Suppose $\mathcal{H}$ is a finite range potential $R \geq 2$, a shift invariant probability measure μ is called a Gibbs measure (in the sense of Bowen) if there are constants $M > 1$ and $\mathcal{P}[\mathcal{H}] \in \mathbb{R}$ s.t.

$$M^{-1} \leq \frac{\mu[\sigma^{1,n}]}{\exp(\sum_{k=1}^{n-R+1} \mathcal{H}(\sigma^{k,k+R-1}) - (n+R-1)\mathcal{P}[\mathcal{H}])} \leq M \tag{17}$$

It is easy to see that the classical form of Boltzmann–Gibbs distributions $\mu[\sigma] = e^{\mathcal{H}(\sigma)}/Z$ is a particular case of (17), when $M = 1$, $\mathcal{H}$ is a potential of range $R = 1$ and $\mathcal{P}[\mathcal{H}] = \log Z$.

Statistical Inference

The functional $\mathcal{P}[\mathcal{H}]$ has the following property:

$$\frac{\partial \mathcal{P}[\mathcal{H}]}{\partial h_k} = \mu[f_k] = C_k, \quad \forall k \in \{1, \dots, K\} \tag{18}$$

where $\mu[f_k]$ is the average of f_k with respect to μ, which is equal to the average value of f_k with respect to the empirical measure from the data C_k, by constraint of the maximization problem. For finite range potentials, $\mathcal{P}(\mathcal{H})$ is a convex function of h_l's. This ensures the uniqueness of the solution of (16). Efficient algorithms exist to estimate the Lagrange multipliers for the maximum entropy problem with non-synchronous constraints [18].

3.4. Ruelle–Perron–Frobenius Transfer Operator

Consider $\mathcal{H}$ to be an arbitrary potential, and w a continuous function on Σ_N. We introduce the *Ruelle–Perron–Frobenius* (R–P–F) transfer operator denoted by $\mathcal{L}_{\mathcal{H}}$, and it is given by,

$$\mathcal{L}_{\mathcal{H}} w(\sigma) = \sum_{\sigma' \in \Sigma_N, \phi(\sigma') = \sigma} e^{\mathcal{H}(\sigma')} w(\sigma').$$

In an analogous way, as it is done for Markov approximations of Gibbs measures [40,41], for a finite range potential $\mathcal{H}$, we introduce the *transfer matrix* $\mathcal{L}_{\mathcal{H}}$,

$$\mathcal{L}_{\mathcal{H}}(\ell, \ell') = \begin{cases} e^{\mathcal{H}(\sigma^{0,L})} & \text{if } \sigma^{0,L} \sim \sigma^{(\ell)} \to \sigma^{(\ell')} \\ 0, & \text{otherwise.} \end{cases} \tag{19}$$

From the assumption $\mathcal{H} > -\infty$, each allowed transition corresponds to a positive entry in the matrix $\mathcal{L}_{\mathcal{H}}$.

3.5. Maximum Entropy Markov Chain for Finite Range Potentials

The matrix (19) is primitive (the matrix A is primitive if there is an $n \in \mathbb{N}$, s.t. A^n has only positive components) by construction, thus it satisfies the Perron–Frobenius theorem [42]. Let $\rho > 0$ be its spectral radius. Because of the irreducibility of the transfer matrix, ρ is an eigenvalue of multiplicity 1 strictly larger in modulus than the other eigenvalues. For every $\sigma^{(\ell)} \in \Sigma_N^L$, let us denote by $L_\ell := L(\sigma^{(\ell)})$ and $R_\ell := R(\sigma^{(\ell)})$, the left and right eigenvectors of $\mathcal{L}_{\mathcal{H}}$ corresponding to the eigenvalue ρ. Notice that $L_\ell > 0$ and $R_\ell > 0$ for all $\sigma^{(\ell)} \in \Sigma_N^L$. Using spectral properties of the transfer matrix, we get the maximum entropy Markov transition probability matrix [39]:

$$P_{\ell,\ell'} := \frac{\mathcal{L}_{\mathcal{H}}(\ell, \ell') R_{\ell'}}{R_\ell \rho}, \quad \forall \sigma^{(\ell)}, \sigma^{(\ell')} \in \Sigma_N^L. \tag{20}$$

The unique stationary probability measure π associated to P is also obtained by the spectral properties of $\mathcal{L}_{\mathcal{H}}$:

$$\pi_\ell := \frac{L_\ell R_\ell}{\langle L, R \rangle}, \quad \forall \sigma^{(\ell)} \in \Sigma_N^L. \tag{21}$$

For a finite range potential $\mathcal{H}$, the unique measure $\mu(\pi, P)$ associated to $\mathcal{H}$, satisfies the variational principle and, furthermore, the topological pressure can be explicitly computed $P[\mathcal{H}] = \ln \rho$.

3.6. IEP of the Inferred Markov Maximum Entropy Process

Consider a potential $\mathcal{H}$ of finite range and the state space Σ_N^L. As we have seen before, using the maximum entropy framework one can build from the transfer matrix $\mathcal{L}_{\mathcal{H}}$, the Markov transition matrix P and its invariant measure π. Furthermore, one can apply straightforwardly (20) and (21) to obtain a formula for the IEP based only on the spectral properties of $\mathcal{L}_{\mathcal{H}}$. After simplifying we get:

$$IEP(\mathcal{L}_{\mathcal{H}}) = \sum_{\sigma^{(\ell)}, \sigma^{(\ell')} \in \Sigma_N^L} \frac{L_\ell}{\langle L, R \rangle} \frac{\mathcal{L}_{\mathcal{H}}(\ell, \ell') R_{\ell'}}{\rho} \log \left[\frac{L_\ell R_{\ell'} \mathcal{L}_{\mathcal{H}}(\ell, \ell')}{L_{\ell'} R_\ell \mathcal{L}_{\mathcal{H}}(\ell', \ell)} \right] \tag{22}$$

This is a quantity of major interest in spike train statistics, as it measures the degree of time irreversibility of the inferred maximum entropy Markov chain. Although it is a straightforward result, it is quite general and of practical use, as we will see in the examples below.

We can apply (20) and (21) to Equation (7), we obtain the detailed balance condition in terms of the transfer matrix and its spectral properties:

$$\frac{L_\ell R_\ell}{\langle L, R \rangle} \frac{\mathcal{L}_{\mathcal{H}}(\ell, \ell') R_{\ell'}}{R_\ell s} = \frac{L_{\ell'} R_{\ell'}}{\langle L, R \rangle} \frac{\mathcal{L}_{\mathcal{H}}(\ell', \ell) R_\ell}{R_{\ell'} s}$$

Simplifying, we obtain:

$$\frac{\mathcal{L}_{\mathcal{H}}(\ell, \ell')}{\mathcal{L}_{\mathcal{H}}(\ell', \ell)} = \frac{R_\ell L_{\ell'}}{R_{\ell'} L_\ell} \tag{23}$$

3.7. Large Deviations for Observables of Maximum Entropy Markov Chains

The goal of large deviations is to compute the asymptotic probability distribution $\mathbb{P}(A_n(f) = s)$ for a given finite range observable f and for $s \neq \mathbb{E}(f)$. More precisely, we say that $\mathbb{P}(A_n(f))$ satisfies a large deviation principle with rate $I_f(s)$ if the following limit exists,

$$\lim_{n \to \infty} -\frac{1}{n} \ln \mathbb{P}(A_n(f) = s) = I_f(s).$$

where the dominant behavior of $\mathbb{P}(A_n(f))$ is decaying exponentially fast with the sample size n, as

$$\mathbb{P}(A_n(f) = s) \approx e^{-n I_f(s)}. \tag{24}$$

We define the *scaled cummulant generating function* (SCGF) associated to the random variable (observable) f denoted by $\lambda_f(k)$ as follows,

$$\lambda_f(k) := \lim_{n \to \infty} \frac{1}{n} \ln \mathbb{E}\left[e^{nk A_n(f)} \right], \quad k \in \mathbb{R}. \tag{25}$$

The n-th cumulant of the random variable f can be obtained by differentiating $\lambda_f(k)$ with respect to k, n times and evaluating the result at $k = 0$. The next theorem by Gärtner–Ellis theorem relates the SCGF and the large deviations rate function. The Gärtner–Ellis theorem relies on the differentiability of $\lambda_f(k)$, which is guaranteed for finite state Markov chains [43]. This theorem has several formulations, which usually require some technical definitions beforehand. Here, we stated it in a simplified form, which is what we need for our purposes.

Gärtner–Ellis theorem: If $\lambda_f(k)$ is differentiable, then there exist a large deviation principle for the average process $A_n(f)$ whose rate function $I_f(s)$ is the Legendre transform of $\lambda_f(k)$:

$$I_f(s) = \max_{k \in \mathbb{R}} \{ks - \lambda_f(k)\} \tag{26}$$

The Gärtner–Ellis Theorem is very useful in our context, because it bypasses the direct calculation of $\mathbb{P}(A_n(f))$ in (24), i.e., having $\lambda_f(k)$ a simple calculation leads to the rate function of f. As we will see in the next section $\lambda_f(k)$ naturally appears in the context of Maximum entropy Markov chains.

3.8. Large Deviations for the IEP

Consider an irreducible Markov chain with transition matrix $P_{\ell,\ell'}$. We define the *tilted transition matrix by f* denoted by $\tilde{P}^{(f)}(k)$, whose elements for a one time step observable are:

$$\tilde{P}^{(f)}_{\ell,\ell'}(k) = P_{\ell,\ell'} e^{kf(\ell')} \tag{27}$$

or for a two time step observable:

$$\tilde{P}^{(f)}_{\ell,\ell'}(k) = P_{\ell,\ell'} e^{kf(\ell,\ell')} \tag{28}$$

For a Markov transition matrix P inferred from the maximum entropy, the tilted transition matrix can be built directly from the transfer matrix and its spectral properties.

$$\tilde{P}^{(f)}_{\ell,\ell'}(k) = \frac{\mathcal{L}_{\mathcal{H}}(\ell,\ell') R_{\ell'}}{R_\ell \rho} e^{kf(\ell,\ell')} \tag{29}$$

The Markov chain structure underlying $A_n(f)$ can be used here to obtain more explicit expressions for $\lambda_f(k)$. In the case of the additive observables, if a Markov chain is homogeneous and ergodic can compute explicitly the SCGF as the logarithm of the maximum eigenvalue of $\tilde{P}^{(f)}$:

$$\lambda_f(k) = \ln(\rho(\tilde{P}^{(f)})) \tag{30}$$

This result is valid if the state-space of the Markov chain is finite, where it can be furthermore proven that $\lambda_f(k)$ is differentiable and $\lambda'_f(0) = \mathbb{E}(f)$.

Remark 1. *The observable f does not need to belong in the set $\{f_k\}_{k=1}^K$ of chosen observables to fit the Markov maximum entropy process. We denote $\rho(\tilde{P}^{(f)})$ the dominant eigenvalue (i.e., with largest magnitude) of the matrix $\tilde{P}^{(f)}$, which is unique by the Perron–Frobenius theorem.*

We are interested in the fluctuations of the IEP. For that purpose, we define the following observable:

$$W_n(\{\sigma^i\}_{i=1}^n) = \ln\left[\frac{\mathbb{P}(\{\sigma^i\}_{i=1}^n)}{\mathbb{P}(\{\sigma^i\}^{(R)})}\right]$$

where $\{\sigma^i\}^{(R)} = \sigma^n, \sigma^{n-1}, \ldots, \sigma^1$ is the temporal inversion of the trajectory $\{\sigma^i\}_{i=1}^n$. It can be shown that for $\mathbb{P}$-almost every trajectory of a stationary ergodic Markov chain (π, P):

$$\lim_{n\to\infty}\frac{W_n(\{\sigma^i\}_{i=1}^n)}{n} = IEP(\pi, P)$$

It can be shown [28] that the SCGF $\lambda_W(k)$ associated to the observable W_n can be found as the logarithm of the maximum eigenvalue $\rho(k)$ of the matrix:

$$\tilde{P}_{\ell,\ell'}^{(W)}(k) = P_{\ell,\ell'}e^{kF_{\ell,\ell'}}$$

where,

$$F_{\ell,\ell'} = \ln\left[\frac{\pi_\ell P_{\ell,\ell'}}{\pi_{\ell'} P_{\ell',\ell}}\right]$$

which is a matrix of positive elements.

Using the Gärtner–Ellis theorem, we obtain the rate function $I_W(s)$ for the IEP observable:

$$I_W(s) = \max_k\{ks - \lambda_W(k)\}$$

The rate function of the IEP observable has the following property:

$$\lambda_W(k) = \lambda_W(-k-1)$$

Since $\lambda'_W(0) = IEP(\pi, P)$ the symmetry implies

$$I_W(s) = I_W(-s) - s$$

Gallavotti–Cohen Fluctuation Theorem

The Gallavotti–Cohen fluctuation theorem refers to a symmetry in the fluctuations of the IEP. It is a statement about the large deviations of $\frac{W_n}{n}$, which is the time-averaged entropy production rate of the sample trajectory $\{\sigma^i\}_{i=1}^n$ of the Markov chain $\mu(\pi, P)$.

$$\frac{P\left[\frac{W_n}{n} \approx s\right]}{P\left[\frac{W_n}{n} \approx -s\right]} \asymp e^{ns}$$

This means that the positive fluctuations of $\frac{W_n}{n}$ are exponentially more probable than negative fluctuations of equal magnitude. This is a universal ratio, i.e., no free parameters are involved and is experimentally observable.

4. Examples

In this section, we provide several examples of applications of our results in the context of spike train statistics. We first provide an example of a discrete time Integrate-and-fire (IF) neuronal network model. This example does not use the MEP as the transition matrix can be explicitly obtained from the dynamics of the model. We then come back to the MEP approach to characterize the spike train statistics and compute the IEP for each example. We finally provide a summary of the results and discuss our findings.

4.1. Example: Discrete Time Spiking Neuronal Network Model

The IF model is one of the most ubiquitous models to simulate and analyze the dynamics of spiking neuronal circuits. This model is the simplest dynamical model that captures the basic properties of neurons, including the temporal integration of noisy sub-threshold inputs and all-or-nothing spiking. At the level of networks postulates a set of equations describing the behavior of the interconnected neurons motivated by the microscopic picture of how the biological neuronal network is supposed to work.

There exist several different versions of this model. Here, we present the discrete time IF model. The model definition follows the presentation given in [44]. Neurons are considered as points, without spatial extension nor biophysical structure (axon, soma, and dendrites). The dynamical system is only ruled by discrete time dynamical variables.

Denote by $V(t)$ the membrane potential vector with entries $V_i(t)$, whose dynamics is defined as follows. Fix a real variable $\theta > 0$ called *firing threshold*. For a fixed discrete time t, we have two possibilities:

- $V_i(t) < \theta$, for all $k = 1, ..., N$. This corresponds to sub-threshold dynamics.
- There exists a k such that, $V_k(t) \geq \theta$. Corresponding to firing dynamics.

The under-threshold dynamics is given by the following equation:

$$V(t+1) = F(V(t)) + \sigma_B B(t) \tag{31}$$

where

$$F_i(V(t)) = \gamma V_i(t)\big(1 - Z[V_i(t)]\big) + \alpha \sum_{j=1}^{N} W_{ij} Z[V_j(t)] + \beta I_i. \tag{32}$$

The function $Z[x] := \mathbb{1}_{x \geq \theta}$ is called the *firing state* of neuron x, where $\mathbb{1}$ is the indicator function. When $Z[V_i(t)] = 1$ one says that neuron i *spike*, otherwise is *silent*. We extend the definition of Z to vectors: $Z[V(t)]$ is the vector with components $Z[V_i(t)], i = 1, ..., N$. The *leak rate* is denoted by $\gamma \in [0, 1]$, and W_{ij} is called the *synaptic weight* from the neuron j to the neuron i. The synaptic weight is said to be *excitatory* if $W_{ij} > 0$ or *inhibitory* if $W_{ij} < 0$. The components of the vector $B(t)$ are independent normalized Gaussian random variables and σ_B is the noise amplitude parameter. The parameters α and β are introduced in order to control the intensity of the synaptic weights and the stimulus, respectively.

From this model, one can obtain a set of conditional probabilities of spike patterns given the network's spike history, allowing a mechanistic and causal interpretation of the origin of correlations (see [44] for details). Here, we consider only one time-step dependence on the past, although in the general approach it is possible to consider infinite memory. The conditional probabilities (transition matrix elements) are given as follows:

$$P[\sigma \mid \sigma'] = \prod_{i=1}^{N} \left[\sigma_i \, \varphi\Big(\frac{\theta - C_i(\alpha, \beta, \sigma')}{\sigma_B}\Big) + (1 - \sigma_i)\left(1 - \varphi\Big(\frac{\theta - C_i(\alpha, \beta, \sigma')}{\sigma_B}\Big)\right) \right], \tag{33}$$

where,

$$C_i(\alpha, \beta, \sigma') = \gamma \, \alpha \sum_{j=1}^{N} W_{ij} \sigma'_j + \beta I_i \tag{34}$$

and

$$\varphi(x) = \int_x^\infty e^{\frac{-u^2}{2}} \, du. \tag{35}$$

The function C takes into account the past and the external stimuli (see [44] for details). These transition probabilities define an ergodic Markov chain specified by the biophysical dynamics of the spiking network. From the transition probabilities (33) and the unique steady state, we compute the IEP of this model using (11) for different values of the parameters α and β (see Figure 1).

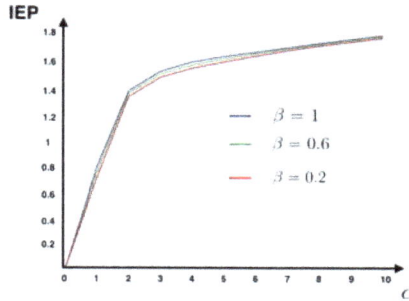

Figure 1. Plot of the average value of IEP for 500 realizations of the synaptic weight matrix for fixed α and β in each case. We fix the following values of the parameters: $N = 6$, $\gamma = 0.2$, $\sigma_b = 1$, $\theta = 1$, $I_i = 1 \; \forall i \in \{1, ..., 6\}$. The components of the synaptic weight matrix W_{ij} were drawn at random from a normalized Gaussian distribution. We plot the average value of IEP for 500 realizations of the synaptic weight matrix for fixed α and β in each case.

Figure 1 shows that for this model the IEP depends mostly on the intensity of the synaptic weights, while the stimulus intensity is playing a minor role. This suggests that IEP (in the stationary case) is essentially a property of the spiking neuronal network structure. The IEP of this neuronal network model is zero only under very restricted and unrealistic biophysical circumstances: when all synaptic weights are identical in amplitude and with the same sign or when they are all zero, i.e., when neurons do not communicate among them. In the first case, spikes play a symmetrical role with respect to time, which cancels out when computing the IEP. In the second case, the associated stochastic process is time-independent (thus time reversible). Therefore, generically this biophysically plausible model of spiking neuronal networks, has positive IEP. This means that the spike dynamics of this model leads to an irreversible Markov process.

4.2. MEMC Example: One Observable

In the previous example, we assume known the transition probabilities i.e., the structure of synaptic connectivity, stimulus and all other parameters defining the spiking neuronal network. Unfortunately, this is not always the case. Alternative approaches based on the MEP are considered when only spiking data are available. Consider a range-2 potential with $N = 2$ neurons:

$$\mathcal{H}(\sigma^{0,1}) = h_1 \sigma_1^1 \sigma_2^0.$$

The transfer matrix (19) associated to $\mathcal{H}$ is in this case a 4×4 matrix:

$$\mathcal{L}_\mathcal{H} = \begin{pmatrix} 1 & 1 & 1 & 1 \\ 1 & 1 & 1 & 1 \\ 1 & e^{h_1} & 1 & e^{h_1} \\ 1 & e^{h_1} & 1 & e^{h_1} \end{pmatrix}.$$

As this matrix is primitive by construction, it satisfies the hypothesis of the Perron–Frobenius theorem. The unique maximum eigenvalue is $\rho = e^{h_1} + 3$. The left and right eigenvectors associated to this largest eigenvalue are respectively:

$$L\begin{pmatrix} 0 \\ 0 \end{pmatrix} = \frac{2}{1 + e^{h_1}}; \ L\begin{pmatrix} 0 \\ 1 \end{pmatrix} = 1; \ L\begin{pmatrix} 1 \\ 0 \end{pmatrix} = \frac{2}{1 + e^{h_1}}; \ L\begin{pmatrix} 1 \\ 1 \end{pmatrix} = 1,$$

$$R\begin{pmatrix} 0 \\ 0 \end{pmatrix} = \frac{2}{1 + e^{h_1}}; \ R\begin{pmatrix} 0 \\ 1 \end{pmatrix} = \frac{2}{1 + e^{h_1}}; \ R\begin{pmatrix} 1 \\ 0 \end{pmatrix} = 1; \ R\begin{pmatrix} 1 \\ 1 \end{pmatrix} = 1.$$

From the spectral properties of $\mathcal{L}_{\mathcal{H}}$, we obtain the Markov transition matrix (20), which reads:

$$P_{\sigma^0,\sigma^1} = \frac{1}{\rho}\begin{pmatrix} 1 & 1 & \frac{1+e^{h_1}}{2} & \frac{1+e^{h_1}}{2} \\ 1 & 1 & \frac{1+e^{h_1}}{2} & \frac{1+e^{h_1}}{2} \\ \frac{2}{1+e^{h_1}} & \frac{2e^{h_1}}{1+e^{h_1}} & 1 & e^{h_1} \\ \frac{2}{1+e^{h_1}} & \frac{2e^{h_1}}{1+e^{h_1}} & 1 & e^{h_1} \end{pmatrix},$$

The unique invariant measure of this irreducible Markov transition matrix is given by Equation (21), and its entries are given by:

$$\pi\begin{pmatrix} 0 \\ 0 \end{pmatrix} = \frac{4}{\rho^2}, \quad \pi\begin{pmatrix} 0 \\ 1 \end{pmatrix} = \frac{2(\rho - 2)}{\rho^2}, \quad \pi\begin{pmatrix} 1 \\ 0 \end{pmatrix} = \frac{2(\rho - 2)}{\rho^2}, \quad \pi\begin{pmatrix} 1 \\ 1 \end{pmatrix} = \frac{(\rho - 2)^2}{\rho^2}.$$

It is easy to check that π is invariant w.r.t. the transition matrix P, that is $\pi P = \pi$.

From this example, we can verify that *generically the detailed balance condition is not satisfied*; for example:

$$P\begin{pmatrix} 0 \\ 1 \end{pmatrix}\begin{pmatrix} 1 \\ 0 \end{pmatrix}\pi\begin{pmatrix} 1 \\ 0 \end{pmatrix} \neq P\begin{pmatrix} 1 \\ 0 \end{pmatrix}\begin{pmatrix} 0 \\ 1 \end{pmatrix}\pi\begin{pmatrix} 0 \\ 1 \end{pmatrix}.$$

As we can see in Figure 2, the maximum entropy measure for the unconstrained problem is attained at the uniform distribution ($h_1 = 0$, eigenvalue $\rho = 4$ assigning probability $\frac{1}{4}$ to each spike pattern).

Let us now consider a constrained version of this problem. Suppose we have a dataset of length T and we measure the average value of the observable considered in this example $f = \sigma_1^1 \sigma_2^0$,

$$A_T(f) = c_1$$

Given this restriction and using the Equation (18), we obtain the following equation:

$$\frac{\partial \log(e^{h_1} + 3)}{\partial h_1} = c_1$$

Solving we find h_1. Among all the Markov chains that match exactly the restriction, the one that maximizes the information entropy is the one obtained by fixing h_1 at the found value. Is easy to check that the variational principle (16) is satisfied.

From the transition probability matrix P and the invariant measure π, we compute the KSE (Equation (9)) and the IEP (Equation (22)) as a function of the parameter h_1 (see Figure 2). Additionally, we fix the value $h_1 = -1$ at which we compute the IEP. We also compute the fluctuations around the mean and the large deviations. The Gallavotti–Cohen theorem applied to this example is illustrated in Figure 3.

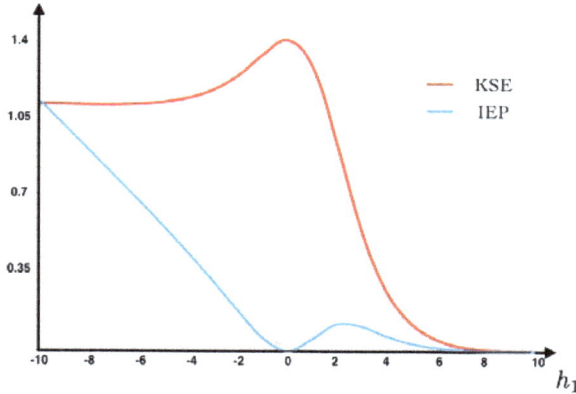

Figure 2. IEP and KSE as a function of h_1. In this example, the detailed balance condition is only satisfied in the trivial case $h_1 = 0$, corresponding to the uniform distribution. In all other cases, we obtain a MEMC with positive IEP, that is a NESS.

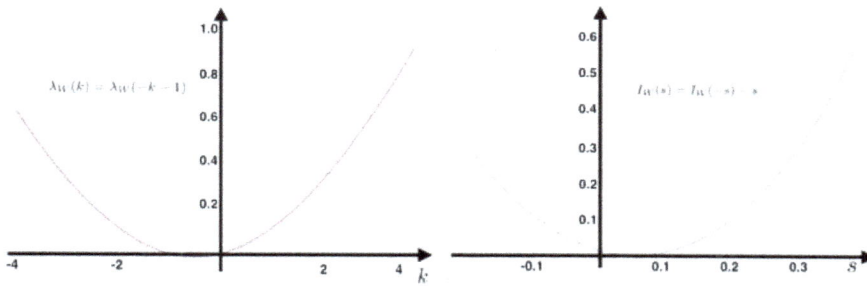

Figure 3. Gallavotti–Cohen fluctuation theorem for the MEMC example with one observable at the parameter value $h_1 = -1$. Left: We show the SCGF associated to W, $\lambda_W(k)$, the derivative at zero is the IEP of the MEMC, which in this case is 0.0557. This value coincides with the minimum of the rate function $I_W(s)$ at the right side of the figure.

4.3. MEMC Example: Two Observables

Consider now a similar neural system of two interacting neurons, but now take into account two observables representing how one neuron influences the other in the next time step.

$$f_1(\sigma^{0,1}) = \sigma_1^1 \sigma_2^0 \qquad \text{and} \qquad f_2(\sigma^{0,1}) = \sigma_2^1 \sigma_1^0$$

From these two features, one can build the corresponding energy function

$$\mathcal{H}(\sigma^{0,1}) = h_1 f_1(\sigma^{0,1}) + h_2 f_2(\sigma^{0,1}), \tag{36}$$

where h_1, h_2 are the parameters.

Given a dataset, let us denote the corresponding empirical averages of both features as $A_T(f_1) = c_1$ and $A_T(f_2) = c_2$. From the energy function (36), we build the transfer matrix and apply the same procedure presented in the previous example to obtain the unique maximum entropy Markov transition matrix and the invariant measure to compute the IEP as a function of c_1 and c_2, as illustrated in Figure 4.

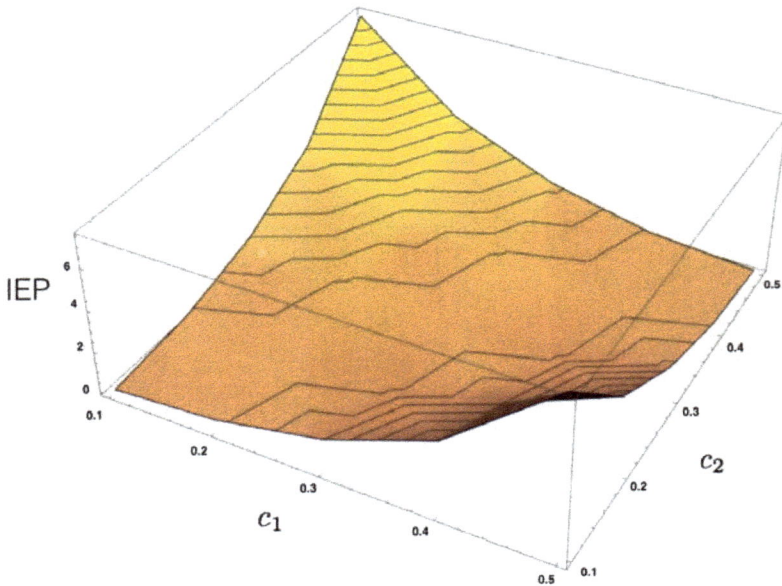

Figure 4. IEP for the MEMC build from each pair of constraints for the example Section 4.3. We use the restrictions on the average values denoted by c_1 and c_2 to build the corresponding MEMC in each case. We compute the IEP for each pair. In this figure, we illustrate that the IEP is zero only when both restrictions are equal and that the IEP increases with the difference in the restrictions.

4.4. Example: Memoryless Potentials

Consider a finite and fix number of neurons N and a potential of range 1. This case includes the Ising model [10], Triplets [15], K-pairwise [25] and all other memoryless potentials that has been used in the context of maximum entropy models of spike train statistics. It represent a limit case in the definition of the transfer matrix. In this case, the potential does not "see" the past, i.e., $\mathcal{L}_{\mathcal{H}}(\sigma, \sigma') = e^{\mathcal{H}(\sigma')}$. The matrix $\mathcal{L}_{\mathcal{H}}$ has a unique maximum eigenvalue:

$$\rho = Z = \sum_{\sigma' \in \Sigma_N^1} e^{\mathcal{H}(\sigma')}$$

and the rest of eigenvalues are equal to 0. The left and right eigenvectors corresponding to ρ are:

$$L(\sigma') = \frac{1}{Z}, \quad R(\sigma') = e^{\mathcal{H}(\sigma')}; \quad \forall \sigma' \in \Sigma_N^1.$$

Note that $\langle L, R \rangle = 1$. We have therefore:

$$P(\sigma' \mid \sigma) = P(\sigma') = \pi(\sigma') = \frac{e^{\mathcal{H}(\sigma')}}{Z}; \quad \forall \sigma, \sigma' \in \Sigma_N^1, \tag{37}$$

In this case, the invariant measure π has the classical Boltzmann–Gibbs form. The associated Markov chain has no memory: successive events are independent.

Taking the formula of IEP (22) we obtain:

$$IEP(\mathcal{L}_{\mathcal{H}}) = \sum_{\sigma,\sigma' \in \Sigma_N^1} \frac{L(\sigma)}{\langle L, R \rangle} \frac{e^{\mathcal{H}(\sigma')} R(\sigma')}{\log(Z)} \left(\mathcal{H}(\sigma') - \mathcal{H}(\sigma) \right) = 0.$$

In the case where only range 1 observables are chosen, the average value of these observables in a given data set is the same as the one taken from another data set where the time indexes have been randomly shuffled or even time reversed. As this is the only information about the process that the maximum entropy principle consider, it is not surprising that the stochastic process associated with the maximum entropy measure is time reversible. Consider a data set consisting in binary patterns $\mathcal{D}^O$. Let $g : \{0, \ldots, T\} \to \{0, \ldots, T\}$ be a function that randomly shuffles the time indexes, we call $\mathcal{D}^{RS}$ the data set obtained after this transformation. Finally consider $\mathcal{D}^I$, the data set with inverted time indexes,

$$\mathcal{D}^O = \{\sigma^0, \sigma^1, \sigma^2, \ldots, \sigma^{T-1}, \sigma^T\}$$
$$\mathcal{D}^{RS} = \{\sigma^{g(0)}, \sigma^{g(1)}, \sigma^{g(2)}, \ldots, \sigma^{g(T-1)}, \sigma^{g(T)}\}$$
$$\mathcal{D}^I = \{\sigma^T, \sigma^{T-1}, \sigma^{T-2}, \ldots, \sigma^1, \sigma^0\}.$$

Observe that, in these three cases (that may correspond to very different biological experiments), the average value of every observable of range one is exactly the same, therefore these data sets are characterized by the same maximum entropy distribution as illustrated in Figure 5.

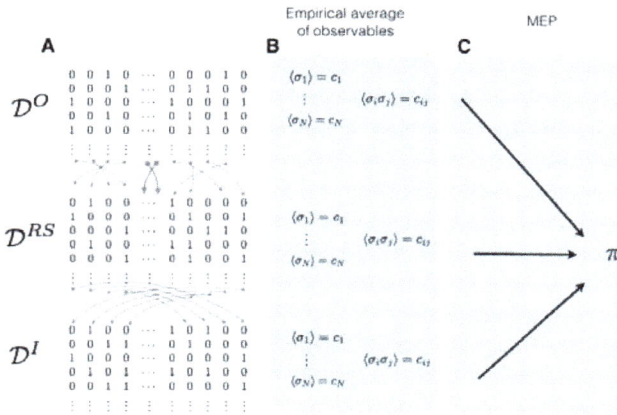

Figure 5. Memoryless potentials do not distinguish shuffled nor time-inverted data sets. Illustrative scheme showing three different data sets sharing the same maximum entropy distribution π. (**A**) We illustrate three data sets: on the top is the original; in the middle is the one obtained by randomly shuffle the time-indexes of the spike patterns; and on the bottom is the data set obtained by inverting the time indexes. (**B**) For each of these datasets, we compute the firing rate of each neuron denoted by $\langle \sigma_i \rangle$ and the pairwise correlations $\langle \sigma_i \sigma_j \rangle$ obtaining for each data set the same average values. (**C**) The spike train statistics of these three data sets are characterized by the same time-independent maximum entropy distribution π.

4.5. Example: 1-Time Step Markov with Random Coefficients

Here, we consider the 1-time step extension of the Ising model, that reads:

$$\mathcal{H}(\sigma^{0,1}) = \sum_{i=1}^{N} h_i\sigma_i + \frac{1}{2}\sum_{i,j=1}^{N} J_{ij}\sigma_i\,\sigma_j + \sum_{i,j=1}^{N} \gamma_{ij}\sigma_i\,\sigma_j^1. \tag{38}$$

This is the potential considered to fit a maximum entropy distribution to spiking data from a mammalian parietal cortex in-vivo in [20]. It is important to notice that in [20], the authors compute the solution of the maximum entropy problem imposing detailed balance condition, so in their case, there is zero IEP by construction. Here, we do not consider a particular data set, instead we investigate the capability of this potential to generate IEP by considering the following scenarios: We consider a network of $N = 10$ neurons, where we draw at random the coefficients h_i and J_{ij} in a range plausible to be the maximum entropy coefficients (or Lagrange multipliers) of an experiment of retinal ganglion cells exposed to natural stimuli (values of from h_i and J_{ij} as in [26]). We generate the matrix γ_{ij} by drawing each component at random from Gaussian distributions with different means and standard deviations. We summarize our results in Figure 6. We observe the following: Independent of h_i and J_{ij} and the parameters of mean and variance from which the matrix of coefficients γ_{ij} is generated, if γ_{ij} is symmetric, the Markov process generated by the potential (38) is reversible in time so the IEP is zero. This includes the limit case when $\gamma_{ij} = 0, \forall i,j \in \{1,\ldots,N\}$, where we recover the Ising model. Next, we fix the values of h_i and J_{ij} (random values), and we generate 100 matrices γ_{ij} by drawing their components from Gaussian distributions $\mathcal{N}(0, e^2)$, another 100 from $\mathcal{N}(1, e^2)$. We also generate 100 anti-symmetric matrices γ_{ij} from $\mathcal{N}(1, e^2)$, that we denote in Figure 6 $\mathcal{N}^A(1, e^2)$. For each realization of γ_{ij}, we generate the transfer matrix and proceed as explained in Section 3 to obtain the IEP in each case.

Figure 6 shows that for fitted data with a maximum entropy 1-time step Markov model, the IEP is zero only when all the measured 1-step correlations between neurons are symmetric, which is very unlikely for an experimental spike train. The degree of symmetry in the matrix of γ's play an important role in the IEP.

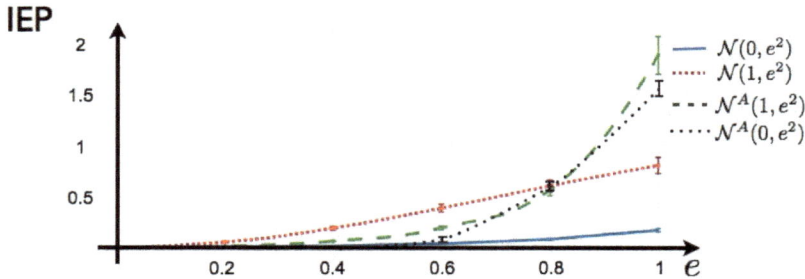

Figure 6. IEP for the 1-time step Markov potential. The parameters h_i and J_{ij} are draw at random one time and remain fixed. We draw at random the components of 100 matrices γ_{ij} from a Gaussian distribution with different values of mean and standard deviation e. We plot the average value of IEP for each case, with the respective error bars.

4.6. Example: Kinetic Ising Model with Random Asymmetric Interactions

This model of spike generation is an example of a non-equilibrium system, which has been used in [45] to approach the question of recovering the interactions of an asymmetrically-coupled Kinetic

Ising model, with a time-independent external field to ensure stationarity. This is a discrete-time, synchronously updated Markov model in Σ_N^1 with transition matrix is given by:

$$P[\sigma' \mid \sigma] = \prod_{i=1}^{N} \frac{\exp[(2\sigma_i' - 1)\theta_i(\sigma)]}{2\cosh[\theta_i(\sigma)]}, \quad \forall \sigma, \sigma' \in \Sigma_N^1 \tag{39}$$

$$\theta_i(\sigma) = \beta\, h_i + \alpha \sum_{j=1}^{N} J_{ij}(2\sigma_i - 1) \quad \forall \sigma \in \Sigma_N^1. \tag{40}$$

The fields h_i and the couplings J_{ij} are independent Gaussian variables and $\alpha, \beta \in \mathbb{R}$. These set of stationary transition probabilities characterize an ergodic Markov chain with a unique invariant measure. With these two quantities at hand, the scene is set to compute information entropy production under different scenarios.

In Figure 7, we recover the same structure found in Figure 1 for the IF model. This fact suggests that in this model the synaptic couplings are playing a major role in IEP, while the intensity of the stimulus is less relevant.

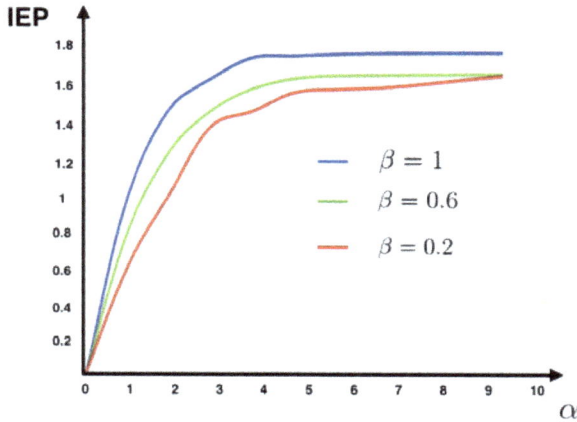

Figure 7. IEP for the Kinetic Ising model with random asymmetric interactions. We consider $N = 6$. The components of field vector were drawn at random from a Gaussian $\mathcal{N}(-3, 1)$ and the coupling matrix J_{ij} were drawn at random from a Gaussian $\mathcal{N}(0, 1)$. We plot the average value of IEP for 500 realizations of the synaptic coupling matrix for fixed α and β in each case.

4.7. Summary

We have shown several examples of applications of our results in the context of spike train statistics. We first provide an example of a discrete time Integrate-and-fire (IF) neuronal network model to illustrate that, in multiple scenarios of synaptic connectivity, even with constant stimulus, we find positive IEP (see Figure 2). This example does not use the MEP as the transition matrix can be explicitly obtained from the dynamics of the model. We use this example to illustrate that time irreversible statistical models arise naturally from biologically realistic spiking neuronal network models and to emphasize that IEP can be obtained from this approach as a benchmark for the MEP approach. We then consider the MEP approach to characterize the spike train statistics. In the second example, we detail the transfer matrix technique to compute the maximum entropy Markov transition matrix and the invariant measure, from these two quantities the IEP is easily computed using (11). We illustrate the Gallavotti–Cohen fluctuation theorem of the IEP for this example (see Figure 3). The third example is used to illustrate how unlikely is to find a reversible MEMC from data as a strong condition $c_1 = c_2$ need to be satisfied as shown in Figure 4. The fourth example is presented to show that our framework is general enough to consider the memoryless scenario as a limit case producing zero IEP. We illustrate in Figure 5 that memoryless maximum entropy models are not capable to distinguish among very different datasets. In the fifth example, we consider a MEMC, which has been considered in the literature of this field imposing detailed balance. We simulate different scenarios for the inter-neuron temporal parameters to illustrate the capability of this approach to capture the time-irreversible character of the underlying spiking network (see Figure 6). In the last example, we consider a popular model in the literature of this field to compute the IEP. Surprisingly, we recover the same structure of IEP (see Figure 7) as in the IF example (see Figure 1).

5. Discussion

The aim of population spike train statistical analysis is to deduce the principles of operation of neuronal populations. When trying to characterize the spike train statistics of networks of spiking neurons using the MEP, one hopes that the fitted parameters shed light on the understanding of some aspects of the population spiking phenomena in all its complexity. Therefore, to include and quantify time order in neuronal populations becomes a compulsory step towards a deeper understanding of the correlations observed in experimental data and consequently to better understand some aspects of the population neural code. The main message of this work is that limiting the complexity of the maximum entropy model using arguments of parsimony may not be appropriate to model a complex underlying stochastic process.

One of the consequences of including non-synchronous constraints in the MEP framework is that opens the possibility to broke the time-reversal symmetry imposed by time-independent models, and consequently to capture the irreversible character of the underlying biological process, allowing in this way to fit statistical models biologically more realistic. We have emphasized that the IEP is zero for time-independent processes (time-reversible) derived from commonly used statistical models in this field, for example, Ising, *K*-pairwise, triplets, among others [10,26]. However, *only time-dependent maximum entropy models induce time irreversible processes*, feature highly expected from biological systems.

While many spiking neuronal network models as the IF or the Generalized Linear Model (GLM) consider the influence of spike events occurred in the past, the most popular maximum entropy models in this field ignore them, causing a clear phenomenological disagreement between these two approaches, which can be corrected including non-synchronous constraints [22]. Leaving aside the fact that biophysical quantities used to fit realistic spiking neuronal network models may be difficult to obtain experimentally, the IEP obtained from both approaches to characterize the same neuronal tissue should be the same, thus the IEP may provide an alternative biologically based measure (going beyond goodness of fit type) of the adequacy of the chosen maximum entropy model.

Unfortunately, we do not yet know how to quantify the spiking activity in ways that yield the most meaningful insights into the relationship between the activity patterns and nervous system function and we are still looking for better conceptual-mathematical frameworks to better describe and understand spiking dynamics. We believe IEP may play an important conceptual role in future studies as help thinking about this dynamics.

However, there are two main drawbacks of our approach, both inherited from the MEP. The first is that the MEP assumes stationarity in the data, which is not a common situation from recordings in neuronal systems, so requires careful experimental control to approach this condition. The second is methodological and related to the fact that the Markov transition matrix as presented here is obtained from the transfer matrix technique, so it may require an important computational effort for large-scale and long memory spiking neuronal networks. Indeed as discussed in [18] this approach can reliably recover Markov transition matrices for systems of *N* neurons and memory $R - 1$ that satisfies $N \times R \leq 20$. However, new methods based on Monte Carlo methods can overcome this limitation [46].

There is a lot of room for progress going beyond the scope of this work, one possibility is to quantify the IEP for different choices of non-synchronous constraints and binning sizes on biological

Entropy **2018**, *20*, 34

spike train recordings. A more ambitious goal would be to link the IEP as a signature of an underlying physiological process depending on time such as adaptation or learning. IEP is a much broader concept which can also be measured along non-stationary trajectories, thus IEP can be measured for time-dependent models where transition probabilities are explicitly given or can be computed (for example the GLM [8]). Previous studies in the context of spike train statistics have measured the dynamical entropy production in spiking neuron networks using a deterministic approach based on the Pesin identity (sum of positive Lyapunov exponents) [47]. There are relationships between the deterministic and stochastic dynamics [48], and some interpretations of deterministic dynamical entropy production with information loss which should be investigated in more detail, in particular, if these relationships bring new knowledge in the field of computational neuroscience.

We have focused on spike train statistics, but our results are not restricted to this field and can be applied wherever Markov maximum entropy measures under constraints have to be inferred from data, especially for irreversible Markov chains arising from stochastic network theory [49], information theory [37], and finance [38], among other disciplines.

Acknowledgments: We thank Jean-Pierre Eckmann and Fernando Rosas for discussions and careful reading of the manuscript. Rodrigo Cofre was supported by an ERC advanced grant "Bridges", CONICYT-PAI Insercion # 79160120 and Proyectos REDES ETAPA INICIAL, Convocatoria 2017 REDI170457. Cesar Maldonado was supported by the CONICYT-FONDECYT Postdoctoral Grant No. 3140572.

Author Contributions: Both authors conceived the algorithm and wrote the and revised manuscript. Both authors have read and approved the final manuscript.

Conflicts of Interest: The authors declare no conflict of interest.

Abbreviations

The following abbreviations are used in this manuscript:

MEP	Maximum entropy principle
MEMC	Maximum entropy Markov chain
IEP	Information entropy production
KSE	Kolmogorov–Sinai entropy
IF	Integrate-and-Fire
GLM	Generalized Linear model
NESS	Non-equilibrium steady states

Symbol List

σ_k^n	Spiking state of neuron k at time n.
σ^n	Spike pattern at time n
σ^{n_1,n_2}	Spike block from time n_1 to n_2.
$A_T(f)$	Empirical Average value of the observable f considering T spike patterns.
Σ_N^L	Set of spike blocks of N neurons and length L.
$\mathcal{S}[\mu]$	Entropy of the probability measure μ.
$\mathcal{H}$	Potential function.
$\mathcal{P}[\mathcal{H}]$	Free energy or topological pressure.

References

1. Lefebvre, B.; Yger, P.; Marre, O. Recent progress in multi-electrode spike sorting methods. *J. Physiol. Paris* **2016**, *4*, 327–335.
2. Schneidman, E.; Freedman, B.; Segev, I. Ion channel stochasticity may be critical in determining the reliability and precision of spike timing. *Neural Comput.* **1998**, *10*, 1679–1703.
3. Rieke, F.; Warland, D.; de Ruyter van Steveninck, R.; Bialek, W. *Spikes, Exploring the Neural Code*; MIT Press: Cambridge, MA, USA, 1996.
4. Faisal, A.; Selen, L.; Wolpert, D. Noise in the nervous system. *Nat. Rev. Neurosci.* **2008**, *9*, 292–303.
5. Borst, A.; Theunissen, F. Information theory and neural coding. *Nat. Neurosci.* **1999**, *2*, 947–957.

6. Rolls, E.; Treves, A. The neuronal encoding of information in the brain. *Prog. Neurobiol.* **2011**, *95*, 448–490.
7. Cafaro, J.; Rieke, F. Noise correlations improve response fidelity and stimulus encoding. *Nature* **2010**, *468*, 964–967.
8. Pillow, J.; Paninski, L.; Uzzell, V.; Simoncelli, E.; Chichilnisky, E. Prediction and decoding of retinal ganglion cell responses with a probabilistic spiking model. *J. Neurosci.* **2005**, *25*, 11003–11013.
9. Shadlen, M.; Newsome, W. The variable discharge of cortical neurons: Implications for connectivity, computation, and information coding. *J. Neurosci.* **1998**, *18*, 3870–3896.
10. Schneidman, E.; Berry , M.J.; Segev, R.; Bialek, W. Weak pairwise correlations imply string correlated network states in a neural population. *Nature* **2006**, *440*, 1007–1012.
11. Nirenberg, S.; Latham, P. Decoding neuronal spike trains: How important are correlations. *Proc. Natl. Acad. Sci. USA* **2003**, *100*, 7348–7353.
12. Nirenberg, S.; Latham, P. Population coding in the retina. *Curr. Opin. Neurobiol.* **1998**, *8*, 488–493.
13. Ohiorhenuan, I.E.; Mechler, F.; Purpura, K.P.; Schmid, A.M.; Hu, Q.; Victor, J.D. Sparse coding and high-order correlations in fine-scale cortical networks. *Nature* **2010**, *466*, 617–621.
14. Panzeri, S.; Schultz, S. A unified approach to the study of temporal, correlational, and rate coding. *Neural Comput.* **2001**, *13*, 1311–1349.
15. Ganmor, E.; Segev, R.; Schneidman, E. The architecture of functional interaction networks in the retina. *J. Neurosci.* **2011**, *31*, 3044–3054.
16. Moore, G.; Segundo, J.; Perkel, D.; Levitan, H. Statistical signs of synaptic interaction in neurons. *Biophys. J.* **1970**, *10*, 876–900.
17. Moran, J.; Desimone, R. Selective attention gates visual processing in the extrastriate cortex. *Science* **1985**, *229*, 782–784.
18. Nasser, H.; Cessac, B. Parameter estimation for spatio-temporal maximum entropy distributions: Application to neural spike trains. *Entropy* **2014**, *16*, 2244–2277.
19. Tang, A.; Jackson, D.; Hobbs, J.; Chen, W.; Smith, J.; Patel, H.; Prieto, A.; Petrusca, D.; Grivich, M.; Sher, A.; et al. A maximum entropy model applied to spatial and temporal correlations from cortical networks in vitro. *J. Neurosci.* **2008**, *28*, 505–518.
20. Marre, O.; El Boustani, S.; Frégnac, Y.; Destexhe, A. Prediction of spatiotemporal patterns of neural activity from pairwise correlations. *Phys. Rev. Lett.* **2009**, *102*, 138101.
21. Vasquez, J.; Palacios, A.; Marre, O.; Berry, M.J.; Cessac, B. Gibbs distribution analysis of temporal correlation structure on multicell spike trains from retina ganglion cells. *J. Physiol. Paris* **2012**, *106*, 120–127.
22. Cofré, R.; Cessac, B. Exact computation of the maximum entropy potential of spiking neural networks models. *Phys. Rev. E* **2014**, *89*, 052117.
23. Jaynes, E. Information theory and statistical mechanics. *Phys. Rev.* **1957**, *106*, 620.
24. Pillow, J.W.; Shlens, J.; Paninski, L.; Sher, A.; Litke, A.M.; Chichilnisky, E.J.; Simoncelli, E.P. Spatio-temporal correlations and visual signaling in a complete neuronal population. *Nature* **2008**, *454*, 995–999.
25. Tkačik, G.; Marre, O.; Amodei, D.; Schneidman, E.; Bialek, W.; Berry, M.J. Searching for collective behavior in a large network of sensory neurons. *PLoS Comput. Biol.* **2013**, *10*, e1003408.
26. Tkačik, G.; Mora, T.; Marre, O.; Amodei, D.; Palmer, S.E.; Berry, M.J.; Bialek, W. Thermodynamics and signatures of criticality in a network of neurons. *Proc. Natl. Acad. Sci. USA* **2015**, *112*, 11508–11513.
27. Shi, P.; Qian, H. Irreversible stochastic processes, coupled diffusions and systems biochemistry. In *Frontiers in Computational and Systems Biology*; Springer: London, UK, 2010; pp. 175–201.
28. Jiang, D.Q.; Qian, M.; Qian, M.P. *Mathematical Theory of Nonequilibrium Steady States*; Springer: Berlin/Heidelberg, Germany, 2004.
29. Bowen, R. *Equilibrium States and the Ergodic Theory of Anosov Diffeomorphisms*, revised edition; Springer: Berlin, Germany, 2008; Volume 470.
30. Fernandez, R.; Maillard, G. Chains with complete connections: General theory, uniqueness, loss of memory and mixing properties. *J. Stat. Phys.* **2005**, *118*, 555–588.
31. Cessac, B.; Cofré, R. Spike train statistics and Gibbs distributions. *J. Physiol. Paris* **2013**, *107*, 260–368.
32. Galves, A.; Löcherbach, E. Infinite systems of interacting chains with memory of variable length—A stochastic model for biological neural nets. *J. Stat. Phys.* **2013**, *151*, 896–921.
33. Gaspard, P. Time-reversed dynamical entropy and irreversibility in Markovian random processes. *J. Stat. Phys.* **2004**, *117*, 599–615.

34. Kitchens, B.P. *Symbolic Dynamics: One-Sided, Two-Sided and Countable State Markov Shifts*; Springer: Berlin/Heidelberg, Germany, 1998.

35. Pollard, B.S. Open Markov processes: A compositional perspective on non-equilibrium steady states in biology. *Entropy* **2016**, *18*, 140.

36. Ruelle, D. *Thermodynamic Formalism*; Addison-Wesley: Reading, MA, USA, 1978.

37. Van der Straeten, E. Maximum entropy estimation of transition probabilities of reversible Markov chains. *Entropy* **2009**, *11*, 867–887.

38. Chliamovitch, G.; Dupuis, A.; Chopard, B. Maximum entropy rate reconstruction of Markov dynamics. *Entropy* **2015**, *17*, 3738–3751.

39. Baladi, V. *Positive Transfer Operators and Decay of Correlations*; World Scientific: Singapore, 2000.

40. Chazottes, J.R.; Ramirez, L.; Ugalde, E. Finite type approximations of Gibbs measures on sofic subshifts. *Nonlinearity* **2005**, *18*, 445–463.

41. Maldonado, C.; Salgado-García, R. Markov approximations of Gibbs measures for long-range interactions on 1D lattices. *J. Stat. Mech. Theory Exp.* **2013**, *8*, P08012.

42. Gantmacher, F.R. *The Theory of Matrices*; AMS Chelsea Publishing: Providence, RI, USA, 1998.

43. Lancaster, P. *Theory of Matrices*; Academic Press: Cambridge, MA, USA, 1969.

44. Cessac, B. A discrete time neural network model with spiking neurons. Rigorous results on the spontaneous dynamics. *J. Math. Biol.* **2008**, *56*, 311–345.

45. Roudi, Y.; Hertz, J. Mean field theory for non-equilibrium network reconstruction. *Phys. Rev. Lett.* **2011**, *106*, 048702.

46. Nasser, H.; Marre, O.; Cessac, B. Spatio-temporal spike trains analysis for large scale networks using maximum entropy principle and Monte-Carlo method. *J. Stat. Mech.* **2013**, *2013*, P03006.

47. Monteforte, M.; Wolf, F. Dynamical entropy production in spiking neuron networks in the balanced state. *Phys. Rev. Lett.* **2010**, *105*, 268104.

48. Gaspard, P. Time asymmetry in nonequilibrium statistical mechanics. *Adv. Chem. Phys.* **2007**, *135*, 83–133.

49. Delvenne, J.; Libert, A. Centrality measures and thermodynamic formalism for complex networks. *Phys. Rev. E* **2011**, *83*, 046117.

![entropy logo] *entropy*

MDPI

Article

A Moment-Based Maximum Entropy Model for Fitting Higher-Order Interactions in Neural Data

N. Alex Cayco-Gajic [1],* [iD], **Joel Zylberberg [2] and Eric Shea-Brown [3]**

[1] Department of Neuroscience, Physiology, and Pharmacology, University College London, London WC1E 6BT, UK
[2] Department of Physiology and Biophysics, University of Colorado School of Medicine, Aurora, CO 80045, USA; joel.zylberberg@ucdenver.edu
[3] Department of Applied Mathematics, University of Washington, Seattle, WA 98195, USA; etsb@uw.edu
* Correspondence: natasha.gajic@ucl.ac.uk

Received: 1 May 2018; Accepted: 19 June 2018; Published: 23 June 2018

Abstract: Correlations in neural activity have been demonstrated to have profound consequences for sensory encoding. To understand how neural populations represent stimulus information, it is therefore necessary to model how pairwise and higher-order spiking correlations between neurons contribute to the collective structure of population-wide spiking patterns. Maximum entropy models are an increasingly popular method for capturing collective neural activity by including successively higher-order interaction terms. However, incorporating higher-order interactions in these models is difficult in practice due to two factors. First, the number of parameters exponentially increases as higher orders are added. Second, because triplet (and higher) spiking events occur infrequently, estimates of higher-order statistics may be contaminated by sampling noise. To address this, we extend previous work on the Reliable Interaction class of models to develop a normalized variant that adaptively identifies the specific pairwise and higher-order moments that can be estimated from a given dataset for a specified confidence level. The resulting "Reliable Moment" model is able to capture cortical-like distributions of population spiking patterns. Finally, we show that, compared with the Reliable Interaction model, the Reliable Moment model infers fewer strong spurious higher-order interactions and is better able to predict the frequencies of previously unobserved spiking patterns.

Keywords: maximum entropy; higher-order correlations; neural population coding; Ising model

---

1. Introduction

An essential step in understanding neural coding is the characterization of the correlated structure of neural activity. Over the past two decades, much theoretical work has clarified the strong impact that correlated variability between pairs of neurons can have on the amount of information that can be encoded in neural circuits [1–6]. Beyond pairs, recent experimental studies have shown evidence of *higher-order* correlations in cortical [7–11] and retinal [12,13] population activity. Depending on their stimulus-dependent structure, these higher-order correlations could also have a strong impact on population coding [14,15]. Moreover, capturing higher-order correlations in neural spiking may be important for identifying functional networks in neural circuits [16], or for characterizing their collective statistical activity [17]. Therefore, to incorporate higher-order spiking statistics into an information theoretic framework, we require flexible modeling tools that can capture the coordinated spiking of arbitrary orders within neural populations.

Maximum entropy models are an increasingly common tool for fitting and analyzing neural population spiking patterns. Intuitively, maximum entropy models fit certain specified features (e.g., firing rates, correlations between cells) while making minimal additional assumptions about the population structure [18]. Several variants of the maximum entropy model have been used to fit the collective activity of spiking patterns in neural data [4,11,13,16,19,20]. However, it is still unclear how to efficiently incorporate higher-order features into maximum entropy models for two reasons. First, the number of parameters (and hence the computational expense of model fitting) increases exponentially as higher-order features are incorporated. Second, because higher-order synchronous spiking occurs infrequently, empirical estimates tend to be noisy; therefore, massive amounts of data may be necessary to create a model with higher-order interactions that can generalize to held-out data. These issues have been addressed by the Reliable Interaction model [12], which uses a maximum entropy inspired model to fit a sparse network of features based on the most "reliable" (i.e., high-frequency) spiking patterns within their data. This approach is extremely efficient numerically and reproduces the frequencies of the most commonly occurring patterns with high accuracy. However, because the model is not a normalized probability distribution, it cannot be used to calculate information theoretic quantities such as the Kullback–Leibler divergence or mutual information.

To address these challenges, we introduce an adaptive maximum entropy model that identifies and fits spiking interactions of all orders, based on the criterion that they can be accurately estimated from the data for a specified confidence level. Towards this end, we adapt the Reliable Interaction model by making two small but critical modifications in the fitting procedure and fitting criterion; these modifications normalize the model, allowing information theoretic quantities to be calculated. The resulting model is able to fit cortical-like distributions of spiking patterns with dense higher-order statistics. Finally, we show that these modifications have two further important consequences: they reduce spurious higher-order interactions, and improve the model's ability to predict the frequencies of previously unseen spiking patterns.

2. Results

2.1. The Reliable Moment Model

To analyze population-level activity in neural recordings, it is often necessary to first model the distribution of spiking patterns. Certain spiking features of neural population activity are likely to be more relevant for modeling than others: for example, each neuron's firing rate and the correlations between pairs of neurons. In general, there may be an infinite family of models that fit these key features in the data, making any particular choice seem potentially arbitrary. One approach is to take the distribution that captures the identified statistical features while making the fewest additional assumptions on the structure of the data. Mathematically, this is equivalent to matching the average values of the features observed in the data while maximizing the statistical entropy [21]. The resulting distribution is called the maximum entropy model and can be derived analytically via Lagrange multipliers [18], resulting in the following probability:

$$P(x) = \frac{1}{Z} \exp\left\{ \sum_i h_i f_i(x) \right\}. \tag{1}$$

Here, x represents a binary spiking pattern across the population in a small time bin (i.e., $x_i = 1$ if neuron i spiked in that time bin, otherwise $x_i = 0$), $f_i(x)$ are the chosen spiking features, and h_i are interaction parameters that are fitted to match the average $f_i(x)$ to the values observed in the data. Z is a normalizing factor, also called the partition function.

The quality of fit of a maximum entropy model relies critically on which features are included. Traditionally, first-order (i.e., firing rate) and second-order features (correlations) are chosen [4] to isolate the effect of pairwise correlations on population activity patterns. However, this may miss important information about higher-order dependencies within the data. In principle, the pairwise maximum entropy model can be generalized by fitting features of up to kth order; but this becomes computationally expensive for large datasets as the number of parameters grows as $O(N^k)$. Moreover, higher-order features are more susceptible to overfitting, because they represent spiking features that occur less frequently in the data (and consequently have noisy empirical estimates). An alternative is to incorporate a limited subset of predetermined phenomenological features that increase the predictive power of the model, such as the spike count distribution [13] or frequency of the quiescent state [11]. While these models have been able to capture the collective activity of populations of neurons (e.g., to determine whether neural activity operates at a critical point [17]), they are not able to dissect how the functional connectivity between specific subgroups of neurons contributes to the population level activity.

To address these challenges, a method is needed for data-driven adaptive identification of relevant spiking features of all orders. The Reliable Interaction (RI) model [12] has previously been used to fit sparse networks of pairwise and higher-order interactions to retinal populations. The RI model fits only the features corresponding to spiking patterns whose observed frequencies are larger than an arbitrary threshold. For example, in a 10-cell population, the fourth-order feature $f_i(x) = x_1 x_3 x_5 x_9$ would be fitted only if the spiking pattern $x = 1010100010$ occurs with frequency above this threshold. Once these features have been identified, the RI model uses an algebraic approximation for rapid parameter fitting by first calculating the partition function Z as the inverse of the frequency of the silent state: $Z = P(00\ldots0)^{-1}$. Subsequently, the interaction parameters can be estimated recursively from the observed frequencies and Z. However, while the RI model has been shown to be able to accurately fit the frequencies of spiking patterns, its fitting procedure does not generate a normalized probability distribution (as originally discussed in [12]; see Appendix A for an intuitive example). This limits certain applications of the model: for example, information theoretic measures such as the Kullback–Leibler divergence and mutual information cannot be calculated. Another limitation (demonstrated below and in Appendix A) is that the RI model often cannot predict the frequencies of rarely occurring spiking patterns.

We propose the Reliable Moment (RM) model, an adaptation of the RI model that makes two key modifications in the fitting procedure and fitting criterion. First, we take advantage of a recently developed method for rapid parameter estimation: Minimum Probability Flow (MPF) learning [22]. While still substantially slower than the algebraic method employed in [12] (which is essentially instantaneous), using a parameter estimation method such as MPF guarantees a probability distribution that, in theory, can be readily normalized. In practice, calculating the partition function (Z in Equation (1)) may be computationally expensive, as it requires summing 2^N probabilities. In this case, the partition function can be quickly estimated using other techniques, such as the Good–Turing estimate [23] (see Methods). As we shall see below, attempting to apply these approaches to the RI model strongly disrupts its predictions.

Second, instead of fitting the features corresponding to the most commonly occurring spiking patterns, we fit the features corresponding to the largest moments. Taking the previous example, feature $f_i(x) = x_1 x_3 x_5 x_9$ would be fitted only if the moment $x_1 x_3 x_5 x_9$ is greater than some threshold. As in the RI model, the threshold parameter p_{min} implicitly determines the number of fitted features. For binary systems, the uncentered moment of a subset of neurons is equal to the marginal probability of those neurons spiking, so that the previous condition is equivalent to:

$$P(x_1 = 1, x_3 = 1, x_5 = 1, x_9 = 1) \geq p_{min}.$$

The choice of p_{min} can be made less arbitrary by choosing its value to bound the 95% confidence interval of the relative error in the sample moments (with some minimal assumptions; [14]):

$$p_{min} = \frac{1}{1 + M\left(\frac{\alpha}{2}\right)^2}. \tag{2}$$

where M is the number of samples and α is the maximum desired relative error. In this way, the RM model can adaptively identify which moments within a specific dataset are large enough to be accurately estimated by the sample frequency.

Unlike the spiking pattern frequencies used in the RI model, these marginal probabilities satisfy an important hierarchy: the moment of any set of neurons is necessarily bounded by the moment of any subset of those neurons, e.g.:

$$x_1 x_3 x_5 x_9 \leq x_1 x_3 x_5 \leq x_3 x_5 \leq x_3$$

This means that for every higher-order interaction fitted by the RM model, all of its corresponding lower-order interactions are automatically fitted as well. Although this may seem to be a minor change from the RI model, we will demonstrate the significance of this change with the following toy model (we later consider larger and more realistic models, see Sections 2.3–2.5).

2.2. Illustration with a Toy Example

Consider $N = 3$ homogeneous neurons with only first and second-order interactions:

$$P(x) = \frac{1}{Z} \exp \left\{ -\alpha \sum_i x_i + \frac{\beta}{2} \sum_{i \neq j} x_i x_j \right\}. \tag{3}$$

The probability of each pattern can be found analytically:

$$P(x) = \begin{cases} \frac{1}{Z} & \text{if 0 spikes} \\ \frac{e^{-\alpha}}{Z} & \text{if 1 spike} \\ \frac{e^{-2\alpha+\beta}}{Z} & \text{if 2 spikes} \\ \frac{e^{-3\alpha+3\beta}}{Z} & \text{if 3 spikes} \end{cases}$$

where $Z = 1 + 3e^{-\alpha} + 3e^{-2\alpha+\beta} + e^{-3\alpha+3\beta}$. In particular, for $\alpha = 1$, $\beta = 1.2$:

$$P(x) \approx \begin{cases} 0.1896 & \text{if 0 spikes} \\ 0.0698 & \text{if 1 spike} \\ 0.0852 & \text{if 2 spikes} \\ 0.3455 & \text{if 3 spikes} \end{cases}$$

To gain intuition on the fundamental differences between the RM and RI models, we will take the "best-case" scenario for the model fits; i.e., assuming infinite data and infinite fitting time. This eliminates any error due to statistical sampling or parameter fitting for this toy example. We will first see that the difference in fitting criterion can lead the RI model to identify spurious higher-order interactions. This can be seen by setting the threshold at $p_{min} = 0.1$. Then, the RI model will only identify the spiking patterns $x = 000$ and 111 as reliable, resulting in the following:

$$P_{RI}(x) = \frac{1}{Z} e^{h_{123} x_1 x_2 x_3}, \tag{4}$$

where $h_{123} = \log(Z * P(111)) = 0.6$. While the ground truth distribution only contains first- and second-order interactions, the RI fitting procedure mistakenly infers a pure triplet model. This happens because the RI model criterion for selection is based on the frequencies of spiking patterns, which (unlike the moments) do not necessarily follow a natural hierarchy. In contrast, because it relies on the frequency of the marginal probabilities, the RM model identifies all first, second, and third order interaction parameters:

$$P_{RM}(x) = \frac{1}{Z} \exp \left\{ \sum_i h_i^{(1)} x_i + \sum_{i \neq j} h_{ij}^{(2)} x_i x_j + h_{123}^{(3)} x_1 x_2 x_3 \right\}. \tag{5}$$

This demonstrates that the RM model cannot infer higher-order interactions without also fitting the corresponding lower-order interactions.

Second, the RI model can fail to predict the frequencies of rare spiking patterns; i.e., those that were not selected as reliable by the model. To see this, consider that the RI model estimates the partition function as $Z = P(000)^{-1}$. While this gives an accurate estimate of the partition function of the true underlying distribution (in this example, the pairwise model; Equation (3)), it may be a poor estimate of the partition function for the model with interactions inferred by the RI fitting criterion (i.e., the pure triplet model). This mismatch between model form and the estimated partition function is the reason the model cannot be normalized. Because the estimated Z is also used to determine the interaction parameters, the RI model frequencies match the true probabilities of the spiking patterns that are used for fitting (i.e., the most common or reliable patterns), but is inaccurate for patterns that are below the threshold frequency (Figure 1). However, naïve renormalization of the model would make all of the probabilities inaccurate.

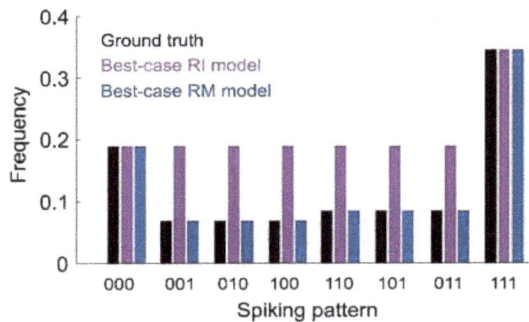

Figure 1. Toy model of $N = 3$ neurons with only first- and second-order interactions. Ground-truth probabilities are shown for each spiking pattern (black). Also shown are the frequencies predicted by the best-case (i.e., assuming infinite data and fitting time) Reliable Interaction (RI, magenta) and Reliable Moment (RM, blue) models (assuming a threshold of 0.1). Under these assumptions, the RM model would fit the ground-truth frequencies exactly. The RI model exactly fits the frequencies for spiking patterns above threshold, but is inaccurate for rare patterns. Note that the RI model cannot be normalized because the fitted partition function does not match fitted interaction terms (see main text and Appendix A for a detailed explanation). Model parameters: $\alpha = 1$, $\beta = 1.2$ (see Equation (3)).

On the other hand, because it falls in the class of maximum entropy distributions, the RM model is guaranteed to converge to the ground-truth solution under the following assumptions: first, assuming that all interaction terms in the ground-truth model are incorporated into the RM model; second, assuming infinite data; and finally, assuming infinite time and a convex iterative fitting procedure such as Iterative Scaling [24]. For this toy example, this means that the "best case" RM model given by Equation (5) will converge to the ground-truth distribution (Equation (3)). However, note that this is not necessarily the case due to sampling noise, unidentified interaction terms, and the necessity for approximate methods due to time limitations. In the latter case, we advocate the use of the approximate MPF learning algorithm as a more practical option than Iterative Scaling, but this choice introduces some error into the fitted model.

Approximate methods are also useful for calculating the partition function. While the partition function can be calculated exactly by brute-force summing all 2^N unnormalized probabilities, this can become prohibitively slow for large populations. We instead approximate the partition function; e.g., by the Good–Turing estimate [23]. Another alternative is to use Gibbs sampling [25] to generate spiking patterns from the inferred interaction parameters, then use the RI estimate of the partition function as the inverse probability of the non-spiking state in the Gibbs sampled "data". Regardless of which of these methods is used, our toy example shows the fundamental differences between the RM and RI models, namely, that the RM model can in principle be normalized without disrupting its predictions of spike pattern probabilities.

2.3. The RM Model Infers Fewer Strong Spurious Higher-Order Interactions

Using this toy model, we have demonstrated that the RM model may be: (1) less likely to infer spurious higher-order interactions, and (2) better able to predict the frequencies of new spiking patterns. Do these improvements hold for more realistic population spiking statistics? To test this, we modeled populations of $N = 20$ neurons using pairwise maximum entropy models. Specifying the desired statistics of a maximum entropy model is a notoriously difficult inverse problem. We therefore tuned the ground-truth interaction parameters to generally give low firing rates (Figure 2a, mean $\pm$ std, 3.3 ± 1.9 Hz) and a broad distribution of correlations (Figure 2b, 0.01 ± 0.05; see Methods). However, we will subsequently test the ability of the RM model to fit a class of models for which we can directly prescribe cortical-like distributions of firing rates and spiking correlations (see Section 2.5).

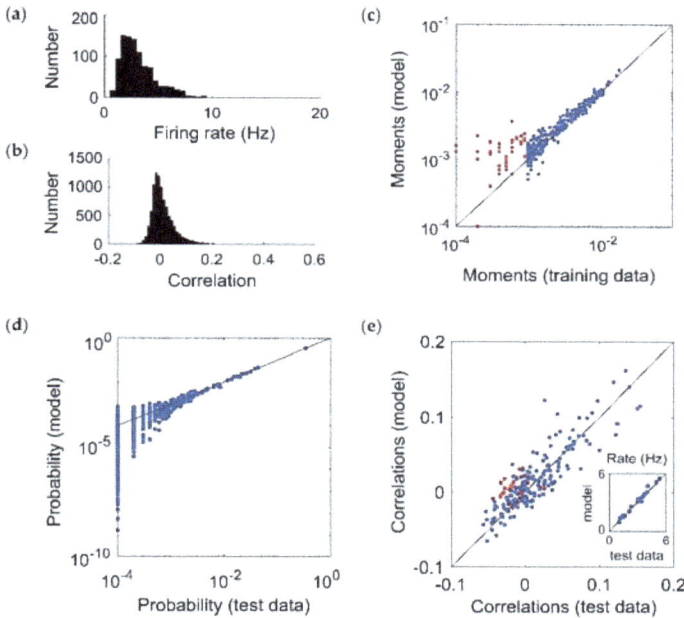

Figure 2. Fitting a ground-truth pairwise maximum entropy model ($N = 20$). (**a**,**b**) Distribution of (**a**) firing rates (assuming a time window of 20 ms) and (**b**) pairwise correlation coefficients generated by the ground truth models. (**c**–**e**) Example of Reliable Moment (RM) model fit to 200 s of a simulated pairwise ground truth model ($p_{min} = 10^{-3}$). In this example, the RM model identified all 20 units, 154 pairs, 103 triplets, and 5 quadruplets. (**c**) Uncentered sample moments in the fitted RM model plotted against the empirical sample moments (estimated from training data) to show quality of model fit. Blue indicates all moments (single, pairwise, and higher-order) that were identified by the RM model. For comparison, red indicates the 36 pairs that were not identified by the RM model (and hence not fitted). (**d**) Cross-validated RM model probabilities versus ground-truth probability (i.e., estimated from held-out "test" data), for an example ground-truth model. Each point represents a different spiking pattern. (**e**) RM model correlations plotted against cross-validated empirical correlations (i.e., sample correlations plotted against empirical sample correlations from test data). Again, red points indicate pairs whose corresponding interaction terms were not identified. Inset shows the same for firing rates.

We generated population spiking patterns under the resulting distribution using Gibbs sampling (equivalent to 200 s worth of data) [22,25]. Figure 2c shows the fitted moments of a RM model for an example simulated population dataset ($p_{min} = 10^{-3}$). This choice of threshold parameter identifies all 20 units, and 154 pairs (out of 190 possible) as having moments above threshold, which are fitted via MPF learning to reproduce the sample moments from the training data (blue; for comparison, the 36 pairs that were not included in the fitting are shown in red). The model is able to reproduce the probability distribution of spiking patterns in a "test" dataset that was not used to fit the model (Figure 1d), as well as the firing rates and correlations (Figure 1e; including the pairs that were not explicitly used for fitting, in red). This choice of model also identifies 108 higher-order moments (103 triplets and 5 quadruplets) as being above threshold. Since the ground truth model is pairwise, ideally their interaction parameters should be zero after fitting. Because of sampling noise in the data, as well as idiosyncrasies of MPF learning (see Discussion), they are nonzero but small on average (magnitude 0.235 ± 0.231).

How does this compare to the RI model? We next systematically tested whether the RM and RI models infer spurious higher-order interactions by simulating 50 random pairwise populations (using the same firing rates and correlations given by the distributions in Figure 2a,b). For each ground-truth model, we fit 20 RM and RI models with varying thresholds (see Methods), and compared the magnitudes of the higher-order interaction parameters. We found that the fitted higher-order interaction terms were smaller for the RM model than the RI model, regardless of the number of inferred parameters (Figure 3). This was true even when correcting for potential differences in the fitted lower-order interaction parameters (see Appendix B). Moreover, for the RM model, the average magnitude of the higher-order interaction terms was nonzero, but small and constant across different thresholds; whereas for the RI model, they increased in both magnitude and in variance. When a sparse subset of triplet interaction terms is added to the ground-truth model, the RM model is also better able to fit the corresponding interaction parameters (see Appendix C). These results reinforce the intuition we developed previously with the toy model (Figure 1) that the RM model finds fewer strong, spurious higher-order interactions, and is better able to fit existing higher-order interactions.

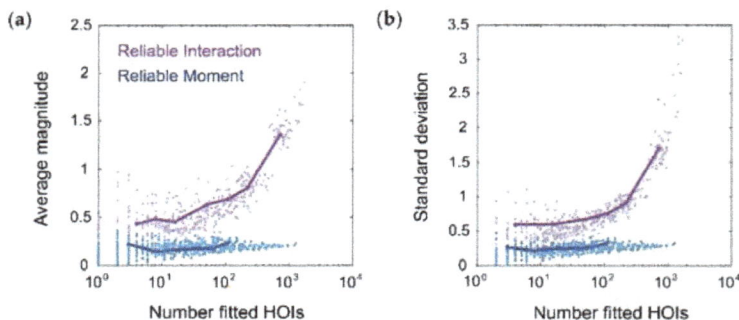

Figure 3. The Reliable Moment (RM) infers fewer strong, spurious higher-order interactions. (a) Average magnitude of all fitted higher-order interaction parameters as a function of the number of fitted higher-order interactions, shown for both the Reliable Interaction (RI; magenta) and RM (blue) models. Note that all higher-order interactions should have magnitude 0. Points represent 50 random ground-truth models (i.e., random interaction parameters), each of which is fitted 20 times with varying threshold parameters (see Methods). Solid lines indicate the RM and RI fits to a specific example ground-truth model. (b) Same as (a) but for standard deviation.

2.4. The RM Model Fits Rare Spiking Patterns

Our toy model also predicted that, while the RI model is very accurate at capturing the frequencies of commonly occurring spiking patterns, it is unable to predict the probabilities of rare patterns. This could be a strong limitation for large population recordings, as the number of previously-unseen spiking patterns grows as $O(2^N)$ assuming fixed recording lengths. We therefore tested this effect by generating a new testing dataset for each ground-truth model, and separating it into "old" spiking patterns (those that also occurred within the training dataset) and "new" spiking patterns (those that only occurred within the test dataset). In order to compare the RM and RI models, we must specify which threshold values to use for each model. Since the RM and RI threshold use different "units" (i.e., the RI threshold is based on the frequencies population spiking patterns, and the RM threshold is based on marginal probabilities or moments), it is difficult to directly compare them. For a fair comparison of the model fits, it is therefore necessary to compare models that have the same number of fitted interaction parameters. Otherwise, any difference in model performance might be attributed to a model having more parameters to fit. We therefore first chose the threshold parameters in this example so that the RM and RI models have exactly the same number of fitted interaction parameters (in this case, 395). Figure 4a shows an example of model vs. empirical frequencies (calculated from held-out test data) for old spiking patterns.

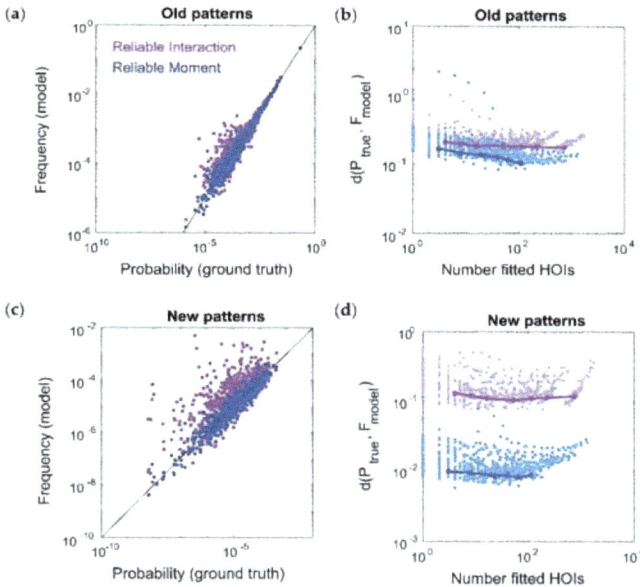

Figure 4. The Reliable Moment (RM) model is able to predict the probabilities of new spiking patterns. (**a**) Reliable interaction (RI; magenta) model frequencies and RM (blue) model probabilities of previously observed spiking patterns plotted against ground-truth probability, for an example ground-truth model. Each point represents a different "old" spiking pattern (i.e., occurring within both test and training datasets). For a fair comparison, we chose an example in which the RM and RI models had the same number of fitted interaction parameters (in this case, 395). (**b**) Dissimilarity (see Methods) between ground-truth distribution and model distribution of spiking patterns over different numbers of fitted higher-order interactions. Points represent 50 random ground-truth models (i.e., random interaction parameters), each of which is fitted 20 times with varying threshold parameters. Solid lines indicate the RM and RI fits to a specific example ground-truth model. (**c**,**d**) Same as (**a**,**b**) for new spiking patterns (i.e., those observed in the test data but not observed in the training data).

Because the RI model is unnormalized, we cannot use the Kullback–Leibler divergence. Instead, we calculated the dissimilarity between the distributions using the weighted average of the magnitude of the log-likelihood (see Methods, [12]). RM and RI model performances were comparable across different ground-truth populations and different threshold parameters (Figure 4b). However, the RI model was much less accurate for predicting the frequencies of new spiking patterns (Figure 4c,d). As discussed for the toy model, this is because the RI fitting procedure is only able to capture data that was used for fitting, which precludes new spiking patterns. Therefore, in both the toy model and the more realistic case here, the RM model is better able to predict the frequencies of the many unobserved spiking patterns that inevitably occur in large array recordings.

2.5. Fitting a Model with Cortical-Like Statistics and Dense Higher-Order Correlations

Thus far we have focused on fitting data generated by ground-truth pairwise maximum entropy models. Therefore, we now test the performance of the RM model on the Dichotomized Gaussian (DG) model, which simulates population spiking activity by generating multivariate Gaussian samples (representing correlated inputs to the population) and thresholding them [26]. The DG model generates dense higher-order statistics and can reproduce higher-order correlations observed in cortical data [8]. Unlike maximum entropy models, we can directly specify the firing rates for the DG model in order to generate cortical-like statistics. We chose log-normal, low-rate (mean 4 Hz) firing rate distributions [27,28] (Figure 5a), and normally distributed (mean 0.1) pairwise correlations [29] (Figure 5b; see Methods). We next compared the ability of the RM and RI models to fit the DG model spike patterns by comparing the dissimilarity between the model frequencies and the empirical probabilities from a held-out test dataset. The RM model was able to fit the DG patterns well, with the classic U-shaped curve with the number of parameters, whereas the RI model had an oscillatory shape (Figure 5c). The oscillations occur due to instabilities in the model's ability to fit rare spiking patterns. To see this, Figure 5d shows an example of cross-validated model vs. empirical frequencies of spiking patterns that occur more than once in the test dataset. This is analogous to comparing the performance of the models for old spiking patterns (as in Figure 4a,b). For a fair comparison, we chose this example so that the RM and RI models had the same number of fitted interaction parameters (in this case, 239). Both models describe the data well, with the RI model performing slightly better because of its more accurate fit to the most common (quiescent) spiking patterns. However, when considering all the spiking patterns that occur in the test dataset, the RI model produces incoherent values for rare spiking patterns (i.e., those that only occur once, analogous to the "new" spiking patterns in Figure 4c,d), with frequencies often far surpassing 1 (Figure 5d, inset). Finally, note that an advantage of the RI model is that its fitting procedure is essentially instantaneous (Figure 5e). We therefore conclude that the RI model is a highly efficient method for capturing the frequencies of observed spiking patterns with relatively few parameters, but is unstable for predicting previously-unseen spiking patterns.

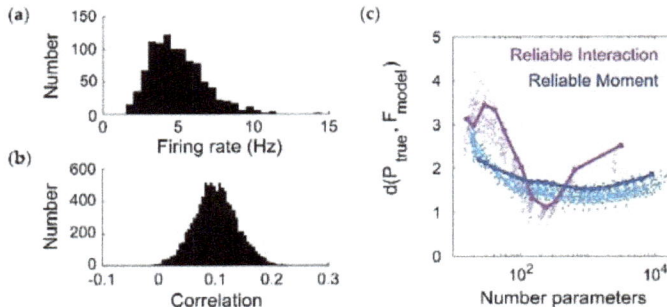

Figure 5. *Cont.*

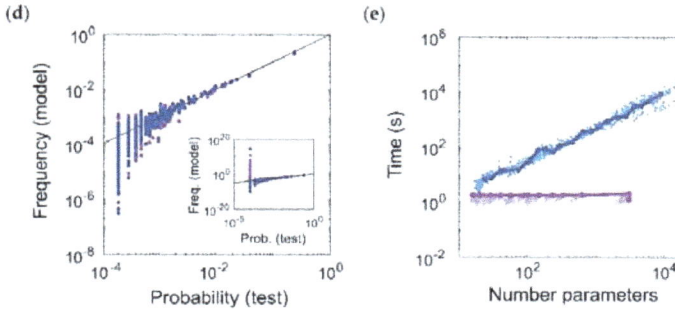

Figure 5. Fitting a Dichotomized Gaussian model with cortical-like statistics ($N = 20$). (**a**,**b**) Distribution of (**a**) firing rates (assuming a time window of 20 ms) and (**b**) pairwise correlation coefficients generated by the model. The Dichotomized Gaussian model is known to generate dense higher-order correlations [9,26]. (**c**) Cross-validated dissimilarity between the empirical and model distributions, for both Reliable Interaction (RI; magenta) and Reliable Moment (RM; blue) models. Points represent 50 random ground-truth models (i.e., random interaction parameters), each of which is fitted 20 times with varying threshold parameters. Solid lines indicate the RM and RI fits to a specific example ground-truth model. (**d**) Cross-validated model frequencies versus empirical probability, for an example ground-truth model. Each point represents a different spiking pattern. Only patterns occurring at least twice in the dataset are shown. Inset shows same plot, including spiking patterns that only occur once. For a fair comparison, we chose an example in which the RM and RI models had the same number of fitted interaction parameters (in this case, 239). (**e**) Time required for fitting RM and RI models.

3. Discussion

We developed the Reliable Moment (RM) model, a novel class of maximum entropy model for adaptively identifying and fitting pairwise and higher-order interactions to neural data. To do this, we extended a previous model [12] by making two key modifications in the fitting criterion and the fitting procedure. First, we include spiking features whose corresponding uncentered moments are above a threshold value. This threshold need not be arbitrary, as it can be used to bound the confidence interval of the relative error (Equation (2)) [14]. Second, we take advantage of recent fast parameter fitting techniques [22], which results in a normalized probability distribution. We show that the RM model is able to fit population spike trains with cortical-like statistics, while inferring fewer strong, spurious higher-order correlations, and more accurately predicting the frequencies of rare spiking patterns.

We extended the intuition of the Reliable Interaction (RI) model [12] by determining which spiking features were most "reliable" as a criterion for inclusion in the model. However, our modifications confer several benefits. First, the RM model is normalized. While this does not necessarily affect the ability of a model to fit spiking pattern frequencies, it means that certain quantities that depend on the full distribution, such as mutual information or specific heat, can be applied to the RM model (although the RI model can be used to decode spiking patterns, as in [12]). This allows the RM model to be used for analyzing population coding or Bayesian inference, or for measuring signatures of criticality [17]. Second, the RM model is better able to predict the frequencies of previously-unseen spiking patterns. This is important for neural data, as the number of unseen spiking patterns increases significantly for large-scale population recordings. On the other hand, as a result of its fitting method, the RI model can be unstable for rare patterns (Figure 5d, inset; although it is able to predict the frequencies of common patterns well). Third, the RM model is less likely to find spurious higher-order interactions in a pairwise model, as compared to the RI model. This is because the hierarchical structure of the uncentered moments guarantees that no higher-order spiking feature can be fitted without also fitting all of its lower-order feature subsets. Finally, the main disadvantage of the RM model is that it is much slower to fit than the RI model, even using Minimum Probability Flow learning [22]. Therefore, the RM model performs better for determining the higher-order statistical structure of the data or predicting the frequencies of new patterns, while the RI model performs better as a fast method for fitting commonly occurring spiking patterns.

Several variants on the RM model are possible. While we chose to use MPF learning due to its speed, there are many alternative methods that are available [24,30–32]. In particular, classic Iterative Scaling [24] finds the interaction parameters that maximize the log-likelihood of the data. This is equivalent to minimizing the Kullback–Leibler divergence between the data and the model, which can be shown to be a convex problem. However, it can be prohibitively slow even for reasonably sized populations. On the other hand, MPF defines a dynamical system that would transform the empirical distribution into the model distribution, then minimizes the Kullback–Leibler divergence between the empirical distribution and the distribution established by this flow (after an infinitesimal time) [22]. While there is no guarantee on convexity, MPF in general works very well in practice (see Figure 2) and is much faster. Another possibility is to add a regularization term to the cost function during fitting to ensure sparse interaction parameters. Moreover, there is some flexibility in choosing the threshold parameter. Here, we advocated determining the threshold parameter to bound the error of the moments (Equation (2)). An alternative option would be to use the Akaike Information Criterion to determine the threshold that results in the optimal number of interaction parameters [33]; however, this would require multiple model fittings for validation. The criterion for inclusion of specific interactions may also be modified, for instance, by requiring that fitted interaction parameters have moments exceeding a threshold based on the empirical values. For each of these variants, the RM model extends the ideas behind the RI model by fitting a sparse subset of the most "relevant" higher-order interactions, while ensuring that the corresponding lower-order interactions are also fit.

We focused on capturing stationary correlations in neural data, while neglecting temporal dependencies between neurons. In principle, temporal correlations could be incorporated into the RM model, and into the maximum entropy models more generally, by fitting concatenated spatiotemporal spiking patterns [34–36]. This dramatically increases the cost of fitting, although emerging techniques are making this problem more tractable [37,38]. Another, more widely-used approach to fitting spatiotemporal models of neural spiking is the generalized linear model (GLM) [39,40]. GLMs and Maximum entropy models are not mutually exclusive; hybrid approaches are possible [41], and maximum entropy models can be interpreted in a linear-nonlinear framework [15,19]. Future work could incorporate higher-order moments into the GLM framework, as has been done for common inputs [42]; indeed, there is a long history of moment-based methods for point process models that could be drawn upon [43–46]. Such an advance could provide a powerful tool for fitting higher-order spatiotemporal statistics to neural circuits, and help to illuminate the structure of collective neural activity.

4. Materials and Methods

4.1. Ground Truth Models

We simulated ground-truth pairwise maximum entropy models for $N = 20$ neurons. Throughout, we assumed time bins of 20 ms for the spiking patterns. To test the performance of the Reliable Moment (RM) model on data without any higher-order interactions, we first assumed a pairwise maximum entropy distribution of spiking patterns with random, normally distributed first and second-order interaction parameters: $h_i \sim \mathcal{N}(3, 0.25)$, $h_{ij} \sim \mathcal{N}\left(0, \frac{2}{N}\right)$. The metaparameters for the distributions were tuned to give low average firing rates and a broad distribution of correlations (Figure 2a). To calculate probabilities from the ground-truth model, we either calculate the empirical frequency of spiking patterns ("empirical probability", Figure 2d) or else we calculate the exact probability from model parameters ("ground-truth probability", Figure 4). The latter requires an expensive calculation of the partition function.

For cortical-like models with higher-order statistics, we used a technique based on the Dichotomized Gaussian (DG) model [47] to generate spike trains with specified firing rates and spiking correlations. In this case, we drew firing rates from a lognormal distribution with a mean of 4 Hz and standard deviation of 2 Hz, and correlations were normally distributed $\rho_{ij} \sim \mathcal{N}(0.1, 0.05)$. In this case, all probabilities are calculated based on empirical frequency (Figure 5).

4.2. Identification of Reliable Moments

To fit the RM model, we must first identify which moments are greater than p_{min}. This process can be made efficient by taking advantage of the hierarchical arrangement of moments. We first find the set of neurons whose mean firing rates in the training data are greater than threshold:

$$S_1 = \{i \,:\, x_i \geq p_{min}\}.$$

S_1 is the set of first-order interaction parameters. Similarly, the set of kth-order interaction parameters is given by:

$$S_k = \left\{ \{s_1 \cdots s_k\} : \prod_{i=1}^{k} x_{s_i} \geq p_{min} \right\}.$$

The RM model fits the interactions corresponding to all elements in S_k, $k = 1, \ldots, N$. Enumerating all S_k can be computationally expensive as the number of possible interactions increases as $O(N^k)$. Because of the hierarchy of moments, this search can be expedited by only considering the kth-order subsets $\{s_1 \cdots s_k\}$ for which all of their $(k-1)$th-order subsets are elements of S_{k-1}. This determines whether the corresponding moment is above threshold. This step is performed iteratively until $S_k = \varnothing$.

4.3. Model Fitting and Sampling

We fit the interaction parameters for the RM model using Minimum Probability Flow learning [22], which we adapted to accommodate arbitrary spiking interactions. After fitting the model, we used the Good–Turing estimate [23] to estimate the partition function empirically. For each ground-truth model, we fit 20 RM models with threshold parameters varying from $p_{min} = 0.05$ to $p_{min} = 0.001$. Because MPF is not convex (and therefore not guaranteed to converge), it is important to check that the model correlations reproduce the data correlations. To do this, we calculate sample correlations via Gibbs sampling.

The Reliable Interaction (RI) models were fit using the procedure described in [12]. Because spiking pattern frequencies are smaller than the marginal frequencies, we used smaller thresholds for the RI model, ranging from $p_{min} = 5 * 10^{-3}$ to 10^{-5}, as these resulted in similar numbers of fitted parameters in the RM and RI models.

4.4. Dissimilarity Between Empirical Data and Models

Since the RI model is not normalized, the Kullback–Leibler divergence returns incongruent (negative) values. We therefore follow [12] in measuring the dissimilarity between the ground-truth and the model frequencies as:

$$d(P, Q) = \sum_{x \in D} P(x) \left| \log_2 \frac{P(x)}{Q(x)} \right|$$

where $\mathcal{D}$ is the set of all observed spiking patterns in the test data (however, in contrast to [1], we do not exclude spiking patterns that only occurred once).

4.5. Code Availability

All relevant code is available at: https://github.com/caycogajic/Reliable-Moment.

Entropy **2018**, *20*, 489

Author Contributions: N.A.C.G., J.Z., and E.S.B. conceived and designed the experiments; N.A.C.G. performed the computational experiments and analysis; N.A.C.G., J.Z., and E.S.B. wrote the paper.

Acknowledgments: J.Z. gratefully acknowledges the following funding: Google Faculty Research Award, Sloan Research Fellowship, and Canadian Institute for Advanced Research (CIFAR) Azrieli Global Scholar Award. This work was also supported by an NSF Grants 1514743 and 1056125 to E.S.B.

Conflicts of Interest: The authors declare no conflict of interest.

Appendix A

Here we demonstrate why the RI model cannot be normalized, using an intuitive example that can only be described in the maximum entropy formulation in the limit that the interaction parameters $\to -\infty$. However, note that the RI model for the toy example discussed in the main text (described by Equations (3) and (4)) is also unnormalizable in its exact form.

Consider N neurons that never spike: then, $P(00\ldots 0) = 1$, and zero for all other spiking patterns. Informally, this distribution can be described by the limit of the following first-order maximum entropy model:

$$P(x) = \frac{1}{Z} \exp\left\{ \sum_i h_i x_i \right\},$$

as $h_i \to -\infty$. Under the RI model, the partition function is estimated as $\hat{Z} = P(00\ldots 0)^{-1} = 1$. Since this is the only occurring pattern, all interactions are set to 0. As a result, the frequency of any spiking pattern is:

$$P_{RI}(x) = \frac{1}{Z} \exp\{0\} = 1,$$

so that the frequencies sum to 2^N. Although the RI model accurately (in this example, perfectly) fits the most common spiking pattern (silence), it is unnormalized. Furthermore, naive renormalization would result in a probability distribution that is inaccurate for every spiking pattern. This dilemma occurs because $\hat{Z}$ is an accurate (in this example, perfect) estimate of the partition function of the underlying distribution, but not for the model defined by the interaction parameters identified by the RI model.

Appendix B

We have shown that the RM model predicts infers smaller higher-order interaction parameters in a ground-truth pairwise model than the RI model. In principle, this fact could be due to changes in the lower-order terms. In other words, it is possible that the RI higher-order interaction terms are larger in absolute magnitude, but not in relative magnitude as compared to the e.g., pairwise terms. We therefore quantified the average magnitude of fitted higher-order interaction parameters, normalized by the average magnitude of all fitted pairwise terms. However, we still found that the normalized higher-order interaction terms were larger in the RI model than in the RM model (Figure A1a). This is due to the fact that the pairwise interaction terms were similar between the RM and RI models (Figure A1b). Therefore, we conclude that the RM model infers fewer strong, spurious higher-order interactions, even when controlling for differences in the lower-order terms.

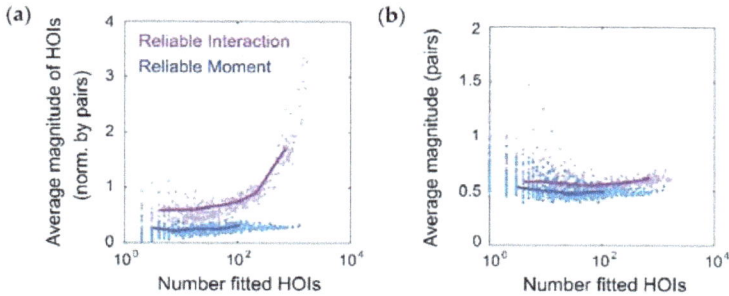

Figure A1. The Reliable Moment (RM) model infers smaller parameters for spurious higher-order interaction terms, relative to lower-order terms. (**a**) Average magnitude of all fitted higher-order interaction parameters normalized by the average magnitude of all pairwise interaction parameters, shown for both the Reliable Interaction (RI; magenta) and RM (blue) models (cf. Figure 3a). (**b**) Average magnitude of all pairwise interaction parameters (RI, magenta; RM, blue).

Appendix C

To test whether the RM model is better able to fit higher-order interactions than the RI model, we augmented the pairwise maximum entropy model with a sparse set of nonzero triplet interaction terms. Specifically, the lower-order terms were generated in the same manner as for the pairwise model (see Methods). Then, each of the ($\begin{smallmatrix} N \\ 3 \end{smallmatrix}$) possible triplet terms was chosen with probability $p = 0.05$, and the corresponding interaction parameters for these triplets (referred to as the "ground-truth triplets") were drawn from a standard normal distribution. We repeated the same fitting protocol as previously described for the pairwise maximum entropy model. The number of ground-truth triplets that were not identified by the RM model was slightly higher than the RI model (mean $\pm$ std, 53.00 ± 0.29, RM model; 46.00 ± 0.37, RI model). However, we found that the inferred interaction parameters for these ground-truth triplets are more accurate for the RM model than the RI model (Figure A2). Therefore, while the RM model misses slightly more ground-truth triplets than the RI model, it more accurately fits their interaction parameters.

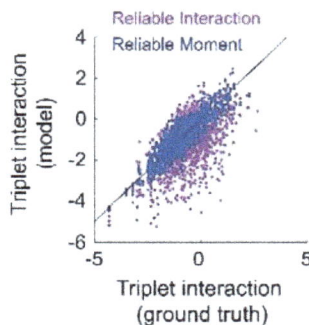

Figure A2. The Reliable Moment (RM) model more accurately fits higher-order interaction parameters of a maximum entropy ground-truth model incorporating a sparse subset of triplet terms. Fitted interaction parameters for ground-truth triplets inferred by the RM model (blue) and the Reliable Interaction model (RI, magenta), plotted against the ground-truth values of the interaction parameters.

References

1. Panzeri, S.; Schultz, S.R.; Treves, A.; Rolls, E.T. Correlations and the encoding of information in the nervous system. *Proc. R. Soc. B Biol. Sci.* **1999**, *266*, 1001–1012. [CrossRef] [PubMed]
2. Nirenberg, S.; Latham, P.E. Decoding neuronal spike trains: How important are correlations? *Proc. Natl. Acad. Sci. USA* **2003**, *100*, 7348–7353. [CrossRef] [PubMed]
3. Averbeck, B.B.; Latham, P.E.; Pouget, A. Neural correlations, population coding and computation. *Nat. Rev. Neurosci.* **2006**, *7*, 358–366. [CrossRef] [PubMed]
4. Schneidman, E.; Berry, M.J.; Segev, R.; Bialek, W. Weak pairwise correlations imply strongly correlated network states in a neural population. *Nature* **2006**, *440*, 1007–1012. [CrossRef] [PubMed]
5. Moreno-Bote, R.; Beck, J.; Kanitscheider, I.; Pitkow, X.; Latham, P.; Pouget, A. Information-limiting correlations. *Nat. Neurosci.* **2014**, *17*, 1410–1417. [CrossRef] [PubMed]
6. Hu, Y.; Zylberberg, J.; Shea-Brown, E. The Sign Rule and Beyond: Boundary Effects, Flexibility, and Noise Correlations in Neural Population Codes. *PLoS Comput. Biol.* **2014**, *10*. [CrossRef] [PubMed]
7. Ohiorhenuan, I.E.; Mechler, F.; Purpura, K.P.; Schmid, A.M.; Hu, Q.; Victor, J.D. Sparse coding and high-order correlations in fine-scale cortical networks. *Nature* **2010**, *466*, 617–621. [CrossRef] [PubMed]
8. Yu, S.; Yang, H.; Nakahara, H.; Santos, G.S.; Nikolic, D.; Plenz, D. Higher-Order Interactions Characterized in Cortical Activity. *J. Neurosci.* **2011**, *31*, 17514–17526. [CrossRef] [PubMed]
9. Shimazaki, H.; Amari, S.; Brown, E.N.; Grün, S. State-space analysis of time-varying higher-order spike correlation for multiple neural spike train data. *PLoS Comput. Biol.* **2012**, *8*. [CrossRef] [PubMed]
10. Köster, U.; Sohl-Dickstein, J.; Gray, C.M.; Olshausen, B.A. Modeling Higher-Order Correlations within Cortical Microcolumns. *PLoS Comput. Biol.* **2014**, *10*. [CrossRef] [PubMed]
11. Shimazaki, H.; Sadeghi, K.; Ishikawa, T.; Ikegaya, Y.; Toyoizumi, T. Simultaneous silence organizes structured higher-order interactions in neural populations. *Sci. Rep.* **2015**, *5*, 9821. [CrossRef] [PubMed]
12. Ganmor, E.; Segev, R.; Schneidman, E. Sparse low-order interaction network underlies a highly correlated and learnable neural population code. *Proc. Natl. Acad. Sci. USA* **2011**, *108*, 9679–9684. [CrossRef] [PubMed]
13. Tkačik, G.; Marre, O.; Amodei, D.; Schneidman, E.; Bialek, W.; Berry, M.J. Searching for Collective Behavior in a Large Network of Sensory Neurons. *PLoS Comput. Biol.* **2014**, *10*. [CrossRef] [PubMed]
14. Cayco-Gajic, N.A.; Zylberberg, J.; Shea-Brown, E. Triplet correlations among similarly tuned cells impact population coding. *Front. Comput. Neurosci.* **2015**, *9*. [CrossRef] [PubMed]
15. Zylberberg, J.; Shea-Brown, E. Input nonlinearities can shape beyond-pairwise correlations and improve information transmission by neural populations. *Phys. Rev. E Stat. Nonlinear Soft Matter Phys.* **2015**, *92*, 062707. [CrossRef] [PubMed]
16. Ganmor, E.; Segev, R.; Schneidman, E. The Architecture of Functional Interaction Networks in the Retina. *J. Neurosci.* **2011**, *31*, 3044–3054. [CrossRef] [PubMed]
17. Tkacik, G.; Mora, T.; Marre, O.; Amodei, D.; Berry, M.J.; Bialek, W. Thermodynamics for a network of neurons: Signatures of criticality. **2014**, *112*, 11508–11513.
18. Berger, A.L.; Pietra, V.J.D.; Pietra, S.A.D. A maximum entropy approach to natural language processing. *Comput. Linguist.* **1996**, *22*, 39–71.
19. Tkacik, G.; Prentice, J.S.; Balasubramanian, V.; Schneidman, E. Optimal population coding by noisy spiking neurons. *Proc. Natl. Acad. Sci. USA* **2010**, *107*, 14419–14424. [CrossRef] [PubMed]
20. Meshulam, L.; Gauthier, J.L.; Brody, C.D.; Tank, D.W.; Bialek, W. Collective Behavior of Place and Non-place Neurons in the Hippocampal Network. *Neuron* **2017**, *96*, 1178–1191. [CrossRef] [PubMed]
21. Jaynes, E.T. Information theory and statistical mechanics. *Phys. Rev.* **1957**, *106*, 620–630. [CrossRef]
22. Sohl-Dickstein, J.; Battaglino, P.B.; Deweese, M.R. New method for parameter estimation in probabilistic models: Minimum probability flow. *Phys. Rev. Lett.* **2011**, *107*, 220601. [CrossRef] [PubMed]
23. Haslinger, R.; Ba, D.; Galuske, R.; Williams, Z.; Pipa, G. Missing mass approximations for the partition function of stimulus driven Ising models. *Front. Comput. Neurosci.* **2013**, *7*, 96. [CrossRef] [PubMed]
24. Darroch, J.N.; Ratcliff, D. Generalized Iterative Scaling for Log-Linear Models. *Ann. Math. Stat.* **1972**, *43*, 1470–1480. [CrossRef]
25. Geman, S.; Geman, D. Stochastic Relaxation, Gibbs Distributions, and the Bayesian Restoration of Images. *IEEE Trans. Pattern Anal. Mach. Intell.* **1984**, *PAMI-6*, 721–741. [CrossRef] [PubMed]

26. Macke, J.H.; Opper, M.; Bethge, M. Common input explains higher-order correlations and entropy in a simple model of neural population activity. *Phys. Rev. Lett.* **2011**, *106*, 208102. [CrossRef] [PubMed]

27. Roxin, A.; Brunel, N.; Hansel, D.; Mongillo, G.; van Vreeswijk, C. On the Distribution of Firing Rates in Networks of Cortical Neurons. *J. Neurosci.* **2011**, *31*, 16217–16226. [CrossRef] [PubMed]

28. Buzsáki, G.; Mizuseki, K. The log-dynamic brain: How skewed distributions affect network operations. *Nat. Rev. Neurosci.* **2014**, *15*, 264–278. [CrossRef] [PubMed]

29. Cohen, M.R.; Kohn, A. Measuring and interpreting neuronal correlations. *Nat. Neurosci.* **2011**, *14*, 811–819. [CrossRef] [PubMed]

30. Ferrari, U. Learning maximum entropy models from finite-size data sets: A fast data-driven algorithm allows sampling from the posterior distribution. *Phys. Rev. E* **2016**, *94*, 023301. [CrossRef] [PubMed]

31. Malouf, R. A comparison of algorithms for maximum entropy parameter estimation. In Proceedings of the 6th Conference on Natural Language Learning, Taipei, Taiwan, 31 August–1 September 2002; Volume 20, pp. 1–7.

32. Broderick, T.; Dudik, M.; Tkacik, G.; Schapire, R.E.; Bialek, W. Faster solutions of the inverse pairwise Ising problem. *arXiv* **2007**, arXiv:0712.2437.

33. Bozdogan, H. Model selection and Akaike's Information Criterion (AIC): The general theory and its analytical extensions. *Psychometrika* **1987**, *52*, 345–370. [CrossRef]

34. Tang, A.; Jackson, D.; Hobbs, J.; Chen, W.; Smith, J.L.; Patel, H.; Prieto, A.; Petrusca, D.; Grivich, M.I.; Sher, A.; et al. A maximum entropy model applied to spatial and temporal correlations from cortical networks in vitro. *J. Neurosci.* **2008**, *28*, 505–518. [CrossRef] [PubMed]

35. Marre, O.; El Boustani, S.; Frégnac, Y.; Destexhe, A. Prediction of spatiotemporal patterns of neural activity from pairwise correlations. *Phys. Rev. Lett.* **2009**, *102*, 138101. [CrossRef] [PubMed]

36. Vasquez, J.C.; Marre, O.; Palacios, A.G.; Berry, M.J.; Cessac, B. Gibbs distribution analysis of temporal correlations structure in retina ganglion cells. *J. Physiol. Paris* **2012**, *106*, 120–127. [CrossRef] [PubMed]

37. Nasser, H.; Cessac, B. Parameter estimation for spatio-temporal maximum entropy distributions application to neural spike trains. *Entropy* **2014**, *16*, 2244–2277. [CrossRef]

38. Herzog, R.; Escobar, M.-J.; Cofre, R.; Palacios, A.G.; Cessac, B. Dimensionality Reduction on Spatio-Temporal Maximum Entropy Models of Spiking Networks. *bioRxiv* **2018**. [CrossRef]

39. Paninski, L.; Pillow, J.; Lewi, J. Statistical models for neural encoding, decoding, and optimal stimulus design. *Prog. Brain Res.* **2007**, *165*, 493–507. [PubMed]

40. Pillow, J.W.; Shlens, J.; Paninski, L.; Sher, A.; Litke, A.M.; Chichilnisky, E.J.; Simoncelli, E.P. Spatio-temporal correlations and visual signalling in a complete neuronal population. *Nature* **2008**, *454*, 995–999. [CrossRef] [PubMed]

41. Granot-Atedgi, E.; Tkačik, G.; Segev, R.; Schneidman, E. Stimulus-dependent Maximum Entropy Models of Neural Population Codes. *PLoS Comput. Biol.* **2013**, *9*. [CrossRef] [PubMed]

42. Vidne, M.; Ahmadian, Y.; Shlens, J.; Pillow, J.W.; Kulkarni, J.; Litke, A.M.; Chichilnisky, E.J.; Simoncelli, E.; Paninski, L. Modeling the impact of common noise inputs on the network activity of retinal ganglion cells. *J. Comput. Neurosci.* **2012**, *33*, 97–121. [CrossRef] [PubMed]

43. Brillinger, D.R.; Bryant, H.L.; Segundo, J.P. Identification of synaptic interactions. *Biol. Cybern.* **1976**, *22*, 213–228. [CrossRef] [PubMed]

44. Krumin, M.; Reutsky, I.; Shoham, S. Correlation-Based Analysis and Generation of Multiple Spike Trains Using Hawkes Models with an Exogenous Input. *Front. Comput. Neurosci.* **2010**, *4*, 147. [CrossRef] [PubMed]

45. Bacry, E.; Muzy, J.F. First- and second-order statistics characterization of hawkes processes and non-parametric estimation. *IEEE Trans. Inf. Theory* **2016**, *62*, 2184–2202. [CrossRef]

46. Etesami, J.; Kiyavash, N.; Zhang, K.; Singhal, K. Learning Network of Multivariate Hawkes Processes: A Time Series Approach. *arXiv* **2016**, arXiv:1603.04319.

47. Macke, J.H.; Berens, P.; Ecker, A.S.; Tolias, A.S.; Bethge, M. Generating spike trains with specified correlation coefficients. *Neural Comput.* **2009**, *21*, 397–423. [CrossRef] [PubMed]

Article

Efficient Algorithms for Searching the Minimum Information Partition in Integrated Information Theory

Jun Kitazono [1,2,*], Ryota Kanai [1] and Masafumi Oizumi [1,3,*]

[1]   Araya, Inc., Toranomon 15 Mori Building, 2-8-10 Toranomon, Minato-ku, Tokyo 105-0001, Japan; kanair@araya.org
[2]   Graduate School of Engineering, Kobe University, 1-1 Rokkodai-cho, Nada-ku, Kobe-shi, Hyogo 657-8501, Japan
[3]   RIKEN Brain Science Institute, 2-1 Hirosawa Wako City, Saitama 351-0198, Japan
*   Correspondence: kitazono@araya.org (J.K.); oizumi@araya.org (M.O.); Tel.: +81-3-6550-9977 (J.K. & M.O.)

Received: 18 December 2017; Accepted: 27 February 2018; Published: 6 March 2018

Abstract: The ability to integrate information in the brain is considered to be an essential property for cognition and consciousness. Integrated Information Theory (IIT) hypothesizes that the amount of integrated information (Φ) in the brain is related to the level of consciousness. IIT proposes that, to quantify information integration in a system as a whole, integrated information should be measured across the partition of the system at which information loss caused by partitioning is minimized, called the Minimum Information Partition (MIP). The computational cost for exhaustively searching for the MIP grows exponentially with system size, making it difficult to apply IIT to real neural data. It has been previously shown that, if a measure of Φ satisfies a mathematical property, submodularity, the MIP can be found in a polynomial order by an optimization algorithm. However, although the first version of Φ is submodular, the later versions are not. In this study, we empirically explore to what extent the algorithm can be applied to the non-submodular measures of Φ by evaluating the accuracy of the algorithm in simulated data and real neural data. We find that the algorithm identifies the MIP in a nearly perfect manner even for the non-submodular measures. Our results show that the algorithm allows us to measure Φ in large systems within a practical amount of time.

Keywords: integrated information theory; integrated information; minimum information partition; submodularity; Queyranne's algorithm; consciousness

1. Introduction

The brain receives various information from the external world. Integrating this information is an essential property for cognition and consciousness [1]. In fact, phenomenologically, our consciousness is unified. For example, when we see an object, we cannot experience only its shape independently of its color. Conversely, we cannot experience only the left half of the visual field independently of the right half. Integrated Information Theory (IIT) of consciousness considers that the unification of consciousness should be realized by the ability of the brain to integrate information [2–4]. That is, the brain has internal mechanisms to integrate information about the shape and color of an object or information of the right and left visual field, and therefore our visual experiences are unified. IIT proposes to quantify the degree of information integration by an information theoretic measure "integrated information" and hypothesizes that integrated information is related to the level of consciousness. Although the hypothesis is indirectly supported by experiments which showed the breakdown of effective connectivity in the brain during loss of consciousness [5,6], only a few studies have directly quantified integrated information in real neural data [7–10] because of the computational difficulties described below.

Conceptually, integrated information quantifies the degree of interaction between parts or equivalently, the amount of information loss caused by splitting a system into parts [11,12]. IIT proposes that integrated information should be quantified between the least interdependent parts so that it quantifies information integration in a system as a whole. For example, if a system consists of two independent subsystems, the two subsystems are the least interdependent parts. In this case, integrated information is 0, because there is no information loss when the system is partitioned into the two independent subsystems. Such a critical partition of the system is called the Minimum Information Partition (MIP), where information is minimally lost, or equivalently where integrated information is minimized. In general, searching for the MIP requires an exponentially large amount of computational time because the number of partitions exponentially grows with the arithmetic growth of system size N. This computational difficulty hinders the application of IIT to experimental data, despite its potential importance in consciousness research and even in broader fields of neuroscience.

In the present study, we exploit a mathematical concept called submodularity to resolve the combinatorial explosion of finding the MIP. Submodularity is an important concept in set functions which is analogous to convexity in continuous functions. It is known that an exponentially large computational cost for minimizing an objective function is reduced to the polynomial order if the objective function satisfies submodularity. Previously, Hidaka and Oizumi showed that the computational cost for finding the MIP is reduced to $O(N^3)$ [13] by utilizing Queyranne's submodular optimization algorithm [14]. They used mutual information as a measure of integrated information that satisfies submodularity. The measure of integrated information used in the first version of IIT (IIT 1.0) [2] is based on mutual information. Thus, if we consider mutual information as a practical approximation of the measure of integrated information in IIT 1.0, Queyranne's algorithm can be utilized for finding the MIP. However, the practical measures of integrated information in the later versions of IIT [12,15–17] are not submodular.

In this paper, we aim to extend the applicability of submodular optimization to non-submodular measures of integrated information. We specifically consider the three measures of integrated information: mutual information Φ_{MI} [2], stochastic interaction Φ_{SI} [15,18,19], and geometric integrated information Φ_G [12]. Mutual information is strictly submodular but the others are not. Oizumi et al. previously showed a close relationship among these three measures [12,20]. From this relationship, we speculate that Queyranne's algorithm might work well for the non-submodular measures. Here, we empirically explore to what extent Queyranne's algorithm can be applied to the two non-submodular measures of integrated information by evaluating the accuracy of the algorithm in simulated data and real neural data. We find that Queyranne's algorithm identifies the MIP in a nearly perfect manner even for the non-submodular measures. Our results show that Queyranne's algorithm can be utilized even for non-submodular measures of integrated information and makes it possible to practically compute integrated information across the MIP in real neural data, such as multi-unit recordings used in Electroencephalography (EEG) and Electrocorticography (ECoG), which typically consist of around 100 channels. Although the MIP was originally proposed in IIT for understanding consciousness, it can be utilized to analyze any system irrespective of consciousness such as biological networks, multi-agent systems, and oscillator networks. Therefore, our work would be beneficial not only for consciousness studies but also to other research fields involving complex networks of random variables.

This paper is organized as follows. We first explain that the three measures of integrated information, Φ_{MI}, Φ_{SI}, and Φ_G, are closely related from a unified theoretical framework [12,20] and there is an order relation among the three measures: $\Phi_{MI} \geq \Phi_{SI} \geq \Phi_G$. Next, we compare the partition found by Queyranne's algorithm with the MIP found by exhaustive search in randomly generated small networks ($N = 14$). We also evaluate the performance of Queyranne's algorithm in larger networks ($N \sim 20$ and 50 for Φ_{SI} and Φ_G, respectively). Since the exhaustive search is intractable, we compare Queyranne's algorithm with a different optimization algorithm called the replica exchange

Markov Chain Monte Carlo (REMCMC) method [21–24]. Finally, we evaluate the performance of Queyranne's algorithm in ECoG data recorded in monkeys and investigate the applicability of the algorithm in real neural data.

2. Measures of Integrated Information

Let us consider a stochastic dynamical system consisting of N elements. We represent the past and present states of the system as $X = (X_1, \ldots, X_N)$ and $X' = (X'_1, \ldots, X'_N)$, respectively. In the case of a neural system, the variable X can be signals of multi-unit recordings, EEG, ECoG, functional magnetic resonance imaging (fMRI), etc. Conceptually, integrated information is designed to quantify the degree of spatio-temporal interactions between subsystems. The previously proposed measures of integrated information are generally expressed as the Kullback–Leibler divergence between the actual probability distribution $p(X, X')$ and a "disconnected" probability distribution $q(X, X')$ where interactions between subsystems are removed [12].

$$\Phi = \min_q D_{KL}\left(p\left(X, X'\right) || q\left(X, X'\right)\right), \tag{1}$$

$$= \min_q \sum_{x,x'} p\left(x, x'\right) \log \frac{p\left(x, x'\right)}{q\left(x, x'\right)}. \tag{2}$$

The Kullback–Leibler divergence measures the difference between the probability distributions, and can be interpreted as the information loss when $q(X, X')$ is used to approximate $p(X, X')$ [25]. Thus, integrated information is interpreted as information loss caused by removing interactions. In Equation (2), the minimum over q should be taken to find the best approximation of p, while satisfying the constraint that the interactions between subsystems are removed [12].

There are many ways of removing interactions between units, which lead to different disconnected probability distributions q, and also different measures of integrated information (Figure 1). The arrows indicate influences across different time points and the lines without arrowheads indicate influences between elements at the same time. Below, we will show that three different measures of integrated information are derived from different probability distributions q.

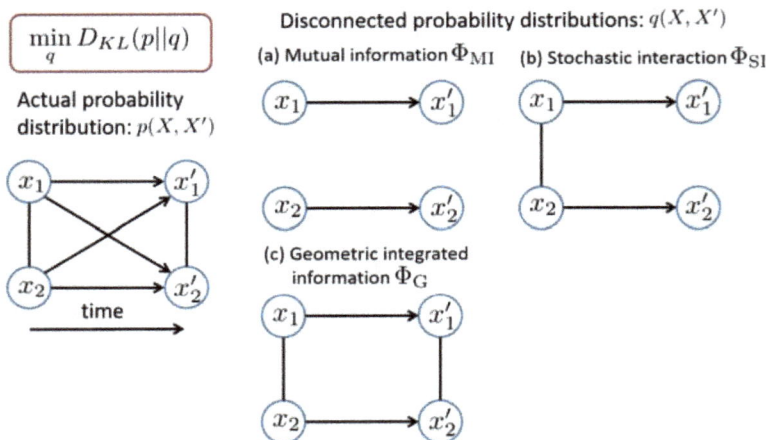

Figure 1. Measures of integrated information represented by the Kullback–Leibler divergence between the actual distribution p and q: (**a**) mutual information; (**b**) stochastic interaction; and (**c**) geometric integrated information. The arrows indicate influences across different time points and the lines without arrowheads indicate influences between elements at the same time. This figure is modified from [12].

2.1. Multi (Mutual) Information Φ_{MI}

First, consider the following partitioned probability distribution q,

$$q\left(X, X'\right) = \prod_{i=1}^{K} q\left(M_i, M_i'\right), \tag{3}$$

where the whole system is partitioned into K subsystems and the past and present states of the i-th subsystem are denoted by M_i and M_i', respectively, i.e., $X = (M_1, \ldots, M_K)$ and $X' = (M_1', \ldots, M_K')$. Each subsystem consists of one or multiple elements. The distribution $q\left(M_i, M_i'\right)$ is the marginalized distribution

$$q\left(M_i, M_i'\right) = \sum_{X \backslash M_i, X' \backslash M_i'} q(X, X'), \tag{4}$$

where $X \backslash M_i$ and $X' \backslash M_i'$ are the complement of M_i and M_i', that is, $(M_1, \ldots, M_{i-1}, M_{i+1}, \ldots, M_K)$ and $(M_1', \ldots, M_{i-1}', M_{i+1}', \ldots, M_K')$, respectively. In this model, all of the interactions between the subsystems are removed, i.e., the subsystems are totally independent (Figure 1a). In this case, the corresponding measure of integrated information is given by

$$\Phi_{MI} = \sum_i H(M_i, M_i') - H(X, X'), \tag{5}$$

where $H(\cdot, \cdot)$ represents the joint entropy. This measure is called total correlation [26] or multi information [27]. As a special case when the number of subsystems is two, this measure is simply equivalent to the mutual information between the two subsystems,

$$\Phi_{MI} = H(M_1, M_1') + H(M_2, M_2') - H(X, X'). \tag{6}$$

The measure of integrated information used in the first version of IIT is based on mutual information but is not identical to mutual information in Equation (6). The critical difference is that the measures in IIT are based on perturbation and those considered in this study are based on observation. In IIT, a perturbational approach is used for evaluating probability distributions, which attempts to quantify actual causation by perturbing a system into all possible states [2,4,11,28]. The perturbational approach requires full knowledge of the physical mechanisms of a system, i.e., how the system behaves in response to all possible perturbations. The measure defined in Equation (6) is based on an observational probability distribution that can be estimated from empirical data. Since we aim for the empirical application of our method, we do not consider the perturbational approach in this study.

2.2. Stochastic Interaction Φ_{SI}

Second, consider the following partitioned probability distribution q,

$$q\left(X'|X\right) = \prod_i q\left(M_i'|M_i\right), \tag{7}$$

which partitions the transition probability from the past X to the present X' in the whole system into the product of the transition probability in each subsystem. This corresponds to removing the causal influences from M_i to M_j' ($j \neq i$) as well as the equal time influences at present between M_i' and M_j' ($j \neq i$) (Figure 1b). In this case, the corresponding measure of integrated information is given by

$$\Phi_{\text{SI}} = \sum_i H(M_i'|M_i) - H(X'|X), \tag{8}$$

where $H(\cdot|\cdot)$ indicates the conditional entropy. This measure was proposed as a practical measure of integrated information by Barrett and Seth [15] following the measure proposed in the second version of IIT (IIT 2.0) [11]. This measure was also independently derived by Ay as a measure of complexity [18,19].

2.3. Geometric Integrated Information Φ_{G}

Aiming at only the causal influences between parts, Oizumi et al. [12] proposed to measure integrated information with the probability distribution that satisfies

$$q\left(M_i'|X\right) = q\left(M_i'|M_i\right), \forall i \tag{9}$$

which means the present state of a subsystem i, M_i' only depends on its past state M_i. This corresponds to removing only the causal influences between subsystems while retaining the equal-time interactions between them (Figure 1c). The constraint Equation (9) is equivalent to the Markov condition

$$q(M_i', M_i^c|M_i) = q(M_i'|M_i)q(M_i^c|M_i), \forall i \tag{10}$$

where M_i^c is the complement of M_i, that is, $M_i^c = X \backslash M_i$. This means when M_i is given, M_i' and M_i^c are independent. In other words, the causal interaction between M_i^c and M_i' is only via M_i.

There is no closed-form expression for this measure in general. However, if the probability distributions are Gaussian, we can analytically solve the minimization over q (see Appendix A).

3. Minimum Information Partition

In this section, we provide the mathematical definition of Minimum Information Partition (MIP). Then, we formulate the search for MIP as an optimization problem of a set function. The MIP is the partition that divides a system into the least interdependent subsystems so that information loss caused by removing interactions among the subsystems is minimized. The information loss is quantified by the measure of integrated information. Thus, the MIP, π_{MIP}, is defined as a partition (since the minimizer is not necessarily unique, strictly speaking, there could be multiple MIPs), where integrated information is minimized:

$$\pi_{\text{MIP}} := \arg\min_{\pi \in \mathcal{P}} \Phi(\pi), \tag{11}$$

where $\mathcal{P}$ is a set of partitions. In general, $\mathcal{P}$ is the universal set of partitions, including bi-partitions, tri-partitions, and so on. In this study, however, we focus only on bi-partitions for simplicity and computational time. Note that, although Queyranne's algorithm [14] is limited to bi-partitions, the algorithm can be extended to higher-order partitions [13]. See Section 7 for more details. By a bi-partition, a whole system Ω is divided into a subset S ($S \subset \Omega, S \neq \varnothing$) and its complement $\bar{S} = \Omega \backslash S$. Since a bi-partition is uniquely determined by specifying a subset S, integrated information can be considered as a function of a set S, $\Phi(S)$. Finding the MIP is equivalent to finding the subset, S_{MIP}, that achieves the minimum of integrated information:

$$S_{\text{MIP}} := \arg\min_{S \subset \Omega, S \neq \varnothing} \Phi(S). \tag{12}$$

In this way, the search of the MIP is formulated as an optimization problem of a set function.

Since the number of bi-partitions for the system with N-elements is $2^{N-1} - 1$, exhaustive search of the MIP in a large system is intractable. However, by formulating the MIP search as an optimization of a set function as above, we can take advantage of a discrete optimization technique and can reduce computational costs to a polynomial order, as described in the next section.

4. Submodular Optimization

The submodularity is an important concept in set functions, which is an analogue of convexity in continuous functions [29]. When objective functions are submodular, efficient algorithms are available for solving optimization problems. In particular, for symmetric submodular functions, there is a well-known algorithm by Queyranne which minimizes them [14]. We utilize this method for finding the MIP in this study.

4.1. Submodularity

Mathematically, the submodularity is defined as follows.

Definition 1 (Submodularrity). *Let Ω be a finite set and 2^Ω its power set. A set function $f : 2^\Omega \to \mathbb{R}$ is submodular if it satisfies the following inequality for any $S, T \subseteq \Omega$:*

$$f(S) + f(T) \geq f(S \cup T) + f(S \cap T).$$

Equivalently, a set function $f : 2^\Omega \to \mathbb{R}$ is submodular if it satisfies the following inequality for any $S, T \subseteq \Omega$ with $S \subseteq T$ and for any $u \in \Omega \setminus T$:

$$f(S \cup \{u\}) - f(S) \geq f(T \cup \{u\}) - f(T).$$

The second inequality means that the function increases more when an element is added to a smaller subset than when the element is added to a bigger subset.

4.2. Queyranne's Algorithm

A set function f is called symmetric if $f(S) = f(\Omega \setminus S)$ for any $S \subseteq \Omega$. Integrated information $\Phi(S)$ computed by bi-partition is a symmetric function, because S and $\Omega \setminus S$ specifies the same bi-partition. If a function is symmetric and submodular, we can find the minimum of the function by Queyranne's algorithm with $O(N^3)$ function calls [14].

4.3. Submodularity in Measures of Integrated Information

In a previous study, Queyranne's algorithm was utilized to find the MIP when Φ_{MI} is used as the measure of integrated information [13]. As shown previously, Φ_{MI} is submodular [13]. However, the other measures of integrated information are not submodular. In this study, we apply Queyranne's algorithm to non-submodular functions, Φ_{SI} and Φ_{G}. When the objective functions are not submodular, Queyranne's algorithm does not necessarily find the MIP. We evaluate how accurately Queyranne's algorithm can find the MIP when it is used for non-submodular measures of integrated information. There is an order relation among the three measures of integrated information [12],

$$\Phi_{\text{G}} \leq \Phi_{\text{SI}} \leq \Phi_{\text{MI}}. \tag{13}$$

This inequality can be graphically understood from Figure 1. The more the connections are removed, the larger the corresponding integrated information (the information loss) is. That is, Φ_{G} measures only the causal influences between subsystems, Φ_{SI} measures the equal-time interactions between the present states as well as the causal influences between subsystems, and Φ_{MI} measures all the interactions between the subsystems. Thus, Φ_{SI} is closer to Φ_{MI} than Φ_{G} is. This relationship implies that Φ_{SI} would behave more similarly to a submodular measure Φ_{MI} than Φ_{G} does. Thus, one may surmise that Queyranne's algorithm would work more accurately for Φ_{SI} than for Φ_{G}. As we will show in Section 6.2, this is indeed the case. However, the difference is rather small because Queyranne's algorithm works almost perfectly for both measures, Φ_{SI} and Φ_{G}.

5. Replica Exchange Markov Chain Monte Carlo Method

To evaluate the accuracy of Queyranne's algorithm, we compare the partition found by Queyranne's algorithm with the MIP found by the exhaustive search when the number of elements n is small enough ($n \lesssim 20$). However, when n is large, we cannot know the MIP because the exhaustive search is unfeasible. To evaluate the performance of Queyranne's algorithm in a large system, we compare it with a different method, the Replica Exchange Markov Chain Monte Carlo (REMCMC) method [21–24]. REMCMC, also known as parallel tempering, is a method to draw samples from probability distributions. REMCMC is an improved version of the MCMC methods. Here, we briefly explain how the MIP search problem is represented as a problem of drawing samples from a probability distribution. Details of the REMCMC method are given in Appendix B.

Let us define a probability distribution $p(S; \beta)$ using integrated information $\Phi(S)$ as follows:

$$p(S; \beta) \propto \exp(-\beta\Phi(S)), \tag{14}$$

where $\beta(> 0)$ is a parameter called inverse temperature. This probability is higher/lower when $\Phi(S)$ is smaller/larger. The MIP gives the highest probability by definition. If we can draw samples from this distribution, we can selectively scan subsets with low integrated information and efficiently find the MIP, compared to randomly exploring partitions independent of the value of integrated information. Simple MCMC methods such as the Metropolis method, which draw samples from Equation (14) with a single value of β, often suffer from the problem of slow convergence. That is, a sample sequence is trapped in a local minimum and the sample distribution takes time to converge to the target distribution. REMCMC aims at overcoming this problem by drawing samples in parallel from distributions with multiple values of β and by continually exchanging the sampled sequences between neighboring β (see Appendix B for more details).

6. Results

We first evaluated the performance of Queyranne's algorithm in simulated networks. Throughout the simulations below, we consider the case where the variable X obeys a Gaussian distribution for the ease of computation. As shown in Appendix A, the measures of integrated information, Φ_{SI} and Φ_{G} can be analytically computed. Note that, although Φ_{SI} and Φ_{G} can be computed in principle even when the distribution of X is not Gaussian, it is practically very hard to compute them in large systems because the computation of Φ involves summation over all possible X. Specifically, we consider the first order autoregressive (AR) model,

$$X' = AX + E, \tag{15}$$

where X and X' are present states and past states of a system, A is the connectivity matrix, and E is Gaussian noise. The stationary distribution of this AR model is considered. The stationary distribution of $p(X, X')$ is a Gaussian distribution. The covariance matrix of $p(X, X')$ consists of covariance of X, $\Sigma(X)$, and cross-covariance of X and X', $\Sigma(X, X')$. $\Sigma(X)$ is computed by solving the following equation,

$$\Sigma(X) = A\Sigma(X)A^T + \Sigma(E). \tag{16}$$

$\Sigma(X, X')$ is given by

$$\Sigma(X, X') = \Sigma(X)A^T. \tag{17}$$

By using these covariance matrices, Φ_{SI} and Φ_{G} are analytically calculated [12] (see Appendix A). The details of the parameter settings are described in each subsection.

6.1. Speed of Queyranne's Algorithm Compared With Exhaustive Search

We first evaluated the computational time of the search using Queyranne's algorithm and compared it with that of the exhaustive search when the number of elements N changed. The connectivity matrices

A were randomly generated. Each element of the connection matrix A was sampled from a normal distribution with mean 0 and variance $0.01/N$. The covariance of Gaussian noise E was generated from a Wishart distribution $\mathcal{W}(\sigma I, 2N)$ with covariance σI and degrees of freedom $2N$, where σ corresponded to the amount of noise E and I was the identity matrix. The Wishart distribution is a standard distribution for symmetric positive-semidefinite matrices [30,31]. Typically, the distribution is used to generate covariance matrices and inverse covariance (precision) matrices. For more practical details, see for example, Ref. [31]. We set σ to 0.1. The number of elements N was changed from 3 to 60. All computation times were measured on a machine with an Intel Xeon Processor E5-2680 at 2.70GHz. All the calculations were implemented in MATLAB R2014b.

We fitted the computational time of the search using Queyranne's algorithm for Φ_{SI} and Φ_G with straight lines, although the computational time for large N is a little deviated from the straight lines (Figure 2a,b). In Figure 2a, the red circles, which indicate the computational time of the search using Queyranne's algorithm for Φ_{SI}, are roughly approximated by the red solid line, $\log_{10} T = 3.066 \log_{10} N - 3.838$. In contrast, the black triangles, which indicate those of the exhaustive search, are fit by the black dashed line, $\log_{10} T = 0.2853N - 3.468$. This means that the computational time of the search using Queyranne's algorithm increases in polynomial order ($T \propto N^{3.066}$), while that of the exhaustive search exponentially increases ($T \propto 1.929^N$). For example, when $N = 100$, Queyranne's algorithm takes $\sim$197 s while the exhaustive search takes 1.16×10^{25} s. This is in practice impossible to compute even with a supercomputer. Similarly, as shown in Figure 2b, when Φ_G is used, the search using Queyranne's algorithm roughly takes $T \propto N^{4.776}$ while the exhaustive search takes $T \propto 2.057^N$. Note that the complexity of the search using Queyranne's algorithm for Φ_G ($O(N^{4.776})$) is much higher than that of Queyranne's algorithm itself ($O(N^3)$). This is because the multi-dimensional equations (Equations (A20) and (A21)) need to be solved by using an iterative method to compute Φ_G (see Appendix A).

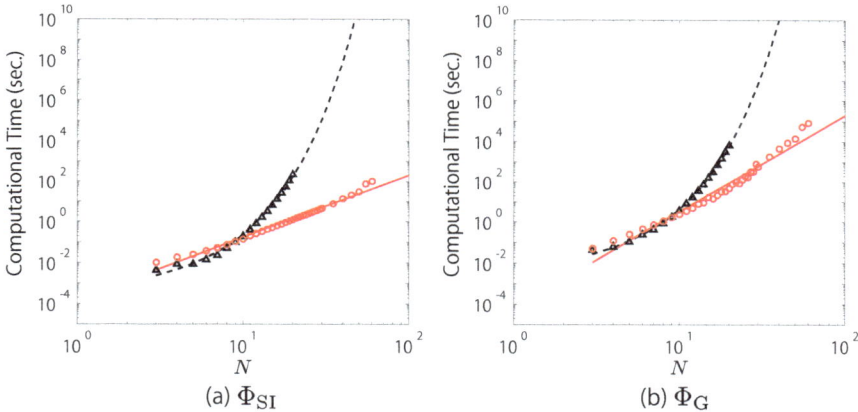

Figure 2. Computational time of the search using Queyranne's algorithm and the exhaustive search. The red circles and the red solid lines indicate the computational time of the search using Queyranne's algorithm and their approximate curves ((**a**) $\log_{10} T = 3.066 \log_{10} N - 3.838$, (**b**) $\log_{10} T = 4.776 \log_{10} N - 4.255$). The black triangles and the black dashed lines indicate the computational time of the exhaustive search and their approximate curves ((**a**) $\log_{10} T = 0.2853N - 3.468$, (**b**) $\log_{10} T = 0.3132N - 2.496$).

6.2. Accuracy of Queyranne's Algorithm

We evaluated the accuracy of Queyranne's algorithm by comparing the partition found by Queyranne's algorithm with the MIP found by exhaustive search. We used Φ_{SI} and Φ_G as the measures of integrated information. We considered two different architectures in connectivity matrix A of AR models. The first one was just a random matrix: Each element of A was randomly sampled from a normal distribution with mean 0 and variance $0.01/N$. The other one was a block matrix consisting of $N/2$ by $N/2$ sub-matrices, $A_{ij}(i, j = 1, 2)$. Each element of diagonal sub-matrices A_{11} and A_{22} was drawn from a normal distribution with mean 0 and variance $0.02/N$. Off-diagonal sub-matrices A_{12} and A_{21} were zero matrices. The covariance of Gaussian noise E in the AR model was generated from a Wishart distribution $\mathcal{W}(\sigma I, 2N)$. The parameter σ was set to 0.1 or 0.01. The number of elements N was set to 14. We randomly generated 100 connectivity matrices A and $\Sigma(E)$ for each setting and evaluated performance using the following four measures. The following measures are averaged over 100 trials:

- **Correct rate (CR):** Correct rate (CR) is the rate of correctly finding the MIP.
- **Rank (RA):** Rank (RA) is the rank of the partition found by Queyranne's algorithm among all possible partitions. The rank is based on the Φ values computed at each partition. The partition that gives the lowest Φ is rank 1. The highest rank is equal to the number of possible bi-partitions, 2^{N-1}.
- **Error ratio (ER):** Error ratio (ER) is the deviation of the value of integrated information computed across the partition found by Queyranne's algorithm from that computed across the MIP, which is normalized by the mean error computed at all possible partitions. Error ratio is defined by

$$\text{Error Ratio} = \frac{\Phi_Q - \Phi_{\text{MIP}}}{\bar{\Phi} - \Phi_{\text{MIP}}}, \tag{18}$$

where Φ_{MIP}, Φ_Q, and $\bar{\Phi}$ are the amount of integrated information computed across the MIP, that computed across the partition found by Queyranne's algorithm, and the mean of the amounts of integrated information computed across all possible partitions, respectively.

- **Correlation (CORR):** Correlation (CORR) is the correlation between the partition found by Queyranne's algorithm and the MIP found by the exhaustive search. Let us represent a bi-partition of N-elements as an N-dimensional vector $\sigma = (\sigma_1, \ldots, \sigma_N) \in \{-1, 1\}^N$, where ± 1 indicates one of the two subgroups. The absolute value of the correlation between the vector given by the MIP (σ^{MIP}) and that given by the partition found by Queyranne's algorithm (σ^Q) is computed:

$$|\text{corr}(\sigma^{\text{MIP}}, \sigma^Q)| = \left| \frac{\sum_{i=1}^N (\sigma_i^{\text{MIP}} - \bar{\sigma}^{\text{MIP}})(\sigma_i^Q - \bar{\sigma}^Q)}{\sqrt{\sum_{i=1}^N (\sigma_i^{\text{MIP}} - \bar{\sigma}^{\text{MIP}})^2 \sum_{i=1}^N (\sigma_i^Q - \bar{\sigma}^Q)^2}} \right|, \tag{19}$$

where $\bar{\sigma}^{\text{MIP}}$ and $\bar{\sigma}^Q$ are the means of σ_i^{MIP} and σ_i^Q, respectively.

The results are summarized in Table 1. This table shows that, when Φ_{SI} was used, Queyranne's algorithm perfectly found the MIPs for all 100 trials, even though Φ_{SI} is not strictly submodular. Similarly, when Φ_G was used, Queyranne's algorithm almost perfectly found the MIPs. The correct rate was 100% for the normal models and 97% for the block structured models. Additionally, even when the algorithm missed the MIP, the rank of the partition found by the algorithm was 2 or 3. The averaged rank over 100 trials were 1.03 and 1.05 for the block structured models. In addition, the error ratio in error trials were around 0.1 and the average error ratios were very small. See Appendix C for box plots of the values of the integrated information at all the partitions. Thus, such miss trials would not affect evaluation of the amount of integrated information in practice. However, in terms of partitions, the partitions found by Queyranne's algorithm in error trials were markedly different from the MIPs. In the block structured model, the MIP for Φ_G was the partition that split the system in halves. In contrast, the partitions found by Queyranne's algorithm were one-vs-all partitions.

Table 1. Accuracy of Queyranne's algorithm.

Model		Φ_{SI}				Φ_{G}			
A	σ	CR	RA	ER	CORR	CR	RA	ER	CORR
Normal	0.01	100%	1	0	1	100%	1	0	1
	0.1	100%	1	0	1	100%	1	0	1
Block	0.01	100%	1	0	1	97%	1.05	2.38×10^{-3}	0.978
	0.1	100%	1	0	1	97%	1.03	9.11×10^{-4}	0.978

In summary, Queyranne's algorithm perfectly worked for Φ_{SI}. With regards to Φ_{G}, although Queyranne's algorithm almost perfectly evaluated the amount of integrated information, we may need to treat partitions found by the algorithm carefully. This slight difference in performance between Φ_{SI} and Φ_{G} can be explained by the order relation in Equation (13). Φ_{SI} is closer to the strictly submodular function Φ_{MI} than Φ_{G} is, which we consider to be why Queyranne's algorithm worked better for Φ_{SI} than Φ_{G}.

6.3. Comparison between Queyranne's Algorithm and REMCMC

We evaluated the performance of Queyranne's algorithm in large systems where an exhaustive search is impossible. We compared it with the Replica Exchange Markov Chain Monte Carlo Method (REMCMC). We applied the two algorithms to AR models generated similarly as in the previous section. The number of elements was 50 for Φ_{SI} and 20 for Φ_{G}, respectively. The reason for the difference in N is because Φ_{G} requires much heavier computation than Φ_{SI} (see Appendix A). We randomly generated 20 connectivity matrices A and $\Sigma(E)$ for each setting. We compared the two algorithms in terms of the amount of integrated information and the number of evaluations of Φ. REMCMC was run until a convergence criterion was satisfied. See Appendix B.3 for details of the convergence criterion.

The results are shown in Tables 2 and 3. "Winning percentage" indicates the fraction of trials each algorithm won in terms of the amount of integrated information at the partition found by each algorithm. We can see that the partitions found by the two algorithms exactly matched for all the trials. We consider that the algorithms probably found the MIPs for the following three reasons. First, it is well known that REMCMC can find a minima if it is run for a sufficiently long time in many applications [24,32–34]. Second, the two algorithms are so different that it is unlikely that they both incorrectly identified the same partitions as the MIPs. Third, Queyranne's algorithm successfully finds the MIPs in smaller systems as shown in the previous section. This fact suggests that Queyranne's algorithm worked well also for the larger systems. Note that, in the case of Φ_{G}, the half-and-half partition is the MIP in the block structured model because $\Phi_{\mathrm{G}} = 0$ under the half-and-half partition. We confirmed that the partitions found by Queyanne's algorithm and REMCMC were both the half-and-half partition for all the 20 trials. Thus, in the block structured case, it is certain that the true MIPs were successfully found by both algorithms.

Table 2. Comparison of Queyranne's algorithm with REMCMC (Φ_{SI}, $N = 50$).

Model		Winning Percentage			Number of Evaluations of Φ		
						REMCMC (Mean $\pm$ std)	
A	σ	Queyranne's	Even	REMCMC	Queyranne's	Converged	Solution Found
Normal	0.01	0%	100%	0%	41,699	$274,257 \pm 107,969$	8172.6 ± 6291.0
	0.1	0%	100%	0%	41,699	$315,050 \pm 112,205$	9084.9 ± 7676.4
Block	0.01	0%	100%	0%	41,699	$308,976 \pm 110,905$	7305.6 ± 6197.0
	0.1	0%	100%	0%	41,699	$339,869 \pm 154,161$	4533.4 ± 3004.8

Table 3. Comparison of Queyranne's algorithm with REMCMC (Φ_G, $N = 20$).

Model		Winning Percentage			Number of Evaluations of Φ		
						REMCMC (Mean ± std)	
A	σ	Queyranne's	Even	REMCMC	Queyranne's	Converged	Solution Found
Normal	0.01	0%	100%	0%	2679	136,271 ± 46,624	862.4 ± 776.3
	0.1	0%	100%	0%	2679	122,202 ± 46,795	894.3 ± 780.2
Block	0.01	0%	100%	0%	2679	129,770 ± 88,483	245.2 ± 194.3
	0.1	0%	100%	0%	2679	146,034 ± 61,880	443.2 ± 642.1

We also evaluated the number of evaluations of Φ in both algorithms before the end of the computational processes. In our simulations, the computational process of Queyranne's algorithm ended much faster than the convergence of REMCMC. Queyranne's algorithm ends at a fixed number of evaluations of Φ depending only on N. In contrast, the number of the evaluations before the convergence of REMCMC depends on many factors such as the network models, the initial conditions, and pseudo random number sequences. Thus, the time of convergence varies among different trials. Note that, by "retrospectively" examining the sequence of the Monte Carlo search, the solutions turned out to be found at earlier points of the Monte Carlo searches than Queyranne's algorithm (which are indicated as "solution found" in Tables 2 and 3). However, it is impossible to stop the REMCMC algorithm at these points where the solutions were found because there is no way to tell whether these points reach the solution until the algorithm is run for enough amount of time.

6.4. Evaluation with Real Neural Data

Finally, to ensure the applicability of Queyranne's algorithm to real neural data, we similarly evaluated the performance with electrocorticography (ECoG) data recorded in a macaque monkey. The dataset is available at an open database, Neurotycho.org (http://neurotycho.org/) [35]. One hundred twenty-eight channel ECoG electrodes were implanted in the left hemisphere. The electrodes were placed at 5 mm intervals, covering the frontal, parietal, temporal, and occipital lobes, and medial frontal and parietal walls. Signals were sampled at a rate of 1k Hz and down-sampled to 100 Hz for the analysis. The monkey "Chibi" was awake with the eyes covered by an eye-mask to restrain visual responses. To remove line noise and artifacts, we performed bipolar re-referencing between nearest neighbor electrode pairs. The number of re-referenced electrodes was 64 in total.

In the first simulation, we evaluated the accuracy. We extracted a 1 min length of the signals of the 64 electrodes. Each 1 min sequence consists of 100 Hz $\times$ 60 s = 6000 samples. Then, we randomly selected 14 electrodes 100 times. We approximated the probability distribution of the signals with multivariate Gaussian distributions. The covariance matrices were computed with a time window of 1 min and a time step of 10 ms. We applied the algorithms to the 100 randomly selected sets of electrodes and measured the accuracy similarly as in Section 6.2. The results are summarized in Table 4. We can see that Queyranne's algorithm worked perfectly for both Φ_{SI} and Φ_G.

Next, we compared Queyranne's algorithm with REMCMC. We applied the two algorithms to the 64 re-referenced signals, and evaluated the performance in terms of the amount of integrated information and the number of evaluations of Φ, as in Section 6.3. We segmented 15 non-overlapping sequences of 1 min each, and computed covariance matrices with a time step of 10 ms. We measured the average performance over the 15 sets. Here, we only used Φ_{SI}, because Φ_G requires heavy computations for 64 dimensional systems. The results are shown in Table 5. We can see that the partitions selected by the two algorithms matched for all 15 sequences. In terms of the amount of computation, Queyranne's algorithm ended much faster than the convergence of REMCMC.

Table 4. Accuracy of Queyranne's algorithm in ECoG data. Randomly-selected 14 electrodes were used.

Φ_{SI}				Φ_G			
CR	RA	ER	CORR	CR	RA	ER	CORR
100%	1	0	1	100%	1	0	1

Table 5. Comparison of Queyranne's algorithm with REMCMC in ECoG data (SI).

Winning Percentage			Number of Evaluations of Φ		
				REMCMC (Mean $\pm$ std)	
Queyranne's	Even	REMCMC	Queyranne's	Converged	Solution Found
0%	100%	0%	87,423	607,797 $\pm$ 410,588	15,859 $\pm$ 10,497

7. Discussion

In this study, we proposed an efficient algorithm for searching for the Minimum Information Partition (MIP) in Integrated Information Theory (IIT). The computational time of an exhaustive search for the MIP grows exponentially with the arithmetic growth of system size, which has been an obstacle to applying IIT to experimental data. We showed here that by using a submodular optimization algorithm called Queyranne's algorithm, the computational time was reduced to $O(N^{3.066})$ and $O(N^{4.776})$ for stochastic interaction Φ_{SI} and geometric integrated information Φ_G, respectively. These two measures of integrated information are non-submodular, and thus it is not theoretically guaranteed that Queyranne's algorithm will find the MIP. We empirically evaluated the accuracy of the algorithm by comparing it with an exhaustive search in simulated data and in ECoG data recorded from monkeys. We found that Queyranne's algorithm worked perfectly for Φ_{SI} and almost perfectly for Φ_G. We also tested the performance of Queyranne's algorithm in larger systems ($N = 20$ and 50 for Φ_{SI} and Φ_G, respectively) where the exhaustive search is intractable by comparing it with the Replica Exchange Markov Chain Monte Carlo method (REMCMC). We found that the partitions found by these two algorithms perfectly matched, which suggests that both algorithms most likely found the MIPs. In terms of the computational time, the number of evaluations of Φ taken by Queyranne's algorithm was much smaller than that taken by REMCMC before the convergence. Our results indicate that Queyranne's algorithm can be utilized to effectively estimate MIP even for non-submodular measures of integrated information. Although the MIP is a concept originally proposed in IIT for understanding consciousness, it can be utilized to general network analysis irrespective of consciousness. Thus, the method for searching MIP proposed in this study will be beneficial not only for consciousness studies but for other research fields.

Here, we discuss the pros and cons of Queyranne's algorithm in comparison with REMCMC. Since the partitions found by both algorithms perfectly matched in our experiments, they were equally good in terms of accuracy. With regards to computational time, Queyranne's algorithm ended much faster than the convergence of REMCMC. Thus, Queyranne's algorithm would be a better choice in rather large systems ($N \sim 20$ and 50 for Φ_{SI} and Φ_G, respectively). Note that, if we retrospectively examine the sampling sequence in REMCMC, we find that REMCMC found the partitions much earlier than its convergence and that the estimated MIPs did not change in the later parts of sampling process. Thus, if we could introduce a heuristic criterion to determine when to stop the sampling based on the time course of the estimated MIPs, REMCMC could be stopped earlier than its convergence. However, setting such a heuristic criterion is a non-trivial problem. Queyranne's algorithm ends within a fixed number of function calls regardless of the properties of data. If the system size is much larger ($N \gtrsim 100$), Queyranne's algorithm will be computationally very demanding because of $O(N^3)$ time complexity and may not practically work. In that case, REMCMC would work better if the above-mentioned heuristics are introduced to stop the algorithm earlier than the convergence.

As an alternative interesting approach for approximately finding the MIP, a graph-based algorithm was proposed by Toker and Sommer [36]. In their method, to reduce the search space, candidate partitions are selected by a spectral clustering method based on correlation. Then Φ_G is calculated for those candidate partitions, and the best partition is selected. A difference between our method and theirs is whether the search method is fully based on the values of integrated information or not. Our method uses no other quantities than Φ for searching the MIP, while their method uses a graph theoretic measure, which may significantly differ from Φ in some cases. It would be an interesting future work to compare our method and the graph-theoretic methods or combine these methods to develop better search algorithms.

In this study, we considered the three different measures of integrated information, Φ_{MI}, Φ_{SI}, and Φ_G. Of these, Φ_{MI} is submodular but the other two measures, while Φ_{SI} and Φ_G, are not. As we described in Section 4.3, there is a clear order relation among them (Equation (13)). Φ_{SI} is closer to a submodular function Φ_{MI} than Φ_G is. This relation implies that Queyranne's algorithm would work better for Φ_{SI} than for Φ_G. We found that it was actually the case in our experiments because there were a few error trials for Φ_G whereas there were no miss trials for Φ_{SI}. For the practical use of these measures, we note that there are two major differences among the three measures. One is what they quantify. As shown in Figure 1, Φ_G measures only causal interactions between units across different time points. In contrast, Φ_{SI} and Φ_{MI} also measure equal time interactions as well as causal interactions. Φ_G best follows the original concept of IIT in the sense that it measures only the "causal" interactions. One needs to acknowledge the theoretical difference whenever applying one of these measures in order to correctly interpret the obtained results. The other difference is in computational costs. The computational costs of Φ_{MI} and Φ_{SI} are almost the same while that of Φ_G is much larger, because it requires multi-dimensional optimization. Thus, Φ_G may not be practical for the analysis of large systems. In that case, Φ_{MI} or Φ_{SI} may be used instead with care taken of the theoretical difference.

Although in this study we focused on bi-partitions, Queyranne's algorithm can be extended to higher-order partitions [13]. However, the algorithm becomes computationally demanding for higher-order partitions, because the computational complexity of the algorithm for K-partitions is $O(N^{3(K-1)})$. This is the main reason why we focused on bi-partitions. Another reason is that there has not been an established way to fairly compare partitions with different K. In IIT 2.0, it was proposed that the integrated information should be normalized by the minimum of the entropy of partitioned subsystems [3], while, in IIT 3.0, it was not normalized [4]. Note that, when integrated information is not normalized, the MIP is always found in bi-partitions because integrated information becomes larger when a system is partitioned into more subsystems.

Whether the integrated information should be normalized and how the integrated information should be normalized are still open questions. In our study, the normalization used in IIT 2.0 is not appropriate, because the entropy can be negative for continuous random variables. Additionally, regardless of whether random variables are continuous or discrete, normalization significantly affects the submodularity of the measures of integrated information. For example, if we use normalization proposed in IIT 2.0, even the submodular measure of integrated information, Φ_{MI}, no longer satisfies submodularity. Thus, Queyranne's algorithm may not work well if Φ is normalized.

Although we resolved one of the major computational difficulties in IIT, an additional issue still remains. Searching for the MIP is an intermediate step in identifying the informational core, called the "complex". The complex is the subnetwork in which integrated information is maximized, and is hypothesized to be the locus of consciousness in IIT. Identifying the complex is also represented as a discrete optimization problem which requires exponentially large computational costs. Queyranne's algorithm cannot be applied to the search for the complex because we cannot formulate it as a submodular optimization. We expect that REMCMC would be efficient in searching for the complex and will investigate its performance in a future study.

An important limitation of this study is that we only showed the nearly perfect performance of Queyranne's algorithm in limited simulated data and real neural data. In general, we cannot tell whether Queyranne's algorithm works well for other data beforehand. For real data analysis, we recommend that the procedure below should be applied. First, as we did in Section 6.2, accuracy should be checked by comparing it with the exhaustive search in small randomly selected subsets. Next, if it works well, the performance should be checked by comparing it with REMCMC in relatively large subsets, as we did in Section 6.3. If Queyranne's algorithm works better than or equally as well as REMCMC, it is reasonable to use Queyranne's algorithm for the analysis. By applying this procedure, we expect that Queyranne's algorithm could be utilized to efficiently find the MIP in a wide range of time series data.

Acknowledgments: We thank Shohei Hidaka, Japan Advanced Institute of Science and Technology, for providing us Queyranne's algorithm codes. This work was partially supported by JST CREST Grant Number JPMJCR15E2, Japan.

Author Contributions: Jun Kitazono and Masafumi Oizumi conceived and designed the experiments; Jun Kitazono performed the experiments; Jun Kitazono and Masafumi Oizumi analyzed the data; and Jun Kitazono, Ryota Kanai and Masafumi Oizumi wrote the paper.

Conflicts of Interest: The authors declare no conflict of interest. The founding sponsors had no role in the design of the study; in the collection, analyses, or interpretation of data; in the writing of the manuscript, or in the decision to publish the results.

Abbreviations

The following abbreviations are used in this manuscript:

IIT	integrated information theory
MIP	minimum information partition
MCMC	Markov chain Monte Carlo
REMCMC	replica exchange Markov chain Monte Carlo
EEG	electroencephalography
ECoG	electrocorticography
AR	autoregressive
CR	correct rate
RA	rank
ER	error ratio
CORR	correlation
MCS	Monte Carlo step

Appendix A. Analytical Formula of Φ for Gaussian Variables

We describe the analytical formula of three measures of integrated information, multi information (Φ_{MI}), stochastic interaction (Φ_{SI}) and geometric integrated information (Φ_{G}), when the probability distribution is Gaussian. For more details about the theoretical background, see [12,15,18,19].

First, let us introduce the notation. We consider a stochastic dynamical system consisting of N elements. We represent the past and present states of the system as $X = (X_1, \ldots, X_N)$ and $X' = (X'_1, \ldots, X'_N)$, respectively, and define a joint vector

$$\tilde{X} = (X, X'). \tag{A1}$$

We assume that the joint probability distribution $p(X, X')$ is Gaussian:

$$p(x, x') = \exp\left\{ -\frac{1}{2}\left(\tilde{x}^T \Sigma(\tilde{X})\tilde{x} - \psi \right) \right\}, \tag{A2}$$

where ψ is the normalizing factor and $\Sigma(\tilde{X})$ is the covariance matrix of $\tilde{X}$. Note that we can assume the mean of the Gaussian distribution is zero without loss of generality because the mean value does not affect the values of integrated information. This covariance matrix $\Sigma(\tilde{X})$ is given by

$$\Sigma(\tilde{X}) = \begin{pmatrix} \Sigma(X) & \Sigma(X, X') \\ \Sigma(X, X')^T & \Sigma(X') \end{pmatrix}, \tag{A3}$$

where $\Sigma(X)$ and $\Sigma(X')$ are the equal time covariance at past and present, respectively, and $\Sigma(X, X')$ is the cross covariance between X and X'. Below we will show the analytical expression of Φ_{MI}, Φ_{SI} and Φ_{G}.

Appendix A.1. Multi Information

Let us consider the following partitioned probability distribution q,

$$q(X, X') = \prod_i q(M_i, M_i'), \tag{A4}$$

where M_i and M_i' are the past and present states of i-th subsystem. Then multi information is defined as

$$\Phi_{\mathrm{MI}} = \sum_i H(M_i, M_i') - H(X, X'). \tag{A5}$$

When the distribution is Gaussian, Equation (A5) is transformed to

$$\Phi_{\mathrm{MI}} = \sum_i \log|\Sigma(\tilde{M}_i)| - \log|\Sigma(\tilde{X})|, \tag{A6}$$

where $\tilde{M}_i = (M_i, M_i')$ and $\Sigma(\tilde{M}_i)$ is the covariance of $\tilde{M}_i$.

Appendix A.2. Stochastic Interaction

We consider the following partitioned probability distribution q,

$$q(X'|X) = \prod_i q(M_i'|M_i). \tag{A7}$$

Then, stochastic interaction [12,15,18,19] is defined as

$$\Phi_{\mathrm{SI}} = \sum_i H(M_i'|M_i) - H(X'|X). \tag{A8}$$

When the distribution is Gaussian, Equation (A8) is transformed to

$$\Phi_{\mathrm{SI}} = \sum_i \log|\Sigma(M_i'|M_i)| - \log|\Sigma(X'|X)|, \tag{A9}$$

where $\Sigma(M_i'|M_i)$ and $\Sigma(X'|X)$ are covariance matrices of conditional distributions. These matrices are represented as

$$\begin{aligned} \Sigma(M_i'|M_i) &= \Sigma(M_i') - \Sigma(M_i, M_i')^T \Sigma(M_i)^{-1} \Sigma(M_i, M_i'), \\ \Sigma(X'|X) &= \Sigma(X') - \Sigma(X, X')^T \Sigma(X)^{-1} \Sigma(X, X'), \end{aligned} \tag{A10}$$

where $\Sigma(M_i)$ and $\Sigma(M_i')$ are the equal time covariance of subsystem i at past and present, respectively, and $\Sigma(M_i, M_i')$ is the cross covariance between M_i and M_i'.

Appendix A.3. Geometric Integrated Information

To calculate the geometric integrated information [12], we first transform Equation (A2). Equation (A2) is equivalently represented as an autoregressive model:

$$X' = AX + E, \tag{A11}$$

where A is the connectivity matrix and E is Gaussian random variables, which are uncorrelated over time. By using this autoregressive model, the joint distribution $p(X, X')$ is expressed as

$$p(x, x') = \exp\left\{-\frac{1}{2}\left(x^T \Sigma(X)x + (x' - Ax)^T \Sigma(E)^{-1}(x' - Ax) - \psi\right)\right\}, \tag{A12}$$

and the covariance matrices as

$$\begin{aligned}
\Sigma(X, X') &= \Sigma(X)A^T, \\
\Sigma(X') &= \Sigma(E) + A\Sigma(X)A^T,
\end{aligned} \tag{A13}$$

where $\Sigma(E)$ is the covariance of E. Similarly, the joint probability distribution in a partitioned model is given by

$$\begin{aligned}
q(x, x') &= \exp\left\{-\frac{1}{2}\left(\tilde{x}^T \Sigma(\tilde{X})_{\mathrm{p}} \tilde{x} - \psi\right)\right\} \\
&= \exp\left\{-\frac{1}{2}\left(x^T \Sigma(X)_{\mathrm{p}} x + (x' - A_{\mathrm{p}}x)^T \Sigma(E)_{\mathrm{p}}^{-1}(x' - A_{\mathrm{p}}x) - \psi\right)\right\},
\end{aligned} \tag{A14}$$

where $\Sigma(X)_{\mathrm{p}}$ and $\Sigma(E)_{\mathrm{p}}$ are the covariance matrices of X and E in the partitioned model, respectively, and A_{p} is the connectivity matrix in the partitioned model.

The geometric integrated information is defined as

$$\Phi_{\mathrm{G}} = \min_{q} D_{\mathrm{KL}}\left(p(X, X') \,||\, q(X, X')\right), \tag{A15}$$

$$D_{\mathrm{KL}}\left(p(X, X') \,||\, q(X, X')\right) = \frac{1}{2}\left(\log \frac{|\Sigma(\tilde{X})_{\mathrm{p}}|}{|\Sigma(\tilde{X})|} + \mathrm{Tr}(\Sigma(\tilde{X})\Sigma(\tilde{X}_{\mathrm{p}})^{-1}) - 2N\right), \tag{A16}$$

such that

$$q\left(M_i'|X\right) = q\left(M_i'|M_i\right), \forall i. \tag{A17}$$

This constraint (Equation (A17)) corresponds to setting the between-subsystem blocks of A_{p} to 0:

$$(A_{\mathrm{p}})_{ij} = 0 \ (i \neq j). \tag{A18}$$

By transforming stationary point conditions, $\partial D_{\mathrm{KL}}/\partial \Sigma(\tilde{X})_{\mathrm{p}}^{-1} = 0$, $\partial D_{\mathrm{KL}}/\partial (A_{\mathrm{p}})_{ii} = 0$, and $\partial D_{\mathrm{KL}}/\partial \Sigma(E)_{\mathrm{p}}^{-1} = 0$, we get

$$\Sigma(X)_{\mathrm{p}} = \Sigma(X), \tag{A19}$$

$$(\Sigma(X)(A - A_{\mathrm{p}})\Sigma(E)_{\mathrm{p}}^{-1})_{ii} = 0, \tag{A20}$$

$$\Sigma(E)_{\mathrm{p}} = \Sigma(E) + (A - A_{\mathrm{p}})\Sigma(X)(A - A_{\mathrm{p}})^T. \tag{A21}$$

By substituting Equations (A19) and (A21) into Equation (A15), Φ_{G} is simplified as

$$\Phi_{\mathrm{G}} = \frac{1}{2}\log \frac{|\Sigma(E)_{\mathrm{p}}|}{|\Sigma(E)|}. \tag{A22}$$

To obtain the value of Equation (A22), we need to find the value of $\Sigma(E)_{\mathrm{p}}$. The computation of $\Sigma(E)_{\mathrm{p}}$ requires solving Equations (A20) and (A21) for $\Sigma(E)_{\mathrm{p}}$ and A_p simultaneously. However, it is difficult to express Equations (A20) and (A21) as closed-form expressions. Therefore, we need to solve the multi-dimensional equations (Equations (A20) and (A21)) using an iterative method. This iterative process increased the complexity of the search using Queyranne's algorithm up to roughly $O(N^{4.776})$ (see Section 6.1). The MALAB codes for this computation of Φ_G are available at [37].

Appendix B. Details of Replica Exchange Markov Chain Monte Carlo Method

The Replica Exchange Markov Chain Monte Carlo (REMCMC) method was originally proposed to investigate physical systems [21–23], and was then rapidly utilized in other applications, including combinatorial optimization problems [32–34,38,39]. For a more detailed history of REMCMC, see, for example, [24].

We first briefly explain how the MIP search problem is dealt with by the Metropolis method. Then, as an improvement of Metropolis method, we introduce REMCMC to more effectively search for the global minimum while avoiding being trapped around at a local minimum. Next, we describe the convergence criterion of MCMC sampling. Finally, we present the parameter settings in our experiments.

Appendix B.1. Metropolis Method

We consider the way to sample subsets from the probability distribution in Equation (14). An initial subset $S^{(0)}$ is randomly selected, and then a sample sequence is drawn as follows.

- **Propose a candidate of the next sample** An element e is randomly selected and if it is in the current subset $S^{(t)}$, the candidate S_c is $S^{(t)} \setminus \{e\}$. If not, the candidate is $S^{(t)} \cup \{e\}$.
- **Determine whether to accept the candidate or not** The candidate S_c is accepted ($S^{(t+1)} = S_c$) or not accepted ($S^{(t+1)} = S^{(t)}$) according to the following probability $a(S^{(t)} \rightarrow S_c)$:

$$a(S^{(t)} \rightarrow S_c) = \min(1, r),$$
$$r = \frac{p(S_c; \beta)}{p(S^{(t)}; \beta)} = \exp\left[\beta \left\{\Phi(S^{(t)}) - \Phi(S_c)\right\}\right]. \tag{A23}$$

This probability means that if the integrated information decreases by stepping from $S^{(t)}$ to S_c, the candidate S_c is always accepted, and otherwise it is accepted with the probability r.

By iterating these two steps with sufficient time, the sample distribution converges to the probability distribution given in Equation (14). N steps of the sampling is referred to as one Monte Carlo step (MCS), where N is the number of elements. In one MCS, each element is attempted to be added or removed once on average.

Depending on the value of β, the behavior of the sample sequence changes. If β is small, the probability distribution given by Equation (14) is close to a uniform distribution and subsets are sampled nearly independently of the value of $\Phi(S)$. If β is large, the candidate is more likely to be accepted when the integrated information decreases. The sample sequence easily falls to a local minimum and cannot explore many subsets. Thus, smaller and larger β have an advantage and a disadvantage: Smaller β is better for exploring around many subsets while larger β is better for finding a (local) minimum. In the Metropolis method, we need to set β to an appropriate value taking account of this trade-off, but it is generally difficult.

Appendix B.2. Replica Exchange Markov Chain Monte Carlo

To overcome the difficulty in setting inverse temperature β, REMCMC samples from distributions at multiple values of β in parallel and the sampled sequences are exchanged between nearby values of β. By this exchange, the sampled sequences at high inverse temperatures can escape from local minima and can explore many subsets.

We consider M-probabilities at different inverse temperatures $\beta_1 > \beta_2 > \cdots > \beta_M$ and introduce the following joint probability:

$$p(S_1, \ldots, S_M; \beta_1, \ldots, \beta_M) = \prod_{m=1}^{M} p(S_m; \beta_m). \tag{A24}$$

Then, the simulation process of the REMCMC consists of the following two steps:

- **Sampling from each distribution:** Samples are drawn from each distribution $p(S_m; \beta_m)$ separately by using the Metropolis method as described in the previous subsection.
- **Exchange between neighboring inverse temperatures:** After a given number of samples are drawn, subsets at neighboring inverse temperatures are swapped, according to the following probability $p(S_m \leftrightarrow S_{m+1})$:

$$p(S_m \leftrightarrow S_{m+1}) = \min(1, r'),$$
$$r' = \frac{p(S_{m+1}; \beta_m)p(S_m; \beta_{m+1})}{p(S_m; \beta_m)p(S_{m+1}; \beta_{m+1})} \tag{A25}$$
$$= \exp\left[(\beta_{m+1} - \beta_m)\left\{\Phi(S_{m+1}) - \Phi(S_m)\right\}\right].$$

This probability indicates that if the integrated information at a higher inverse temperature is larger than that at a lower inverse temperature, subsets are always swapped; otherwise, they are swapped with the probability r'.

By iterating these two steps for sufficient time, the sample distribution converges to the joint distribution in Equation (A24).

To maximize the efficiency of the REMCMC, it is important to appropriately set the multiple inverse temperatures. If the neighboring temperatures are far apart, the acceptance ratio of exchange (Equation (A25)) becomes too small. The REMCMC is then reduced to just separately simulating distributions at different temperatures without any exchange. In a previous study [40], it was recommended to keep the average ratio higher than 0.2 for every temperature pair. At the same time, the highest/lowest inverse temperatures should be high/low enough so that sample sequence at the highest inverse temperature can reach the tips of (local) minima and that at the lowest one can search around many subsets. To satisfy these constraints, a sufficient number M of inverse temperatures are accommodated and the inverse temperatures are optimized to equalize the average of the acceptance ratio of exchanges at all temperature pairs [40–43]. Details of temperature setting are described below.

Appendix B.2.1. Initial Setting

Inverse temperatures $\beta_m (m = 1, \ldots, M)$ are initially set as follows. First, a subset is randomly selected for each m. Then, a randomly chosen element is added to or eliminated from each subset, and the absolute value of the change $\Delta\Phi_m$ in the amount of integrated information is taken. By using these absolute values, the highest and lowest inverse temperatures are determined by a bisection method so that the respective averages of the acceptance ratio $\exp(-\beta\Delta\Phi_1)$ and $\exp(-\beta\Delta\Phi_M)$ match the predefined values. The intermediate inverse temperatures are set to be a geometric progression: $\beta_m = \beta_1 \left(\frac{\beta_M}{\beta_1}\right)^{\frac{m-1}{M-1}}$.

Appendix B.2.2. Updating

The difference in the amount of integrated information between the candidate subset $\Phi(S_c)$ and the current subset $\Phi(S^{(t)})$ is stored when the difference is positive ($\Phi(S_c) - \Phi(S^{(t)}) \geq 0$). Then, by using the stored values at all the inverse temperatures, the highest and lowest inverse temperatures are determined by a bisection method so that the average of the acceptance ratio $\exp\left[\beta\left\{\Phi(S^{(t)}) - \Phi(S_c)\right\}\right]$ matches the predefined value, as in the initial setting. The intermediate inverse temperatures are set to approximately equalize the expected values of acceptance ratio of the exchange at all temperature pairs [40–43]. The expected value is represented as a sum of two probabilities:

$$
\mathbb{E}\left[p(S_m \leftrightarrow S_{m+1})\right] = \int_{-\infty}^{\infty} \int_{-\infty}^{\infty} \left\{ p(\Phi_m \geq \Phi_{m+1}) \right.
$$
$$
\left. + p(\Phi_m < \Phi_{m+1}) e^{(\beta_m - \beta_{m+1})(\Phi_m - \Phi_{m+1})} \right\} d\Phi_m d\Phi_{m+1}.
$$
(A26)

In [43], this expected value is approximated as

$$
\mathbb{E}\left[p(S_m \leftrightarrow S_{m+1})\right] \approx \frac{1}{2} \mathrm{erfc}\left(\frac{\mu(T_{m+1}) - \mu(T_m)}{\sqrt{2\left\{\sigma^2(T_{m+1}) + \sigma^2(T_m)\right\}}} \right)
$$
$$
+ \left\{ 1 - \frac{1}{2} \mathrm{erfc}\left(\frac{\mu(T_{m+1}) - \mu(T_m)}{\sqrt{2\left\{\sigma^2(T_{m+1}) + \sigma^2(T_m)\right\}}} \right) \right\} e^{(\beta_m - \beta_{m+1})(\mu(T_m) - \mu(T_{m+1}))},
$$
(A27)

where $\mu(T)$ and $\sigma^2(T)$ are the mean and variance of Φ, represented as functions of temperature T. In [43], these functions are given by interpolating the sample mean and variance. In this study, these functions are estimated using regression, because the sample mean and variance are highly variable. The mean and variance at each temperature are computed at every update, and these means and variances are regressed on temperature using a continuous piecewise linear function, the T-axis of anchor points of which are current temperatures. The anchor points are interpolated using piecewise cubic Hermite interpolating polynomials. Then, to roughly equalize the expected values of the acceptance ratio of the exchange at all temperature pairs, we minimize the following cost function by varying temperatures [43]:

$$
\mathrm{Cost} = \sum_{m=1}^{M-1} \mathbb{E}\left[p(S_m \leftrightarrow S_{m+1})\right]^{-4}.
$$
(A28)

The minimization is performed by a line-search method.

Appendix B.3. Convergence Criterion

One of the most commonly used MCMC convergence criteria is potential scale reduction factor (PSRF), which was proposed by Gelman and Rubin (1992) [44], and modified by Brooks and Gelman (1998) [45]. In this criterion, multiple MCMC sequences are run. If all of them converge, statistics of the sequences must be about the same. This is assessed by comparing between-sequence variance and within-sequence variance of a random variable and calculating the PSRF, $\hat{R}_c$. Large $\hat{R}_c$ suggests that some of the sequences do not converge yet. If $\hat{R}_c$ is close to 1, we can diagnose them as converged. In this study, we cut the sequence at each inverse temperature into the former and the latter halves, and applied the criterion to these two half sequences. If $\hat{R}_c$ of all the temperatures were below a predefined threshold, we regarded the sequences as converged.

Appendix B.4. Parameter Settings

The number of inverse temperatures M was fixed at 6 throughout out the experiments. The highest/lowest inverse temperatures were set so that the averages of acceptance ratio become 0.01 and 0.5, respectively. The exchange process was done every 5 MCSs. The update of inverse temperatures was performed every 5 MCSs for the 200 initial MCSs. The threshold of $\hat{R}_c$ was set to 1.01. When computing $\hat{R}_c$, we discarded the first 200 MCSs as a burn-in period and started to computing it after 300 MCSs.

Appendix C. Values of Φ

We show some examples of the distributions of the values of Φ in the experiments in Section 6.2. Figure A1a,b are the box plots of Φ_{SI} and Φ_G for the block-structured models at $\sigma = 0.01$, respectively. We can see that in Figure A1a, Φ_{SI} computed at the partition found by Queyranne's algorithm perfectly matched with that at the MIPs. In Figure A1b, Φ_G computed at the partition found by Queyeranne's algorithm did not match that at the MIPs in 3 trials (the trial numbers 11, 54 and 83) but the deviations were very small.

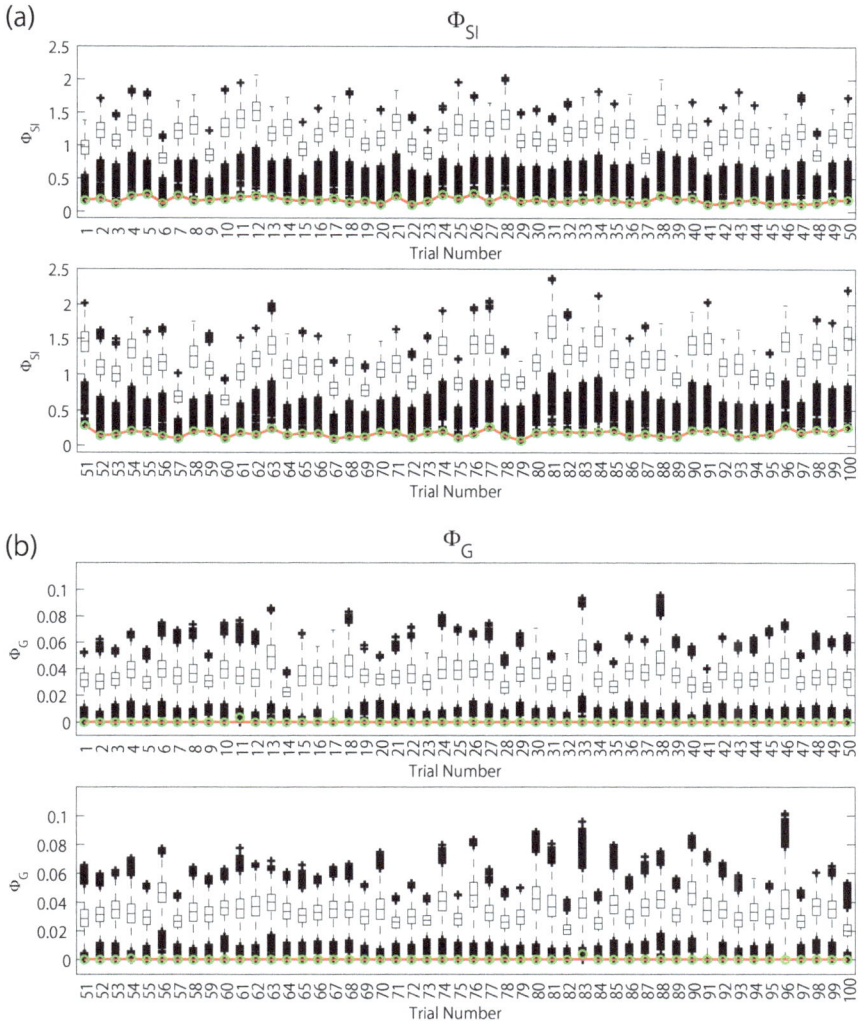

Figure A1. The values of Φ for the block-structured models at $\sigma = 0.01$. The box plots represent the distribution of Φ at all the partitions. The red solid line indicates Φ at the MIP. The green circles indicate Φ at the partitions found by Queyranne's algorithm. (a) Φ_{SI}, (b) Φ_G.

References

1. Tononi, G.; Sporns, O.; Edelman, G.M. A measure for brain complexity: Relating functional segregation and integration in the nervous system. *Proc. Natl. Acad. Sci. USA* **1994**, *91*, 5033–5037.
2. Tononi, G. An information integration theory of consciousness. *BMC Neurosci.* **2004**, *5*, 42, doi:10.1186/1471-2202-5-42.
3. Tononi, G. Consciousness as integrated information: A provisional manifesto. *Biol. Bull.* **2008**, *215*, 216–242.
4. Oizumi, M.; Albantakis, L.; Tononi, G. From the phenomenology to the mechanisms of consciousness: Integrated information theory 3.0. *PLoS Comput. Biol.* **2014**, *10*, e1003588, doi:10.1371/journal.pcbi.1003588.
5. Massimini, M.; Ferrarelli, F.; Huber, R.; Esser, S.K.; Singh, H.; Tononi, G. Breakdown of cortical effective connectivity during sleep. *Science* **2005**, *309*, 2228–2232.
6. Casali, A.G.; Gosseries, O.; Rosanova, M.; Boly, M.; Sarasso, S.; Casali, K.R.; Casarotto, S.; Bruno, M.A.; Laureys, S.; Tononi, G.; et al. A theoretically based index of consciousness independent of sensory processing and behavior. *Sci. Transl. Med.* **2013**, *5*, 198ra105, doi:10.1126/scitranslmed.3006294.
7. Lee, U.; Mashour, G.A.; Kim, S.; Noh, G.J.; Choi, B.M. Propofol induction reduces the capacity for neural information integration: Implications for the mechanism of consciousness and general anesthesia. *Conscious. Cogn.* **2009**, *18*, 56–64.
8. Chang, J.Y.; Pigorini, A.; Massimini, M.; Tononi, G.; Nobili, L.; Van Veen, B.D. Multivariate autoregressive models with exogenous inputs for intracerebral responses to direct electrical stimulation of the human brain. *Front. Hum. Neurosci.* **2012**, *6*, 317, doi:10.3389/fnhum.2012.00317.
9. Boly, M.; Sasai, S.; Gosseries, O.; Oizumi, M.; Casali, A.; Massimini, M.; Tononi, G. Stimulus set meaningfulness and neurophysiological differentiation: A functional magnetic resonance imaging study. *PLoS ONE* **2015**, *10*, e0125337, doi:10.1371/journal.pone.0125337.
10. Haun, A.M.; Oizumi, M.; Kovach, C.K.; Kawasaki, H.; Oya, H.; Howard, M.A.; Adolphs, R.; Tsuchiya, N. Conscious Perception as Integrated Information Patterns in Human Electrocorticography. *eNeuro* **2017**, *4*, 1–18, doi:10.1523/ENEURO.0085-17.2017.
11. Balduzzi, D.; Tononi, G. Integrated information in discrete dynamical systems: Motivation and theoretical framework. *PLoS Comput. Biol.* **2008**, *4*, e1000091, doi:10.1371/journal.pcbi.1000091.
12. Oizumi, M.; Tsuchiya, N.; Amari, S.i. Unified framework for information integration based on information geometry. *Proc. Natl. Acad. Sci. USA* **2016**, *113*, 14817–14822.
13. Hidaka, S.; Oizumi, M. Fast and exact search for the partition with minimal information loss. *arXiv* **2017**, arXiv:1708.01444.
14. Queyranne, M. Minimizing symmetric submodular functions. *Math. Program.* **1998**, *82*, 3–12.
15. Barrett, A.B.; Barnett, L.; Seth, A.K. Multivariate Granger causality and generalized variance. *Phys. Rev. E* **2010**, *81*, 041907, doi:10.1103/PhysRevE.81.041907.
16. Oizumi, M.; Amari, S.; Yanagawa, T.; Fujii, N.; Tsuchiya, N. Measuring integrated information from the decoding perspective. *PLoS Comput. Biol.* **2016**, *12*, e1004654, doi:10.1371/journal.pcbi.1004654.
17. Tegmark, M. Improved measures of integrated information. *PLoS Comput. Biol.* **2016**, *12*, e1005123, doi:10.1371/journal.pcbi.1005123.
18. Ay, N. Information geometry on complexity and stochastic interaction. *MIP MIS Preprint 95* **2001**. Available online: http://www.mis.mpg.de/publications/preprints/2001/prepr2001-95.html (accessed on 6 March 2018).
19. Ay, N. Information geometry on complexity and stochastic interaction. *Entropy* **2015**, *17*, 2432–2458.
20. Amari, S.; Tsuchiya, N.; Oizumi, M. Geometry of information integration. *arXiv* **2017**, arXiv:1709.02050.
21. Swendsen, R.H.; Wang, J.S. Replica Monte Carlo simulation of spin-glasses. *Phys. Rev. Lett.* **1986**, *57*, 2607–2609.
22. Geyer, C.J. Markov chain Monte Carlo maximum likelihood. In Proceedings of the 23rd Symposium on the Interface, Seattle, WA, USA, 21–24 April 1991; Interface Foundation of North America: Fairfax Station, VA, USA, 1991; pp. 156–163.
23. Hukushima, K.; Nemoto, K. Exchange Monte Carlo method and application to spin glass simulations. *J. Phys. Soc. Jpn.* **1996**, *65*, 1604–1608.
24. Earl, D.J.; Deem, M.W. Parallel tempering: Theory, applications, and new perspectives. *Phys. Chem. Chem. Phys.* **2005**, *7*, 3910–3916.
25. Burnham, K.P.; Anderson, D.R. *Model Selection and Multimodel Inference: A Practical Information-Theoretic Approach*; Springer: New York, NY, USA, 2003.

26. Watanabe, S. Information theoretical analysis of multivariate correlation. *IBM J. Res. Dev.* **1960**, *4*, 66–82.
27. Studený, M.; Vejnarová, J. *The Multiinformation Function as a Tool For Measuring Stochastic Dependence*; MIT Press: Cambridge, MA, USA, 1999.
28. Pearl, J. *Causality*; Cambridge University Press: Cambridge, UK, 2009.
29. Iwata, S. Submodular function minimization. *Math. Program.* **2008**, *112*, 45–64.
30. Wishart, J. The generalised product moment distribution in samples from a normal multivariate population. *Biometrika* **1928**, *20A*, 32–52.
31. Bishop, C.M. *Pattern Recognition and Machine Learning*; Springer: New York, NY, USA, 2006.
32. Pinn, K.; Wieczerkowski, C. Number of magic squares from parallel tempering Monte Carlo. *Int. J. Mod. Phys. C* **1998**, *9*, 541–546.
33. Hukushima, K. Extended ensemble Monte Carlo approach to hardly relaxing problems. *Computer Phys. Commun.* **2002**, *147*, 77–82.
34. Nagata, K.; Kitazono, J.; Nakajima, S.; Eifuku, S.; Tamura, R.; Okada, M. An Exhaustive Search and Stability of Sparse Estimation for Feature Selection Problem. *IPSJ Online Trans.* **2015**, *8*, 25–32.
35. Nagasaka, Y.; Shimoda, K.; Fujii, N. Multidimensional recording (MDR) and data sharing: an ecological open research and educational platform for neuroscience. *PLoS ONE* **2011**, *6*, e22561, doi:10.1371/journal.pone.0022561.
36. Toker, D.; Sommer, F. Information Integration in Large Brain Networks. *arXiv* **2017**, arXiv:1708.02967.
37. Kitazono, J.; Oizumi, M. *phi_toolbox.zip*, version 6; Figshare, 6 September 2017. Available online: https://figshare.com/articles/phi_toolbox_zip/3203326/6 (accessed on 6 March 2018).
38. Barthel, W.; Hartmann, A.K. Clustering analysis of the ground-state structure of the vertex-cover problem. *Phys. Rev. E* **2004**, *70*, 066120, doi:10.1103/PhysRevE.70.066120.
39. Wang, C.; Hyman, J.D.; Percus, A.; Caflisch, R. Parallel tempering for the traveling salesman problem. *Int. J. Mod. Phys. C* **2009**, *20*, 539–556, doi:10.1142/S0129183109013893.
40. Rathore, N.; Chopra, M.; de Pablo, J.J. Optimal allocation of replicas in parallel tempering simulations. *J. Chem. Phys.* **2005**, *122*, 024111, doi:10.1063/1.1831273.
41. Sugita, Y.; Okamoto, Y. Replica-exchange molecular dynamics method for protein folding. *Chem. Phys. Lett.* **1999**, *314*, 141–151.
42. Kofke, D.A. On the acceptance probability of replica-exchange Monte Carlo trials. *J. Chem. Phys.* **2002**, *117*, 6911–6914; Erratum in **2004**, *120*, 10852, doi:10.1063/1.1738103.
43. Lee, M.S.; Olson, M.A. Comparison of two adaptive temperature-based replica exchange methods applied to a sharp phase transition of protein unfolding-folding. *J. Chem. Phys.* **2011**, *134*, 244111, doi:10.1063/1.3603964.
44. Gelman, A.; Rubin, D.B. Inference from iterative simulation using multiple sequences. *Stat. Sci.* **1992**, *7*, 457–472.
45. Brooks, S.P.; Gelman, A. General methods for monitoring convergence of iterative simulations. *J. Comput. Graph. Stat.* **1998**, *7*, 434–455.

entropy

MDPI

Article

Novel Brain Complexity Measures Based on Information Theory

Ester Bonmati * , Anton Bardera * , Miquel Feixas and Imma Boada

Graphics and Imaging Laboratory, University of Girona, 17003 Girona, Spain; feixas@ima.udg.edu (M.F.); imma.boada@udg.edu (I.B.)
* Correspondence: ester.bonmati@imae.udg.edu (E.B.); anton.bardera@imae.udg.edu (A.B.); Tel.: +34-638-222-355 (A.B.)

Received: 26 April 2018; Accepted: 19 June 2018; Published: 25 June 2018

check for updates

Abstract: Brain networks are widely used models to understand the topology and organization of the brain. These networks can be represented by a graph, where nodes correspond to brain regions and edges to structural or functional connections. Several measures have been proposed to describe the topological features of these networks, but unfortunately, it is still unclear which measures give the best representation of the brain. In this paper, we propose a new set of measures based on information theory. Our approach interprets the brain network as a stochastic process where impulses are modeled as a random walk on the graph nodes. This new interpretation provides a solid theoretical framework from which several global and local measures are derived. Global measures provide quantitative values for the whole brain network characterization and include entropy, mutual information, and erasure mutual information. The latter is a new measure based on mutual information and erasure entropy. On the other hand, local measures are based on different decompositions of the global measures and provide different properties of the nodes. Local measures include entropic surprise, mutual surprise, mutual predictability, and erasure surprise. The proposed approach is evaluated using synthetic model networks and structural and functional human networks at different scales. Results demonstrate that the global measures can characterize new properties of the topology of a brain network and, in addition, for a given number of nodes, an optimal number of edges is found for small-world networks. Local measures show different properties of the nodes such as the uncertainty associated to the node, or the uniqueness of the path that the node belongs. Finally, the consistency of the results across healthy subjects demonstrates the robustness of the proposed measures.

Keywords: brain network; complex networks; connectome; information theory; graph theory

1. Introduction

The human brain is a complex system composed of a set of regions, which are segregated in order to perform specific tasks and are also efficiently integrated in order to share information [1]. The mapping of the structure and the functionality of brain networks is therefore a main challenge in understanding the functioning, as it cannot be studied as a group of independent elements. An important first step to understand how the information is shared, is the generation of a comprehensive map. Felleman and Essen [2] represented the connections of different regions of the human brain by defining a connectivity matrix. Later, the idea of a *connectome* [1,3] was introduced, which mapped the neural connections in the brain using networks and graph theory [4–6].

In a brain network or graph, nodes correspond to brain regions and edges to structural or functional connections [7–9]. To model the brain, different graphs can be used: un-directed binary graphs which are the most popular; weighted graphs that assign weights to the edges according to the degree of connectivity between the nodes; and directed graphs that take the influence of one region in another [7]. Once the graph is built, it needs to be analyzed to describe the hidden information in this dense network.

From the connectome, it has been shown that each brain region has a unique pattern of connections (known as *connectional fingerprint*) [10] that varies across subjects [6] but preserves a similar structure. Different techniques have been applied to describe the topological features of brain networks [11–13]. For instance, the independence of large areas, denoted as *integration*, has been studied by the *path length* measure, the *characteristic path length* [14], or the *global efficiency* [15]. The independence of small subsets, defined as *segregation*, can be analyzed by the *clustering coefficient* [14], the *transitivity* [16] or the *modularity* [17]. The importance of individual nodes can be defined with centrality measures such as the *degree* [18], or the *density*. A good summary of these measures can be found in [12].

Global measures have also been proposed to describe the overall network structure of the brain. Studies such as Kennedy et al. [19] suggested that a functional and structural central circuit with different areas acting as a cluster governed the information distribution and integration in the brain. Clusters are densely interconnected areas and are defined as a *rich-club* [20–22]. Sporns et al. [23] evidenced small-world properties of human brain networks. Small-world networks are systems with a high level of clusterization, like lattice networks, and with small path lengths, like random graphs.

Information theory has been previously used to study the integration and segregation of brain networks [24–26]. For instance, the *neuronal complexity* measure (C_N) showed a balance between segregation and integration [24,27,28]. Additionally, other measures were proposed such as the *matching complexity* measure (C_M) that shows the change in C_N after receiving signals from the environment [29], the *functional clustering*, which finds groups of regions that are more connected among themselves than with the rest [30], the *degeneracy* measure (D_N) that describes how structurally different elements are able to perform the same function, and the *redundancy* measure (R), that describes how identical elements perform the same function [31,32].

Brain network measures are also able to associate different diseases with disruptions [6,33–35]. As an example, Sato et al. [36] used the assessment of the graphs entropy to distinguish subjects with and without hyperactivity [36]. Unfortunately, the measures that best describe a brain network are still unknown. Therefore, new network measures showing new properties are required to better understand brain networks and their functioning [37].

In this paper, we use a brain network model, where regions correspond to states of a Markov process, to model impulses as random walks on the brain network [38]. Please note that this model differs from the previous ones [24,27,28], where correlations between subsets are used to study the centrality and segregation. This Markov process-based interpretation provides a solid theoretical framework from which global and local measures can be derived. Global measures provide quantitative values to characterize the whole brain network while local measures, which are based on different decompositions of the global measures, are used to quantify the informativeness associated to each node. To evaluate the proposed measures different synthetic model networks, and structural and functional human networks at different scales are considered.

2. Method

2.1. Information Theory Basis

Let the alphabet $\mathcal{X}$ be a finite set and X a random variable taking values x in $\mathcal{X}$. The *Shannon entropy* $H(X)$ of a random variable X is defined by

$$H(X) = - \sum_{x \in \mathcal{X}} p(x) \log p(x), \tag{1}$$

where $p(x) = Pr[X = x]$ is the probability of the value x. Entropy measures the average uncertainty of a random variable X. All logarithms are base 2 and entropy is expressed in bits. In this paper, the convention $0 \log 0 = 0$ is used.

Likewise, let Y be a random variable taking values y in $\mathcal{Y}$. The *conditional entropy* is defined by

$$H(Y|X) = - \sum_{x \in \mathcal{X}} p(x) \sum_{y \in \mathcal{Y}} p(y|x) \log p(y|x), \tag{2}$$

where $p(y|x) = Pr[Y = y|X = x]$ is the conditional probability. The conditional entropy $H(Y|X)$ measures the average uncertainty associated with Y if we know the outcome of X. In general, $H(Y|X) \neq H(X|Y)$, and $H(X) \geq H(X|Y) \geq 0$.

The *mutual information* (MI) between X and Y is defined by

$$I(X;Y) = H(X) - H(X|Y) = \sum_{x \in \mathcal{X}} \sum_{y \in \mathcal{Y}} p(x,y) \log \frac{p(x,y)}{p(x)p(y)}$$

$$= \sum_{x \in \mathcal{X}} p(x) \sum_{y \in \mathcal{Y}} p(y|x) \log \frac{p(y|x)}{p(y)}, \tag{3}$$

where $p(x,y) = Pr[X = x, Y = y]$ is the joint probability. MI measures the shared information between X and Y. It can be seen that $I(X;Y) = I(Y;X) \geq 0$ [39].

The relative entropy or Kullback-Leibler distance, $D_{KL}(p,q)$, between two probability distributions p and q, that are defined over the same alphabet $\mathcal{X}$, is defined by

$$D_{KL}(p,q) = \sum_{x \in \mathcal{X}} p(x) \log \frac{p(x)}{q(x)}. \tag{4}$$

The relative entropy satisfies that $D_{KL}(p,q) \geq 0$, with equality if and only if $p = q$. Kullback-Leibler distance is a basic information theory measure to quantify the dissimilarity between two probability distributions, and other measures, like entropy or mutual information, can be reformulated in terms of this.

A *stochastic process* or a discrete-time information source **X** is an indexed sequence of random variables characterized by the joint probability distribution $p(x_1, x_2, \ldots, x_L) = Pr\{(X_1, X_2, \ldots, X_L) = (x_1, x_2, \ldots, x_L)\}$ with $(x_1, x_2, \ldots, x_L) \in \mathcal{X}^L$ for $L \geq 1$ [39,40]. The *entropy rate* or entropy density h of a stochastic process **X** is defined by

$$h = \lim_{L \to \infty} \frac{1}{L} H(X_1, X_2, \ldots, X_L) \tag{5}$$

when the limit exists. The entropy rate represents the average information content per symbol in a stochastic process. It is the "uncertainty associated with a given symbol if all the preceding symbols are known" and can be viewed as "the intrinsic *unpredictability*" or "the irreducible *randomness*" associated with the chain [41].

A *stochastic Markov process* [39], or *Markov chain*, is a discrete stochastic process defined over a set of states $\mathcal{X}$ which is described by a *transition probability matrix* P. In each step, the process makes a transition from its current state i to a new state j with *transition probability* $P_{ij} = p(x_j^{t+1}|x_i^t) = Pr[X_{t+1} = x_j|X_t = x_i]$.

For a *stationary Markov process* (that is, a Markov process whose statistical properties are invariant to a shift in time), the probability of each state i converge to a *stationary distribution* $\mu = \{\mu_1, \ldots, \mu_n\}$ after several steps. The stationary or equilibrium probabilities μ_i fulfill the relation $\mu_i = \sum_{j=1}^{n} \mu_j P_{ji}$ and also the reciprocity relation $\mu_i P_{ij} = \mu_j P_{ji}$.

In particular, a Markov process can be considered as a chain of random variables complying with

$$H(X_L|X_1, X_2, \ldots, X_{L-1}) = H(X_L|X_{L-1}). \tag{6}$$

An important result is the following theorem: for a stationary Markov chain with stationary distribution μ_i, the *entropy rate* or information content is given by

$$h \quad = \quad H(X_{t+1}|X_t) = -\sum_{i=1}^{n} \mu_i \sum_{j=1}^{n} P_{ij} \log P_{ij}, \tag{7}$$

where μ_i is the stationary distribution and P_{ij} is the transition probability from state i to state j.

The *excess entropy* [42–45] of an infinite chain is defined by

$$E \quad = \quad \lim_{L \to \infty} (H(X_1, X_2, \ldots, X_L) - Lh) \tag{8}$$

$$= \quad \sum_{L=1}^{\infty} (H(L) - h), \tag{9}$$

where h is the entropy rate of the chain, L is the length of this chain, and $H(L) = H(X_L|X_{L-1}, \ldots, X_1)$. The excess entropy can be interpreted as the mutual information between two semi-infinite halves of the chain. Another way of viewing this is that excess entropy is a measure of the apparent memory or *structure* in the system, that is, the excess entropy measures how much more random the system would become if we suddenly forgot all information about the left half of the string [46]. For a stationary Markov process, excess entropy coincides with mutual information, and, hence, in this case, mutual information can be seen as a measure of the system structure.

The *erasure entropy* [47] measures the information content of each symbol knowing its context, i.e., the previous and posterior samples. For any stationary process, the erasure entropy is given by

$$H^- \quad = \quad \lim_{L \to \infty} H(X_0|X_{-L}^{-1}, X_1^L), \tag{10}$$

where X_{-L}^{-1} symbolizes the previous samples (past) and X_1^L the posterior samples (future).

2.2. Markov Process-Based Brain Model

A brain graph can be defined as a pair of sets $G = (N, E)$, where N is a brain parcellation of n nodes labelled $\{N_1, \ldots, N_n\}$, and E is a set of m edges between two nodes of N. This graph can be represented by a connectivity matrix C with $n \times n$ elements, where C_{ij} gives the connectivity weight between node x_i and node x_j. Please note that for undirected graphs $C_{ij} = C_{ji}$.

In this work, brain functions are modeled as a random walk of a particle on the connectivity graph, where the particle randomly goes from node to node defining a path or a sequence of nodes. From node x_i, the next node x_j is chosen among all nodes connected to node x_i, with a probability proportional to the weight C_{ij}. By introducing this model, we are assuming that the next step in the random walk of a neural impulse is determined only by the region and its connections, but not by previous steps of the random walk.

This model leads to a conditional probability $P_{ij} = p(x_j^{t+1}|x_i^t)$ given by $C_{ij}/\sum_i C_{ij}$. The stationary distribution for this Markov chain assigns probability to node x_i proportional to the total weight of the edges emanating from node x_i [39]. Thus, the stationary distribution of a node x_i is given by

$$\mu_i = p(x_i) \quad = \quad \frac{C_i}{C_T}, \tag{11}$$

where $C_i = \sum_j C_{ij}$ is the total weight of the edges emanating from node i and $C_T = \sum_i \sum_j C_{ij}$ is the sum of the weights of all the edges. Observe that this stationary distribution has an interesting property of locality: it depends only on the total weight and the weight of edges connected to the node and hence, it does not change if the weights in some other part of the graph are changed while keeping constant the total weight.

The definition of this model allows to propose new global and local measures to characterize brain networks. *Global measures* describe by a single value the whole connectivity of the brain, while *local measures* assign a value to each brain region, by considering the contribution of the region to the corresponding global measure. In this work, we propose new measures in three different levels: stationary measures, causal measures, and contextual measures. *Stationary measures* are based on the stationary distribution (i.e., current state). *Causal measures* are based on how the previous states influence the current state in the random walk. Finally, *contextual measures* describe how the context (i.e., the previous and future states) is related to the current state. Table 1 summarizes these measures, which are described in the next subsections.

Table 1. Summary of the proposed set of measures.

	Global	Local
Stationary	Entropy	Entropic surprise
Causal	Mutual Information	Mutual surprise Mutual predictability
Contextual	Erasure Mutual Information	Erasure surprise

2.3. Global Informativeness Measures

Global measures provide quantitative values to typify the brain connectome as a whole. Depending on which level is considered (stationary, causal, or contextual), three different measures are given: entropy, mutual information, and erasure mutual information.

2.3.1. Entropy

From the stationary distribution μ Equation (11), the Shannon entropy $H(\mu)$ Equation (1) measures the average uncertainty of the stationary distribution:

$$H(\mu) = -\sum_{i=1}^{n} \mu_i \log \mu_i. \tag{12}$$

Since the probability of each region depends on the weight of their edges, this measure will take high values when all nodes in a network have similar connectivity (weights) and will take low values when there is large variability in terms of number of connections or weights. For instance, in the graph shown in Figure 1a, all nodes have the same number of connections (in this case, each connection has the same weight). Thus, the entropy takes the maximum value given by $\log_2 N = \log_2 4 = 2$, where N is the number of nodes. For the graphs of Figure 1b,c, the value of the entropy decreases since the connectivity of the nodes is not equal for all nodes.

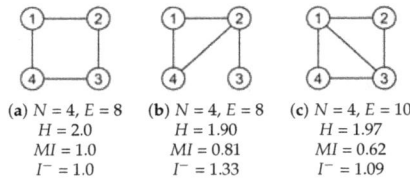

(a) $N = 4, E = 8$ (b) $N = 4, E = 8$ (c) $N = 4, E = 10$
$H = 2.0$ $H = 1.90$ $H = 1.97$
$MI = 1.0$ $MI = 0.81$ $MI = 0.62$
$I^- = 1.0$ $I^- = 1.33$ $I^- = 1.09$

Figure 1. Example values of the entropy (H), mutual information (MI) and erasure (I^-) measures for simple networks (**a**–**c**), where N corresponds to the number of nodes and E to the number of edges. Networks are weighted and undirected, therefore each edge is counted twice.

2.3.2. Mutual Information

As we have previously mentioned, *mutual information* measures the shared information between two random variables. From our Markov process-based brain model, we propose as a global connectivity measure the mutual information between two consecutive states of the process:

$$
\begin{aligned}
I(X_t; X_{t+1}) &= H(X_{t+1}) - H(X_{t+1}|X_t) \\
&= \sum_{x_i^t \in \mathcal{X}} \sum_{x_j^{t+1} \in \mathcal{X}} p(x_i^t, x_j^{t+1}) \log \frac{p(x_i^t, x_j^{t+1})}{p(x_i^t)p(x_j^{t+1})} \\
&= \sum_{i=1}^{n} \mu_i \sum_{j=1}^{n} P_{ij} \log \frac{P_{ij}}{\mu_j}.
\end{aligned}
\tag{13}
$$

From Equations (3) and (17), MI can also be seen as the difference between the uncertainty of the states without any knowledge ($H(X_{t+1})$) and the uncertainty of the states when the past is known ($H(X_{t+1}|X_t)$). In other words, MI measures the information gained when the previous node is known. The higher the MI, the less random the connections. MI can be seen as a measure of brain structure, since it coincides with excess entropy [48].

In the graph of Figure 1a, the fact of knowing the state at a given time t (present) leads to the states for the time $t+1$ (future). For instance, if a given time t the random walk is on the state 1, for the next time $t+1$, the random walk would be either in state 2 or in state 4. Thus, the conditional entropy $H(X_{t+1}|X_t)$ is $\log_2 2 = 1$. The mutual information, given by the entropy (which corresponds to $\log_2 4 = 2$) minus the conditional entropy, is also 1. In the other graphs of Figure 1, the conditional entropy is higher, since there are multiple paths, so the uncertainty of the future step is higher. This fact leads to lower values of MI.

2.3.3. Erasure Mutual Information

The idea of the mutual information measure can be extended by considering not only past states, but also future states. Erasure entropy [47] measures the uncertainty of a system when past and future is known. For a Markov process, this measure can be simplified as

$$
H^-(\mathbf{X}) = H(X_t|X_{t-1}, X_{t+1}),
\tag{14}
$$

where $\mathbf{X}$ symbolizes the whole process. Please note that, in this case, X_{t-1} symbolizes the past, X_t the present, and X_{t+1} the future. From this measure and Equation (3), we can extend mutual information as a measure of the decrease of information when the context (i.e., past and future) is known. Thus, we propose a new global measure, called *erasure mutual information*, defined as

$$
\begin{aligned}
I^-(\mathbf{X}) &= I(X_t; X_{t-1}, X_{t+1}) \\
&= H(X_t) - H(X_t|X_{t-1}, X_{t+1}).
\end{aligned}
\tag{15}
$$

While the mutual information of Equation (13) measures the loss of information taking into account only the previous node in the random walk (past), erasure mutual information measures the loss of information taking into account the previous node (past) and the next node (future). High values of this measure will show a network with predictable paths to go from node to node, and low values will define a network with several possible paths to go from node to node.

In the graph of Figure 1a, the fact of knowing both past and future states does not reduce the uncertainty of the present (compared with only knowing the past). For instance, if in $t-1$ the random walk is the state 1 and in time $t+1$ in 3, the state in t can be either 2 or 4. This uncertainty is the same for all possible pairs of past and future states. Thus, the conditional entropy $H(X_t|X_{t-1}, X_{t+1})$ is $\log_2 2 = 1$ and erasure mutual information is also 1. In the other graphs of Figure 1, the knowledge of future and past states, reduces the uncertainty compared with only knowing the past. For instance, in the graph of Figure 1b, if the past state is 4, the present state can be either 1 or 2, but, if the future state is 3, therefore, the present state is, without uncertainty, state 2. Thus, erasure mutual information takes higher values than mutual information.

2.4. Local Informativeness Measures

In this section, we describe how global measures can be decomposed in order to characterize the degree of informativeness of each state i. When applied to the human connectome, since each state corresponds to an anatomical or functional region, these measures can be seen as the contribution of each node to the whole graph structure, thus, they can describe specific topology of brain areas.

2.4.1. Entropic Surprise

The entropy of X can also be interpreted as the expected value of $-\log(p(x))$, where X is drawn according to probability mass function $p(x)$. Then, in our Markov process-based brain model, the *entropic surprise* value associated to a brain region x_i is defined as

$$E(x_i) = -\log(\mu_i), \qquad (16)$$

where μ_i is the stationary probability of the region x_i.

This measure uses the stationary probability of a node without taking into account the previous or the next node in the random walk. Nodes with a low value will be nodes with a large number of connections or weights in its connections. Therefore, high values will define nodes with a low number of connections to other brain regions. This measure is inversely proportional to the logarithm of the well known strength measure, which is defined as the sum of the edge weights emanating from the node.

Some examples are shown in Figure 2. In the left graph, all nodes have the same entropic surprise value, which is given by $-\log_2 \frac{2}{8} = 2$. Please note that each node has 2 edges and there are 4 bidirectional edges (remember that bidirectional edges are counted twice). In the graphs of Figure 2b,c, it can be seen that nodes with high connectivity take lower entropic values.

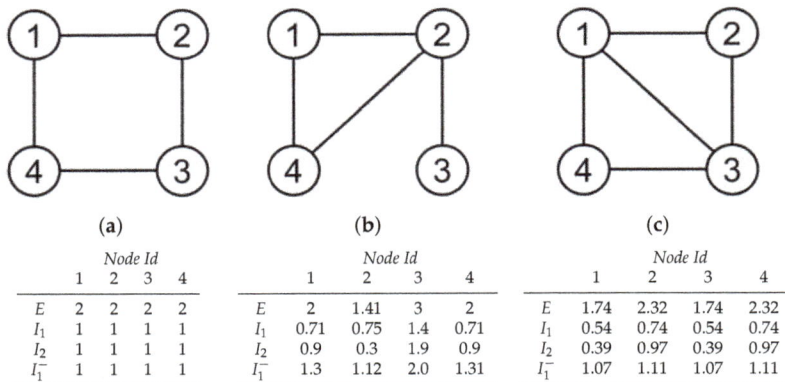

	Node Id			
	1	2	3	4
E	2	2	2	2
I_1	1	1	1	1
I_2	1	1	1	1
I_1^-	1	1	1	1

(a)

	Node Id			
	1	2	3	4
E	2	1.41	3	2
I_1	0.71	0.75	1.4	0.71
I_2	0.9	0.3	1.9	0.9
I_1^-	1.3	1.12	2.0	1.31

(b)

	Node Id			
	1	2	3	4
E	1.74	2.32	1.74	2.32
I_1	0.54	0.74	0.54	0.74
I_2	0.39	0.97	0.39	0.97
I_1^-	1.07	1.11	1.07	1.11

(c)

Figure 2. Example values of entropic surprise (E), mutual surprise (I_1), mutual predictability (I_2) and erasure surprise (I_1^-) measures for simple networks (**a**–**c**). Networks are weighted and undirected, therefore each edge is counted twice.

2.4.2. Mutual Surprise

The interpretation of mutual information explained in Section 2.3 can be extended to define the information associated with a single node $x_i \in \mathcal{X}$, that is, the information gained on X_{t+1} by knowing the original node x_i of the impulse. The definition of *mutual surprise*, denoted by I_1, can be directly derived from the formula of mutual information Equation (3), taking the contribution of a node x_i to $I(X_t; X_{t+1})$, as follows:

$$
\begin{aligned}
I(X;Y) &= H(X) - H(X|Y) \\
&= \sum_{x \in \mathcal{X}} p(x) \sum_{y \in \mathcal{Y}} p(y|x) \log \frac{p(y|x)}{p(y)}, \\
&= \sum_{x \in \mathcal{X}} p(x) I_1(x; Y).
\end{aligned}
\tag{17}
$$

Mutual surprise was used by DeWeese and Meister [49] to emphasize the fact that the observation of x has moved the estimate of another variable Y towards values that seemed very unlikely prior to the observation. I_1 always takes positive values and it can be shown that I_1 is the only positive decomposition of MI [49].

We reformulate Equation (17) in the framework of the Markov process as follows:

$$
\begin{aligned}
I(X_t; X_{t+1}) &= \sum_{x_i^t \in \mathcal{X}} p(x_i^t) \sum_{x_j^{t+1} \in \mathcal{X}} p(x_j^{t+1}|x_i^t) \log \frac{p(x_j^{t+1}|x_i^t)}{p(x_j^{t+1})} \\
&= \sum_{x_i^t \in \mathcal{X}} p(x_i^t) I_1(x_i^t; X_{t+1}),
\end{aligned}
\tag{18}
$$

where

$$
\begin{aligned}
I_1(x_i^t; X_{t+1}) &= \sum_{x_j^{t+1} \in \mathcal{X}} p(x_j^{t+1}|x_i^t) \log \frac{p(x_j^{t+1}|x_i^t)}{p(x_j^{t+1})} \\
&= \sum_{j=1}^{n} P_{ij} \log \frac{P_{ij}}{\mu_j}
\end{aligned}
\tag{19}
$$

expresses the surprise about X_{t+1} from observing x_i^t, i.e., how "surprising" are the nodes connected with the original node. Observe that surprise $I_1(x_i^t; X_{t+1})$ is high when $p(X_{t+1}|x_i^t)$ is very different from $p(X_{t+1})$ (i.e., the stationary distribution), thus, the region x_i is connected with regions which are less connected considering all the connections.

I_1 can be seen as the Kullback-Leibler distance see Equation (4) between $p(X_{t+1})$ (i.e., the stationary distribution) and $p(X_{t+1}|x_i^t)$ (i.e., the distribution of future states if, in the present state, the random walk is on node x_i). Thus, those nodes that are connected with more likely nodes (in terms of the stationary distribution) will lead to low values of I_1, while those with very specific connections or connected with few unlikely nodes will have high I_1 values. This can be seen, for instance, in node 3 of graph shown in Figure 2b.

2.4.3. Mutual Predictability

DeWeese and Meister [49] defined the specific information I_2, which we call *mutual predictability*, using another decomposition of mutual information obtained from Equation (3):

$$
\begin{aligned}
I(X;Y) &= H(Y) - H(Y|X) \\
&= \sum_{x \in \mathcal{X}} p(x) H(Y) - \sum_{x \in \mathcal{X}} p(x) H(Y|x) \\
&= \sum_{x \in \mathcal{X}} p(x) I_2(x; Y),
\end{aligned}
\tag{20}
$$

where

$$I_2(x; Y) = H(Y) - H(Y|x)$$
$$= -\sum_{y \in \mathcal{Y}} p(y) \log p(y) + \sum_{y \in \mathcal{Y}} p(y|x) \log p(y|x) \tag{21}$$

expresses the change in uncertainty of Y when x is observed. In our case, we reformulate I_2 in the framework of the Markov process as follows

$$I_2(x_i^t; X_{t+1}) = H(X_{t+1}) - H(X_{t+1}|x_i^t)$$
$$= -\sum_{x_j^{t+1} \in \mathcal{X}} p(x_j^{t+1}) \log p(x_j^{t+1}) + \sum_{x_j^{t+1} \in \mathcal{X}} p(x_j^{t+1}|x_i^t) \log p(x_j^{t+1}|x_i^t) \tag{22}$$
$$= H(\mu) + \sum_{j=1}^{n} P_{ij} \log P_{ij}.$$

Observe that this measure expresses the difference between global entropy of the graph (i.e., the entropy of the stationary distribution) and entropy of future states of the random walk from node x_i. So, this comparison is done globally and, contrarily to the I_1 measure, it is not affected by the stationary probability of the nodes that is connected to. Another property that fulfills I_2 is additivity, i.e., the information obtained about X from two observations, $y \in Y$ and $z \in Z$, is equal to that obtained from y plus that obtained from z when y is known. Additivity is a desirable property that responds to the intuitive notion that information accumulates additively over a sequence of observations. Because of the additivity property, DeWeese and Meister [49] prefer I_2 against I_1.

Please note that $I_2(x_i^t; X_{t+1})$ can take negative values. In this case, this means that a certain region x_i is connected with more uncertainty than the mean connectivity of the whole brain. Regions with high values of I_2 (like node 3 in the graph of Figure 2b) greatly reduce the uncertainty in X_{t+1} and, thus, they are very significant in the relationship between two consecutive steps in the random walk, X_t and X_{t+1}. Regions with low values of I_2 (like node 2 in the graph of Figure 2b) are assumed to be broadly connected with other brain regions. From this interpretation, we can say that I_2 expresses the capacity of prediction for a given brain region.

2.4.4. Erasure Surprise

In this section, we propose a novel measure based on the decomposition of the erasure mutual information Equation (15) measure. Remember that erasure mutual information represents the reduction of uncertainty when the context (i.e., both past and future) is known.

Then, we can decompose the erasure mutual information measure as:

$$I^-(\mathbf{X}) = H(X_t) - H(X_t|X_{t-1}, X_{t+1})$$
$$= \sum_{x_i^t \in \mathcal{X}} p(x_i^t) I_1^-(x_i^t; \mathbf{X}), \tag{23}$$

where

$$I_1^-(x_i^t; \mathbf{X}) = \sum_{x_j^{t-1} \in \mathcal{X}} \sum_{x_k^{t+1} \in \mathcal{X}} p(x_j^{t-1}, x_k^{t+1}|x_i^t) \log \frac{p(x_j^{t-1}, x_k^{t+1}|x_i^t)}{p(x_j^{t-1}, x_k^{t+1})} \tag{24}$$

$$= \sum_{j=1}^{n} \mu_j P_{ji} \sum_{k=1}^{n} \frac{P_{ik}}{\mu_i} \log \frac{P_{ji} P_{ik}}{\mu_i Q_{jk}} \tag{25}$$

and $Q_{jk} = \sum_{i=1}^{n} P_{ji} P_{ik}$. I_1^- is the *erasure surprise* associated to the region x_i and it always takes positive values. Observe that I_1^- can be seen as the Kullback-Leibler distance see Equation (4) between $p(X_{t-1}, X_{t+1})$ (i.e., joint probability of being at $t-1$ on node x_j and at $t+1$ on node k)

and $p(X_{t-1}, X_{t+1}|x_i^t)$ (i.e., the same as the latter but conditioned to the fact that at t the random walk is on node x_i. Thus, those nodes that connect brain regions which are already connected will lead to low values of I_1^-, and are likely to belong to the same cluster. This can be seen, for instance, in node 2 of the graph shown in Figure 2b. Instead, nodes that connect nodes which would not be connected otherwise (unique paths), will have high values (node 3 of graph shown in Figure 2b).

3. Material

3.1. Synthetic Network Models

The human connectome has been defined as a network with an average short path length which gives a high efficiency in transferring information, a high clustering which provides robustness to random errors, a degree distribution similar to networks with hubs, and a modular community structure [18]. According to these properties, random, lattice, and small-world networks are models that can represent the human connectome. If efficiency was the only property used in the network design, the network would be random [18], with low clustering, short path length [50], and all connections equally probable. However, it is clear that the cortex is not just a uniform system of random connected neurons since random graphs cannot encode and process information [19]. If wiring cost was the priority, the network would be similar to a lattice graph with long paths and high clustering. If we aim for a balance between high clustering and average short path length, then small-world networks are the more accurate representation for both structural and functional networks. For this reason, to illustrate features of the proposed measures, we created three datasets containing random, lattice, ring lattice, and small-world networks.

The first dataset contained random, lattice, ring lattice, and small-world networks with 128 nodes and different number of edges ranging from 128 to 8192 with a step of 128 edges. The second dataset contained the same network models with 256 nodes and edges ranging from 256 to 8192 with a step of 128 edges. Please note that these two datasets provide equivalent networks but with different densities, since the number of nodes was fixed and the number of edges varied. Additionally, a third dataset was created with nodes ranging from 32 to 512 with a step of 32 and a fixed density of 0.4 (varied number of edges). For all graphs, a random weight ranging from 0 to 1 was assigned to all the edges.

The network models were created using the Brain Connectivity Toolbox (BCT) [12]. This toolbox contains a large selection of reference network models and measures that have been previously used in several studies [51–53]. To create the undirected random networks, we used the function *makerandCIJ_und* which generates graphs with no connections on the diagonal (see Figure 3a). The directed non-ring lattice networks were created with the function *makelatticeCIJ*. This lattice is made by placing connections as close as possible to the main diagonal, without wrapping around, and with no connections on the diagonal (see Figure 3b). The ring directed lattice networks were created with the function *makeringlatticeCIJ*. In this case, the lattice is also made by placing connections close to the diagonal, but wrapping around (see Figure 3c). Finally, directed small-world networks were created with the function *makeevenCIJ*. These networks have a specific number of fully connected nodes linked together but with a balanced random connections (see Figure 3d). To transform directed graphs to undirected graphs, all values above diagonal were copied below the diagonal, therefore, all synthetic networks used in this work are weighted and undirected.

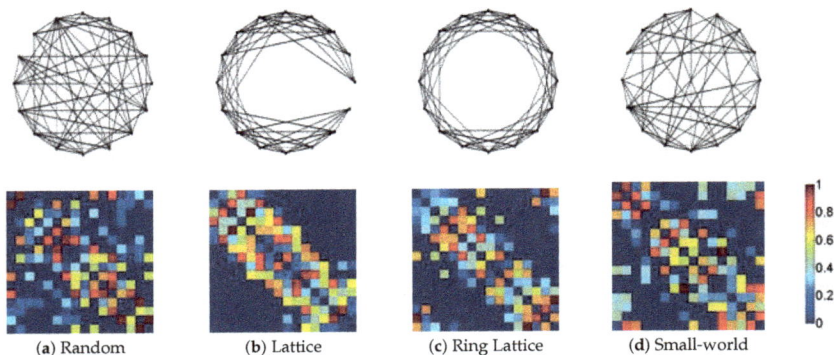

Figure 3. Example of network models (synthetic dataset) used in this work. Each model has the corresponding connectivity matrix illustrated at the bottom. (**a**) Non directed random network (16 nodes, 120 edges); (**b**) Non directed lattice network (16 nodes, 118 edges); (**c**) Non directed ring lattice network (16 nodes, 122 edges); (**d**) Non directed small-world network (16 nodes, 116 edges and cluster size 2).

3.2. Human Datasets

Human datasets were used to test the proposed measures with real data. To show the applicability of our method, we considered both functional and structural brain networks.

3.2.1. Anatomic Dataset

To study the human structural network, we used normalized connection matrices created from MRI tractography [54]. The connectivity matrices were from 10 different subjects at 5 different scales, corresponding to 83, 129, 234, 463 and 1015 cortical and subcortical ROIs. Subjects were all males aged 22 ± 1.3 years old. Edge weights were given by the connectivity density which corresponds to the number of fibers divided by the average of the region surface and by the average length of the fibers. All values were positive, and values on the diagonal were eliminated. The average matrices of the 10 patients for each scale were also created. Figure 4 shows the averaged matrices for the 5 different scales. Edges were resorted to place more edges closer to the diagonal for visualization purposes only.

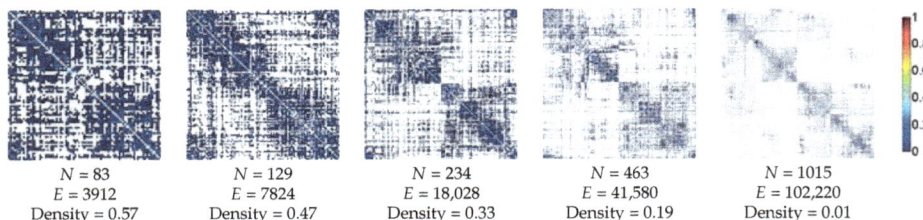

Figure 4. Illustration of the averaged structural connectivity matrices of the anatomic dataset with the corresponding number of nodes (*N*), edges (*E*) and density. 0 values are represented in white. Edges were resorted to place more edges closer to the diagonal for visualization purposes only.

3.2.2. Functional Dataset

Independent component analysis (ICA) is a widely used method to generate functional brain networks of the brain during rest and task. For our analysis, we used the HCP500-PTN functional dataset which belongs to the Human Connectome Project (HCP) beta-release of group-ICA maps [55–57].

This dataset contains functional network matrices of 461 subjects at 5 different scales (25, 50, 100, 200 and 300). For our experiment, we used the approach where the principal eigen-timeseries are estimated and a full normalized temporal correlation has been used. The original matrices contain positive and negative values and no values on the diagonal, but for our experiments, the matrices were thresholded ($Z > 5$) and the negative values were eliminated. The averaged networks were also used. Figure 5 shows the averaged functional matrices at different scales.

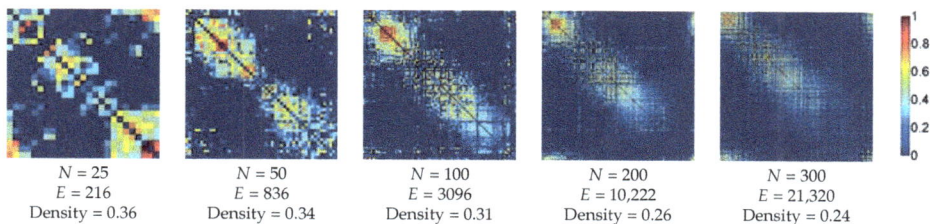

$N = 25$	$N = 50$	$N = 100$	$N = 200$	$N = 300$
$E = 216$	$E = 836$	$E = 3096$	$E = 10,222$	$E = 21,320$
Density = 0.36	Density = 0.34	Density = 0.31	Density = 0.26	Density = 0.24

Figure 5. Illustration of the averaged functional connectivity matrices of the functional dataset with the corresponding number of nodes (N), edges (E) and density.

3.3. Standard Network Measures

The BCT toolbox [12] provides different complex network measures to describe either structural or functional brain connectivity. To evaluate the proposed approach we compared our measures with standard measures included in the BCT. The clustering coefficient is a measure of segregation and expresses the fraction of triangles around a node. The node eccentricity is a measure of distance defined as the maximal shortest path length between a node and any other node. Finally, the node strength is a measure of similarity defined as the sum of weights of links connected to the node.

4. Results and Discussion

In this section, we apply the proposed measures to the synthetic network models and to the human structural and functional connectomes. The results with the global measures and local measures are shown and a comparison with standard measures is presented.

4.1. Global Measures

Firstly, to show the behavior of the global measures, we characterize the synthetic network models (random, lattice, ring lattice and small-world) from the first and second dataset defined in Section 3.1, with 128 and 256 nodes. We apply the global measures (entropy, mutual information, and erasure mutual information) which give a single value per graph.

The first column of Figure 6 presents the entropy measure results. Observe that, when the number of edges increases, the entropy measure tends to a constant value for all types of graphs. This is due to the fact that the higher the number of edges, the more similar the node probability. Thus, the entropy tends to $\log_2 N$, where N is the number of nodes (i.e., for 128 nodes the entropy tends to 7 and for 256 nodes to 8). The slightly decreasing tendency of high values in lattice networks is due to the boundary conditions of extreme nodes which have a lower number of connections which leads to an entropy drop.

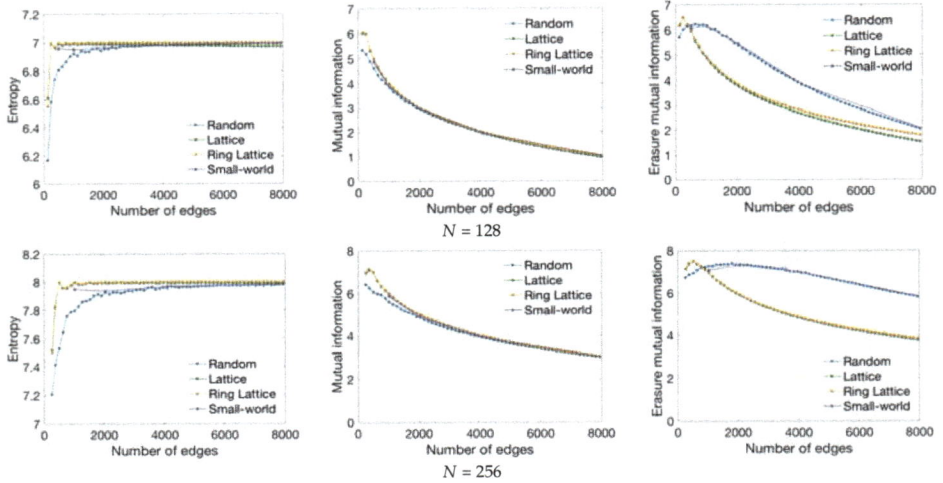

Figure 6. Behavior of the entropy, mutual information and erasure mutual information measures for each network model when the number of edges is increased (from 0 to 8192) and the number of nodes (N) is kept constant (128 nodes on top row and 256 nodes on bottom row).

The second column of Figure 6 shows the behavior of the mutual information measure. In this case, when the number of edges increases, the mutual information of the graph decreases for all types of networks. This is due to the fact that the higher the number of connections, the lower the correlation between consecutive states. For a very low number of edges, we can see that first, the mutual information increases and then decreases. This is due to the fact that for low densities, there are nodes not connected with any node, leading to a decrease of the overall mutual information. Since different tracking methods may provide different number of fibers for a given parcellation [58], the optimal point found with the mutual information measure may allow to find the minimum number of fibers needed for a given brain parcellation to study ring lattice and lattice properties. For a low number of edges, we can also observe that lattice and ring lattice graphs have a slightly higher mutual information than random and small-world graphs. This is due to the higher degree of structure of these kind of graphs, which is what the mutual information measure quantifies.

The third column of Figure 6 presents the values of the erasure mutual information measure. In this case, when the number of edges increases, for all networks, the measure tends to decrease. Note that when there are only a few edges, the uncertainty when past and future states are known is very low ($H(X_t|X_{t-1}, X_{t+1})$, the second term of Equation (15)), leading to high $I^-(\mathbf{X})$ values. When the number of edges increases, the uncertainty tends to increase, thus, the $I^-(\mathbf{X})$ tends to decrease. For this measure, different behaviors can be observed depending on the graph type. For instance, the lattice and ring lattice graphs have a lower erasure mutual information compared to random and small-world graphs. This is due to the fact that the erasure mutual information takes into account the previous node and next node, and for lattice networks, nodes tend to be connected with the closest ones, thus, globally there is more uncertainty. An interesting behavior can be observed for the random and small-world graphs where the measure reaches an optimal point with a larger number of connections compared to lattice and ring lattice networks. In this case, for a low number of edges, there are nodes which are not connected or only connected with intra-module nodes. Thus, all the paths are within the same module. When the number of edges slightly increases, there are more paths that connect different modules but the probability of these paths is very low. Therefore, the erasure mutual information slightly increases. After the optimal point, the erasure mutual information decreases due to the larger number of connections between different modules that increase the uncertainty.

Secondly, we generated different graphs, in this case, modifying the number of nodes but preserving the density (number of edges divided by the number of edges of the complete graph) to 0.4, which is the third dataset described in Section 3.1. Figure 7 shows the behavior of the global measures when the number of nodes increases. As it can be seen, the entropy value increases with the logarithm of the size for a constant edge density. This is consistent with the results of the first experiment where entropy tends to $\log_2 N$, being N the number of nodes. On the contrary, mutual information is not very sensitive for random networks since its connections are randomly placed, so fixing the graph density, the structure of the networks remains similar. A comparable behavior can be observed for the small-world networks. In this case, graphs with a low number of nodes, have a higher mutual information due to more intra-module connections, and, as a consequence, if we increase the number of nodes, the number of edges also increases. On the other side, we can observe that while ring lattice network have a high value, lattice network have a very small value. This is because two nodes of the lattice network are not connected and, the rest of the nodes, have a higher degree compared to ring lattice. Consequently, there are less unique paths. Finally, erasure mutual information is not very sensitive to the graph size but to the graph topology. Random and small world have higher values compared to ring lattice and lattice. This is due to the existence of a large amount of connecting paths for neighbor nodes in ring and lattice networks, so paths are not unique. Since for lattice graphs two nodes are not connected, the rest of the nodes have a slightly higher degree, and, as a result the overall predictability is lower. If we increase the number of nodes we have to increase also the number of edges, thus, as a result, the degree of the nodes increases. Because of this, if we focus on the values for a low number of nodes, we can observe that the erasure mutual information for lattice and ring lattice slightly decreases, and, for random and small-world, increases. The erasure mutual information measure takes into account the next node but also the previous one. Therefore, increasing the degree in the ring lattice and the lattice networks, the overall uncertainty increases. On the contrary, for random and small-world networks with a low node degree, paths are more unique for a low number of nodes. Increasing the number of nodes while keeping the density the same, the erasure mutual information tends to stabilize.

To evaluate the global measures with anatomical data, we applied the global measures to the anatomic and functional datasets at different scales described in Section 3.2. Figure 8 shows the result of entropy, mutual information, and erasure mutual information for the 10 structural networks with 83, 129, 234 and 1015 partitions. Observe that all measures have a similar behavior for all the patients which demonstrates that the measures are consistent among all patients. Figure 9 shows the result of the global measures applied to 468 functional networks with 25, 50, 100, 200 and 300 nodes. In this case, the entropy measure has the same behavior as the structural network. Moreover, since the density is similar between different partitions, the mutual information and the erasure measures have a more uniform value. The same effect has been shown in the behavior of the mutual information and the erasure for model graphs with a constant density.

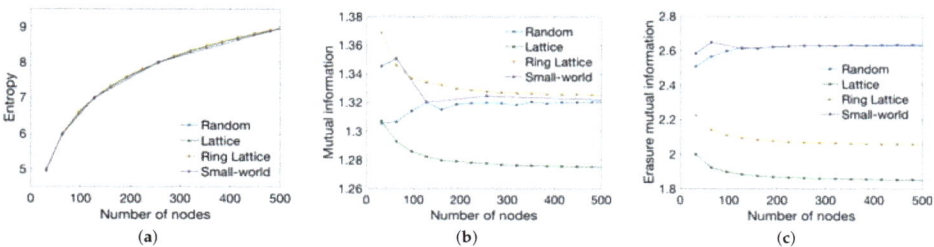

Figure 7. Behavior of the (**a**) entropy; (**b**) mutual information; (**c**) and erasure mutual information measures when the number of nodes is increased (from 0 to 500) while the density is kept constant to 0.4.

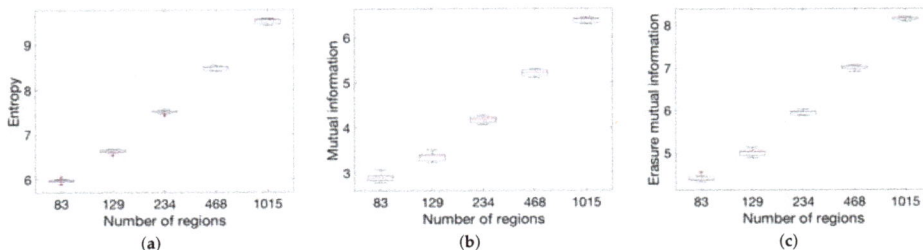

Figure 8. Box-plots showing median, 25th and 75th percentiles for global measures ((a) entropy; (b) mutual information; (c) and erasure mutual information) when applied to the 10 structural connectomes with 83, 129, 254, 463 and 1015 partitions.

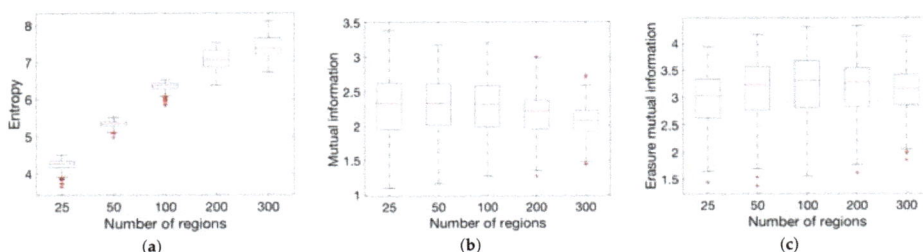

Figure 9. Box-plots showing median, 25th and 75th percentiles for global measures ((a) entropy; (b) mutual information; (c) and erasure mutual information) when applied to 463 functional connectomes with 25, 50, 100, 200 and 300 partitions.

4.2. Local Measures

In this section, we compare local measures (entropic surprise, mutual surprise, mutual predictability and erasure surprise) with standard measures provided in the BCT. In addition, we show the result of the measures applied to the structural and functional human connectivity networks.

First of all, we provide a comparison of the proposed local measures with *strength*, *eccentricity* and *clustering* measures. Strength measures the sum of the weights for each node, eccentricity the maximal shortest path length between a node and any other node, and clustering the fraction of triangles in the node. To carry out this experiment, we have considered the averaged connectivity matrix created from the 10 structural networks with 1015 nodes of the anatomic dataset described in Section 3.2.1. The purpose of this experiment is to demonstrate the feasibility and application of the method in a real clinical scenario. Results are shown in Figure 10. From left to right, columns represent entropic surprise (E), mutual surprise (I_1), mutual predictability (I_2), and erasure surprise (I_1^-), and, from top to down, rows represent the value of our measure with respect to strength, eccentricity, and clustering, respectively. For each scatter plot, x-axis represents the standard measure value and the y-axis the value of our measure. In each plot, the logarithmic curve ($f(x) = a + b \log_2(x)$) that best fits to the data and the determination coefficient, R^2, of the data model are also shown. We can see that the surprise measure is directly related to the strength since both depend on the weight of the node and the surprise measure is mathematically defined as minus the logarithm of the strength see Equation (16). As it can be seen, the other measures are moderately correlated (mainly I_1 and I_1^-) to the strength. This is not directly related to their mathematical definition, but by the fact that those nodes with more connections (high strength) tend to have more uncertainty on their connections and, thus, lower measure values. Comparing with the eccentricity measure, we can observe that nodes with a high maximal shortest path length (high eccentricity) tend to not be highly connected (low E value). On the other side, nodes with a low eccentricity are highly connected. The other measures do not demonstrate significant correlation with eccentricity. With respect to the clustering measure, I_2 is the only measure that slightly correlates with it. This can be explained by the fact that those nodes with a high clustering coefficient will tend to have less uncertainty on their connections.

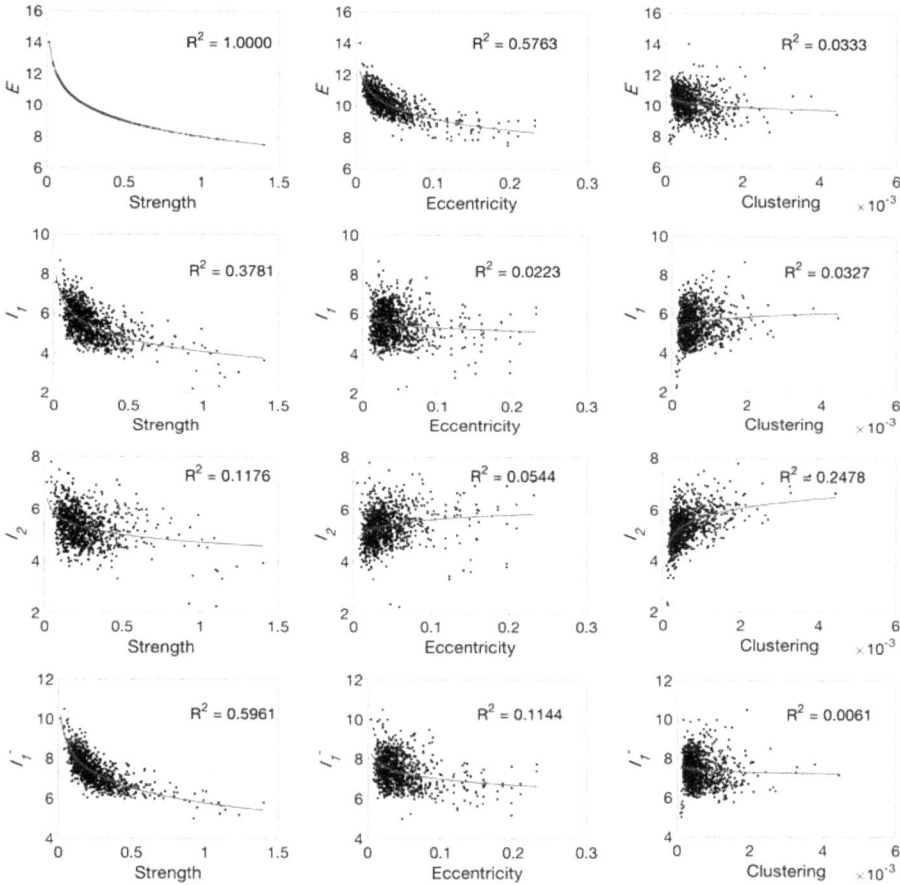

Figure 10. Relationship between the proposed local measures (entropic surprise (E), mutual surprise (I_1), mutual predictability (I_2) and erasure surprise (I_1^-) and standard measures (strength, eccentricity and clustering) using the structural averaged connectivity matrix network with 1015 nodes.

Finally, we show the value of each measure for each node of the human structural and functional averaged networks, with 83 and 25 partitions, respectively. Figure 11a shows all the nodes for the structural network in yellow and the connections between nodes in black.

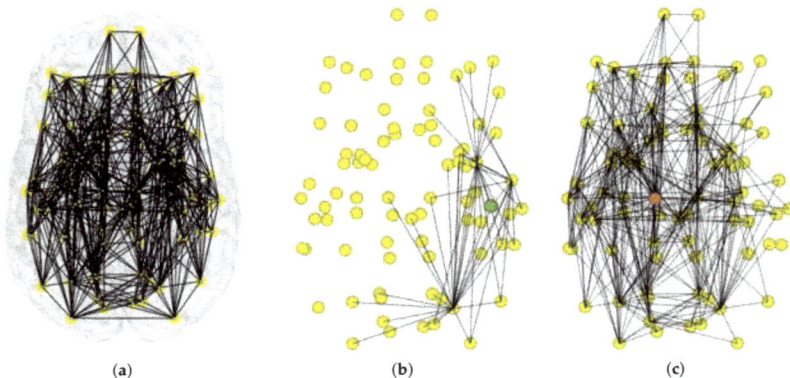

Figure 11. (**a**) Illustration of all the connections in the structural dataset; (**b**) Right hemisphere transverse temporal region (green) connections including its neighbors connections; (**c**) Left hemisphere thalamus proper (orange) connections including its neighbors connections. This figure has been generated using the VisualConnectome software [59].

The value of the entropic surprise E for each node of the human structural averaged network, with 83 partitions is shown on the left histogram of Figure 12. This measure is equivalent to the classic strength measure, where nodes with high values are nodes not highly connected or with low weights, which lead to a low stationary probability. The maximum and minimum E values corresponds to the right hemisphere transverse temporal and the brain stem, respectively. These nodes have been represented on the brain network in green and orange (see Figure 12 top image of the central column). The value of the mutual surprise I_1 for each node is shown on the right histogram of Figure 12. High values correspond to nodes connected to poorly connected nodes (nodes with a low number of connections), while low values correspond to nodes connected to highly connected nodes. This fact is illustrated on the bottom image of the central column of Figure 12 where the right hemisphere transverse temporal, represented as a green node, has the maximum value and the left hemisphere thalamus proper, represented as an orange node, has the minimum one. Comparing entropic surprise and mutual surprise for the structural connectome, we observe an organization, where nodes highly connected are also nodes connected to similar nodes in terms of probability, and nodes not highly connected are connected to nodes with a very different probability compared to them.

The value of the mutual predictability I_2 for each node of the human structural averaged network, with 83 partitions is shown on left histogram of Figure 13. Remember that, for nodes with a high mutual predictability, the distribution of connections with other nodes have a low entropy. For instance, observe the first image of the central column in Figure 13, the green node, which corresponds to the right hemisphere temporal pole, has the highest I_2 value. On the other hand, nodes with low values have more uncertainty in predicting the next node. In this case, the lowest I_2 value corresponds to the right hemisphere putamen, represented as an orange node. The value of the erasure surprise I_1^- is shown on the right histogram of Figure 13. Nodes with high values are nodes that connect different areas otherwise not connected or less connected, like a bridge or a hub. For example, the right hemisphere transverse temporal, shown in Figure 11b together with its neighbor connections, is the region with a higher value in the bottom image of the central column in Figure 13. On the other side, nodes with low values, are nodes that belong to a cluster since there are multiple paths connecting its neighbors. In this case, the lower value of the histogram corresponds to the left hemisphere thalamus proper, which is shown in Figure 11c together with its neighbor connections.

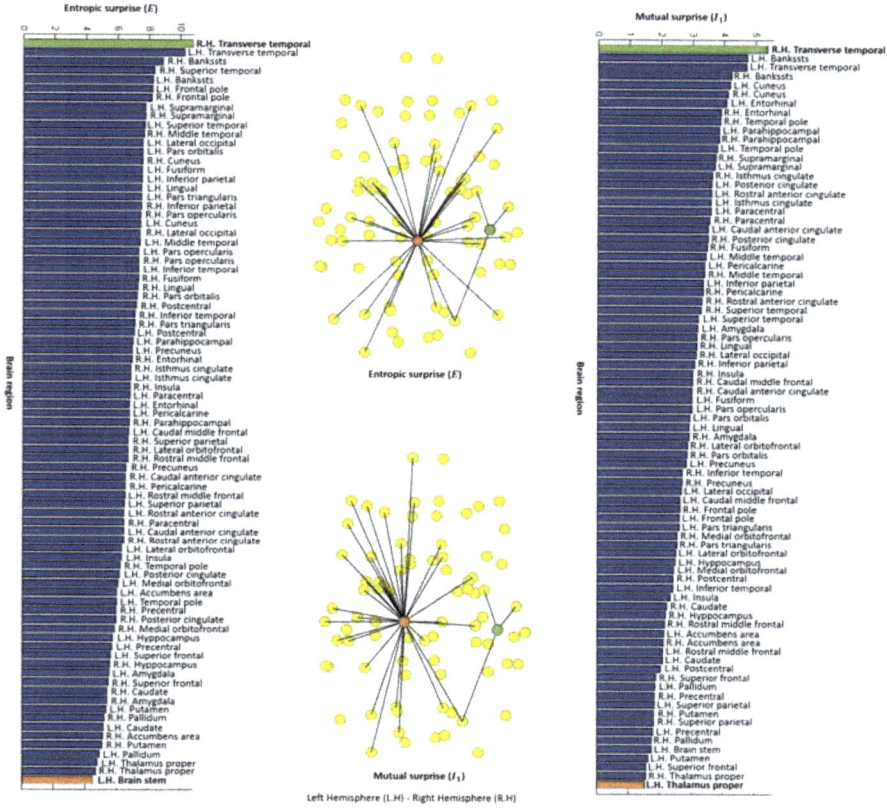

Figure 12. On the left, entropic surprise values obtained with the averaged structural network with 83 partitions. The maximum and minimum values have been represented on the brain network (first image of the central column). The green node corresponds to the right hemisphere transverse temporal area and the orange to the brain stem. On the right, mutual surprise values obtained with the same network. The maximum and minimum values have been represented on the brain network (second image of the central column). The green node corresponds to the right hemisphere transverse temporal area and the orange to the thalamus proper.

Figure 14 shows the results for the entropic surprise, the mutual surprise, the mutual predictability, and the erasure surprise applied to the human functional network with 25 partitions. An illustrative image of each partition is shown in Figure 15. Analyzing independently the measures, we found a behavior similar to the structural networks. However, evaluating all the measures and comparing them, we can observe interesting properties. For instance, regions 14 and 19 have both a high erasure surprise value, while mutual surprise is high for region 14 and low for region 19. Thus, these two regions belong to a unique path (due to a high erasure surprise value) but region 14 connects regions highly connected while region 19 connects regions poorly connected (due to the mutual information value). On the other side, region 1, which has a high sum of weights, is also connected to regions similar to itself, so regions with also a big amount of connections. Region 19 has also a low mutual predictability, which means that there is a high capacity to predict the regions which is connected to, on the contrary, region 1 has a lower mutual predictability, so even if it is highly connected to similar nodes it is difficult to predict which are the nodes. Finally, region 1 has a low erasure surprise, which indicates that is likely to belong to a cluster, and region 14 has a high erasure surprise, so it acts more as a bridge of areas which are not strongly connected with other areas.

These results show a proof of principle of the proposed brain model and the suggested set of measures, that provide robust results using structural or functional data. Prior to a further investigation with more clinical data, the proposed approach provides new insights into the brain complexity which may be of interest in studying the functioning of the brain and the connections between regions.

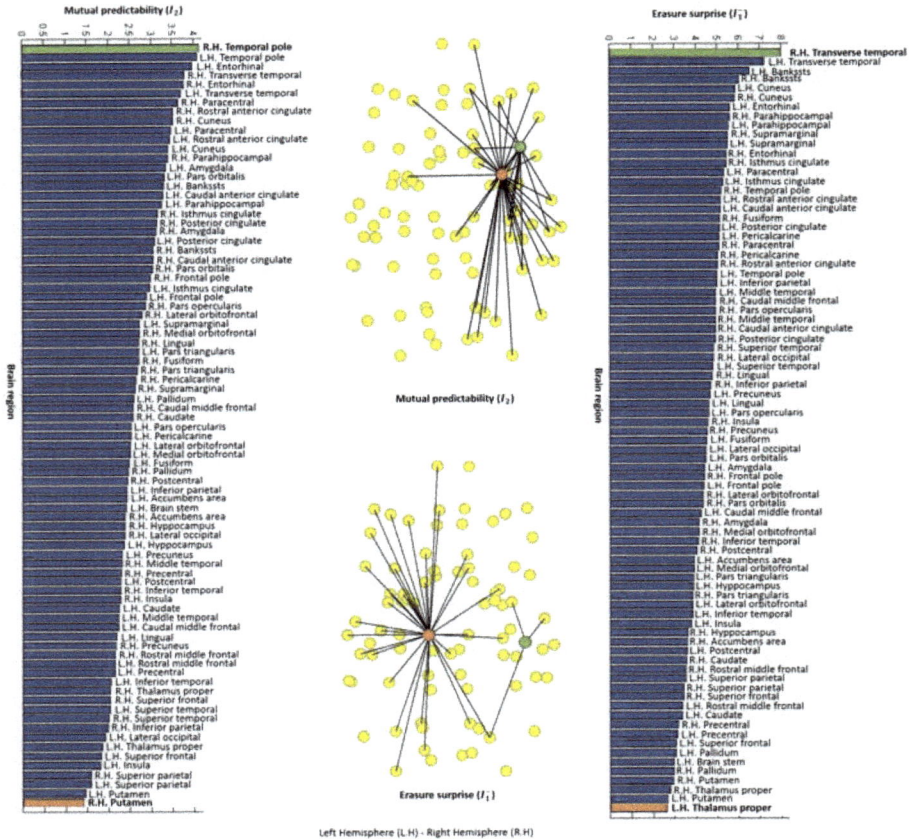

Figure 13. On the left, mutual predictability values obtained with the averaged structural network with 83 partitions. The maximum and minimum values have been represented on the brain network (first image of the central column). The green node corresponds to the right hemisphere temporal pole area and the orange to the putamen. On the right, erasure surprise values obtained with the same network. The maximum and minimum values have been represented on the brain network (second image of the central column). The green node corresponds to the right hemisphere transverse temporal area and the orange to the thalamus proper.

(**a**) Entropic surprise

(**b**) Mutual surprise

(**c**) Mutual predictability

(**d**) Erasure surprise

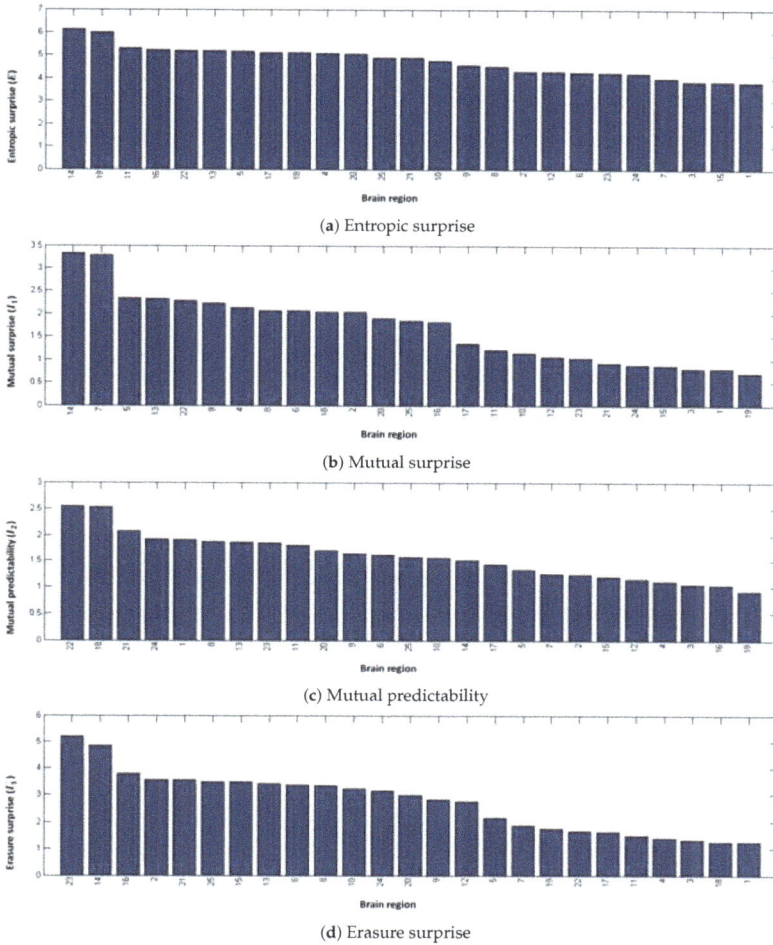

Figure 14. Local measures values ((**a**) entropic surprise; (**b**) mutual surprise (**c**) mutual predictability and (**d**) erasure surprise) obtained with the averaged functional dataset with 25 partitions. An illustrative image of each partition is shown in Figure 15.

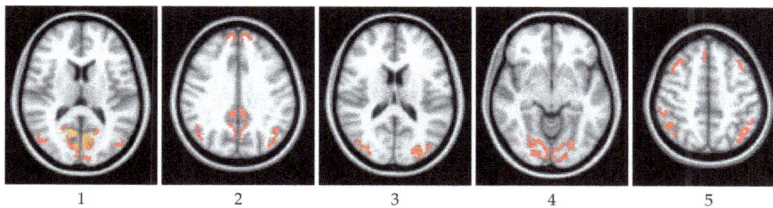

Figure 15. *Cont.*

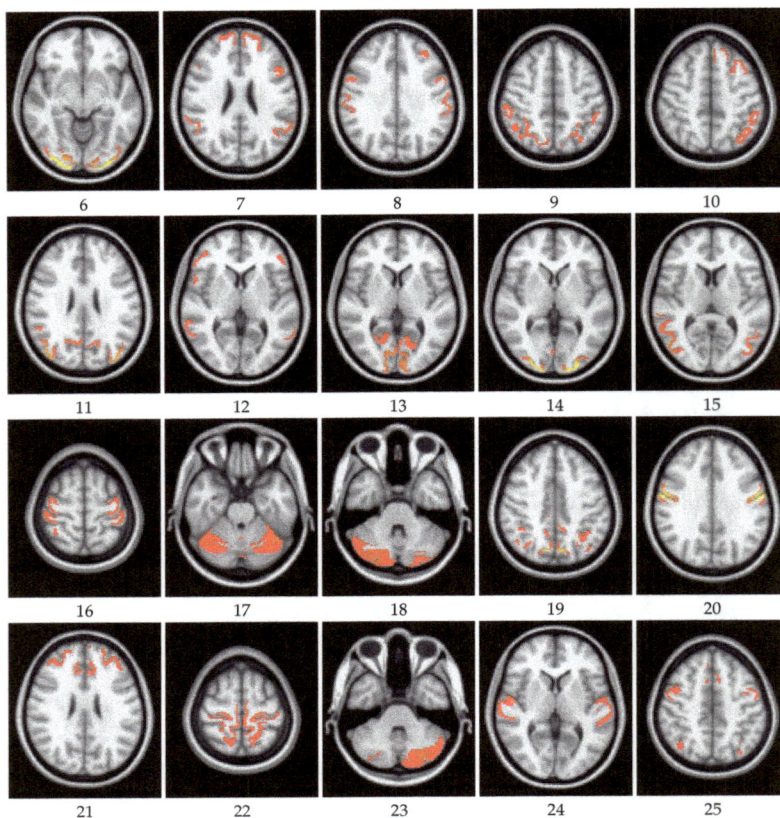

Figure 15. Illustrative images of the 25 regions from the averaged functional dataset [55–57].

5. Conclusions

In this paper, we have used a Markov process-based brain model in which we apply existent and novel information theory-based measures to characterize new properties of complex brain networks. The main contribution of the paper is the proposal of new local and global measures to describe new properties of brain networks in terms of topology and organization, with the main novelty being the definition of erasure mutual information and the erasure surprise. The proposed measures have been tested on synthetic model networks, increasing the number of nodes and the number of edges, and with structural and functional human networks at different scales.

From our experiments and focusing on global measures, we showed that, given a network, the entropy, describes the overall uncertainty of the nodes connectivity. In addition, mutual information, which is a measure of structure, is able to differentiate the topology of network models. Finally, the erasure mutual information, which is a new measure defined by extending the mutual information, describes how unique the paths for a given network are. With this measure, we show an optimal point for small-world networks.

Focusing on local measures, we observed that the entropic surprise, which describes how connected is a node taking into account all the connections in the network, is inversely proportional to the logarithm of the standard strength measure. The mutual surprise, which defines the connectivity of the neighbor nodes for a given node, allows to identify nodes whose nodes have a high connectivity taking into account all connections. The mutual predictability, which given a node, determines the uncertainty associated to a node in predicting the next node, shows that regions with a high clustering tend to be more predictable. Finally, the erasure surprise, which takes into account previous and next nodes, defines how unique the path is which the node belongs to. Results show that regions with a high strength belong to a module where all nodes are strongly connected. The consistency of the results for structural and functional human networks demonstrates the robustness of the proposed measures.

In future work, we will analyze in detail the properties of specific anatomical areas of the human brain and we will study how it can help to detect different diseases. Furthermore, we will investigate clinically informative visualizations using the presented measures.

Author Contributions: E.B., A.B., M.F. and I.B. conceived and designed the experiments; E.B. performed the experiments; E.B. and A.B. analyzed the data; E.B., A.B. and I.B. wrote the paper.

Funding: This work was supported by the Spanish Government (Grant No. TIN2016-75866-C3-3-R) and by the Catalan Government (Grant No. 2017-SGR-1101). Data were provided, in part, by the Human Connectome Project, WU-Minn Consortium (Principal Investigators: David Van Essen and Kamil Ugurbil; 1U54MH091657) funded by the 16 NIH Institutes and Centers that support the NIH Blueprint for Neuroscience Research; and by the McDonnell Center for Systems Neuroscience at Washington University.

Conflicts of Interest: The authors declare no conflict of interest.

References

1. Sporns, O.; Tononi, G.; Kötter, R. The human connectome: A structural description of the human brain. *PLoS Comput. Biol.* **2005**, *1*, e42. [CrossRef] [PubMed]
2. Felleman, D.J.; Essen, D.C.V. Distributed hierarchical processing in the primate cerebral cortex. *Cereb. Cortex* **1991**, *1*, 1–47. [CrossRef] [PubMed]
3. Hagmann, P. From Diffusion MRI to Brain Connectomics. Ph.D. Thesis, École polytechnique fédérale de Lausanne (EPFL), Lausanne, Switzerland, 2005.
4. Hagmann, P.; Kurant, M.; Gigandet, X.; Thiran, P.; Wedeen, V.J.; Meuli, R.; Thiran, J.P. Mapping human whole-brain structural networks with diffusion MRI. *PLoS ONE* **2007**, *2*, e597. [CrossRef] [PubMed]
5. Hagmann, P.; Cammoun, L.; Gigandet, X.; Gerhard, S.; Grant, P.E.; Wedeen, V.; Meuli, R.; Thiran, J.P.; Honey, C.J.; Sporns, O. MR connectomics: Principles and challenges. *J. Neurosci. Methods* **2010**, *194*, 34–45. [CrossRef] [PubMed]
6. Sporns, O. The human connectome: Origins and challenges. *NeuroImage* **2013**, *80*, 53–61. [CrossRef] [PubMed]
7. Bullmore, E.T.; Bassett, D.S. Brain graphs: Graphical models of the human brain connectome. *Ann. Rev. Clin. Psychol.* **2011**, *7*, 113–140. [CrossRef] [PubMed]
8. Sporns, O. The human connectome: A complex network. *Ann. New York Acad. Sci.* **2011**, *1224*, 109–125. [CrossRef] [PubMed]
9. Wu, G.R.; Liao, W.; Stramaglia, S.; Ding, J.R.; Chen, H.; Marinazzo, D. A blind deconvolution approach to recover effective connectivity brain networks from resting state fMRI data. *Med. Image Anal.* **2013**, *17*, 365–374. [CrossRef] [PubMed]
10. Passingham, R.E.; Stephan, K.E.; Kotter, R. The anatomical basis of functional localization in the cortex. *Nat. Rev. Neurosci.* **2002**, *3*, 606–616. [CrossRef] [PubMed]
11. Kaiser, M. A tutorial in connectome analysis: Topological and spatial features of brain networks. *NeuroImage* **2011**, *57*, 892–907. [CrossRef] [PubMed]
12. Rubinov, M.; Sporns, O. Complex network measures of brain connectivity: Uses and interpretations. *NeuroImage* **2010**, *52*, 1059–1069. [CrossRef] [PubMed]
13. Stam, C.; Reijneveld, J. Graph theoretical analysis of complex networks in the brain. *Nonlinear Biomed. Phys.* **2007**, *1*, 3. [CrossRef] [PubMed]
14. Watts, D.J.; Strogatz, S.H. Collective dynamics of 'small-world' networks. *Nature* **1998**, *393*, 440–442. [CrossRef] [PubMed]
15. Latora, V.; Marchiori, M. Efficient behavior of small-world networks. *Phys. Rev. Lett.* **2001**, *87*, 198701. [CrossRef] [PubMed]
16. Newman, M. The structure and function of complex networks. *SIAM* **2003**, *45*, 167–256. [CrossRef]

17. Newman, M. Fast algorithm for detecting community structure in networks. *Phys. Rev. E* **2003**, *69*, 066133. [CrossRef] [PubMed]

18. Bullmore, E.; Sporns, O. Complex brain networks: Graph theoretical analysis of structural and functional systems. *Nat. Rev. Neurosci.* **2009**, *10*, 186–198. [CrossRef] [PubMed]

19. Kennedy, H.; Knoblauch, K.; Toroczkai, Z. Why data coherence and quality is critical for understanding interareal cortical networks. *NeuroImage* **2013**, *80*, 37–45. [CrossRef] [PubMed]

20. Colizza, V.; Flammini, A.; Serrano, M.A.; Vespignani, A. Detecting rich-club ordering in complex networks. *Nat. Phys.* **2006**, *2*, 110–115. [CrossRef]

21. Harriger, L.; van den Heuvel, M.P.; Sporns, O. Rich club organization of macaque cerebral cortex and its role in network communication. *PLoS ONE* **2012**, *7*, e46497. [CrossRef] [PubMed]

22. Van Den Heuvel, M.; Kahn, R.; Goñi, J.; Sporns, O. High-cost, high-capacity backbone for global brain communication. *Proc. Natl. Acad. Sci. USA* **2012**, *109*, 11372–11377. [CrossRef] [PubMed]

23. Sporns, O.; Chialvo, D.R.; Kaiser, M.; Hilgetag, C.C. Organization, development and function of complex brain networks. *Trends Cognit. Sci.* **2004**, *8*, 418–425. [CrossRef] [PubMed]

24. Tononi, G.; Sporns, O.; Edelman, G.M. A measure for brain complexity: Relating functional segregation and integration in the nervous system. *Proc. Natl. Acad. Sci. USA* **1994**, *91*, 5033–5037. [CrossRef] [PubMed]

25. Marrelec, G.; Bellec, P.; Krainik, A.; Duffau, H.; Pélégrini-Issac, M.; Lehéricy, S.; Benali, H.; Doyon, J. Regions, systems, and the brain: Hierarchical measures of functional integration in fMRI. *Med. Image Anal.* **2008**, *12*, 484–496. [CrossRef] [PubMed]

26. Kitazono, J.; Kanai, R.; Oizumi, M. Efficient algorithms for searching the minimum information partition in integrated information theory. *Entropy* **2018**, *20*, 173. [CrossRef]

27. Tononi, G.; Edelman, G.M.; Sporns, O. Complexity and coherency: Integrating information in the brain. *Trends Cognit. Sci.* **1998**, *2*, 474–484. [CrossRef]

28. Sporns, O.; Tononi, G.; Edelman, G.M. Connectivity and complexity: The relationship between neuroanatomy and brain dynamics. *Neural Netw.* **2000**, *13*, 909–922. [CrossRef]

29. Tononi, G.; Sporns, O.; Edelman, G.M. A complexity measure for selective matching of signals by the brain. *Proc. Natl. Acad. Sci. USA* **1996**, *93*, 3422–3427. [CrossRef] [PubMed]

30. Tononi, G.; McIntosh, A.R.; Russell, D.P.; Edelman, G.M. Functional clustering: Identifying strongly interactive brain regions in neuroimaging data. *NeuroImage* **1998**, *7*, 133–149. [CrossRef] [PubMed]

31. Edelman, G.M.; Gally, J.A. Degeneracy and complexity in biological systems. *Proc. Natl. Acad. Sci. USA* **2001**, *98*, 13763–13768. [CrossRef] [PubMed]

32. Tononi, G.; Sporns, O.; Edelman, G.M. Measures of degeneracy and redundancy in biological networks. *Proc. Natl. Acad. Sci. USA* **1999**, *96*, 3257–3262. [CrossRef] [PubMed]

33. Crossley, N.; Mechelli, A.; Scott, J.; Carletti, F.; Fox, P.; McGuire, P.; Bullmore, E. The hubs of the human connectome are generally implicated in the anatomy of brain disorders. *Brain* **2014**, *137*, 2382–2395. [CrossRef] [PubMed]

34. Meskaldji, D.E.; Fischi-Gomez, E.; Griffa, A.; Hagmann, P.; Morgenthaler, S.; Thiran, J.P. Comparing connectomes across subjects and populations at different scales. *NeuroImage* **2013**, *80*, 416–425. [CrossRef] [PubMed]

35. van den Heuvel, M.P.; Pol, H.E.H. Exploring the brain network: A review on resting-state fMRI functional connectivity. *Eur. Neuropsychopharmacol.* **2010**, *20*, 519–534. [CrossRef] [PubMed]

36. Sato, J.R.; Takahashi, D.Y.; Hoexter, M.Q.; Massirer, K.B.; Fujita, A. Measuring network's entropy in ADHD: A new approach to investigate neuropsychiatric disorders. *NeuroImage* **2013**, *77*, 44–51. [CrossRef] [PubMed]

37. Papo, D.; Buldú, J.M.; Boccaletti, S.; Bullmore, E.T. Complex network theory and the brain. *Phil. Trans. R. Soc. B* **2014**, *369*, 20130520. [CrossRef] [PubMed]

38. Bonmati, E.; Bardera, A.; Boada, I. Brain parcellation based on information theory. *Comput. Methods Programs Biomed.* **2017**, *151*, 203–212. [CrossRef] [PubMed]

39. Cover, T.M.; Thomas, J.A. *Elements of Information Theory*; Wiley: Hoboken, NJ, USA, 2006.

40. Yeung, R.W. *A First Course in Information Theory*; Springer Science & Business Media: New York, NY, USA, 2002.

41. Feldman, D.P.; Crutchfield, J.P. *Discovering Noncritical Organization: Statistical Mechanical, Information Theoreticand Computational Views of Patterns in One-Dimensional Spin Systems*; Working Paper 98-04-026; Santa Fe Institute: Santa Fe, NM, USA, 1998.

42. Crutchfield, J.P.; Packard, N. Symbolic dynamics of noisy chaos. *Physica D* **1983**, *7*, 201–223. [CrossRef]
43. Grassberger, P. Toward a quantitative theory of self-generated complexity. *Int. J. Theor. Phys.* **1986**, *25*, 907–938. [CrossRef]
44. Shaw, R. *The Dripping Faucet as a Model Chaotic System*; Aerial Press: Santa Cruz, CA, USA, 1984.
45. Szépfalusy, P.; Györgyi, G. Entropy decay as a measure of stochasticity in chaotic systems. *Phys. Rev. A* **1986**, *33*, 2852. [CrossRef]
46. Feldman, D.P. *A Brief Introduction to: Information Theory, Excess Entropy and Computational Mechanics*; Lecture notes; Department of Physics, University of California: Berkeley, CA, USA, 1997.
47. Verdú, S.; Weissman, T. The information lost in erasures. *IEEE Trans. Inf. Theory* **2008**, *54*, 5030–5058. [CrossRef]
48. Feldman, D.; Crutchfield, J. Structural information in two-dimensional patterns: Entropy convergence and excess entropy. *Phys. Rev. E* **2003**, *67*, 051104. [CrossRef] [PubMed]
49. DeWeese, M.R.; Meister, M. How to measure the information gained from one symbol. *Network Comput. Neural Syst.* **1999**, *10*, 325–340. [CrossRef]
50. Fornito, A.; Zalesky, A.; Breakspear, M. Graph analysis of the human connectome: Promise, progress, and pitfalls. *NeuroImage* **2013**, *80*, 426–444. [CrossRef] [PubMed]
51. Dennis, E.L.; Jahanshad, N.; Toga, A.W.; McMahon, K.; de Zubicaray, G.I.; Martin, N.G.; Wright, M.J.; Thompson, P.M. Test-retest reliability of graph theory measures of structural brain connectivity. In Proceedings of the International Conference on Medical Image Computing and Computer-Assisted Intervention, Nice, France, 1–5 October 2012; Ayache, N., Delingette, H., Golland, P., Mori, K., Eds.; Springer: Berlin, Germany, 2012; Volume 7512, pp. 305–312.
52. Messé, A.; Rudrauf, D.; Giron, A.; Marrelec, G. Predicting functional connectivity from structural connectivity via computational models using MRI: An extensive comparison study. *NeuroImage* **2015**, *111*, 65–75. [CrossRef] [PubMed]
53. Santos Ribeiro, A.; Miguel Lacerda, L.; Ferreira, H.A. Multimodal imaging brain connectivity analysis toolbox (MIBCA). *PeerJ PrePrints* **2014**, *2*, e699v1.
54. Cammoun, L.; Gigandet, X.; Sporns, O.; Thiran, J.; Do, K.; Maeder, P.; Meuli, R.; Hagmann, P.; Bovet, P.; Do, K. Mapping the human connectome at multiple scales with diffusion spectrum MRI. *J. Neurosci. Methods* **2012**, *203*, 386–397. [CrossRef] [PubMed]
55. Essen, D.V.; Ugurbil, K.; Auerbach, E.; Barch, D.; Behrens, T.; Bucholz, R.; Chang, A.; Chen, L.; Corbetta, M.; Curtiss, S.; et al. The Human Connectome Project: A data acquisition perspective. *NeuroImage* **2012**, *62*, 2222–2231. [CrossRef] [PubMed]
56. Glasser, M.F.; Sotiropoulos, S.N.; Wilson, J.A.; Coalson, T.S.; Fischl, B.; Andersson, J.L.; Xu, J.; Jbabdi, S.; Webster, M.; Polimeni, J.R.; et al. The minimal preprocessing pipelines for the Human Connectome Project. *NeuroImage* **2013**, *80*, 105–124. [CrossRef] [PubMed]
57. Hodge, M.R.; Horton, W.; Brown, T.; Herrick, R.; Olsen, T.; Hileman, M.E.; McKay, M.; Archie, K.A.; Cler, E.; Harms, M.P.; et al. ConnectomeDB—Sharing human brain connectivity data. *NeuroImage* **2016**, *124*, 1102–1107. [CrossRef] [PubMed]
58. Christidi, F.; Karavasilis, E.; Samiotis, K.; Bisdas, S.; Papanikolaou, N. Fiber tracking: A qualitative and quantitative comparison between four different software tools on the reconstruction of major white matter tracts. *Eur. J. Radiol. Open* **2016**, *3*, 153–161. [CrossRef] [PubMed]
59. Dai, D.; He, H. VisualConnectome: Toolbox for brain network visualization and analysis. In Proceedings of the Organization on human Brain Mapping, 2011, Québec City, QC, Canada, 26–30 June 2011.

![entropy logo] *entropy*

MDPI

Article

Life on the Edge: Latching Dynamics in a Potts Neural Network

Chol Jun Kang [1,2] (iD) , Michelangelo Naim [1,3] (iD) , Vezha Boboeva [1] and Alessandro Treves [1,4,*] (iD)

1   Cognitive Neuroscience, SISSA—International School for Advanced Studies, Via Bonomea 265, 34136 Trieste, Italy; ckang@sissa.it (C.J.K.); michelangelonaim@gmail.com (M.N.); vboboeva@sissa.it (V.B.)
2   The Abdus Salam International Centre for Theoretical Physics, Strada Costiera 11, 34151 Trieste, Italy
3   Department of Physics, La Sapienza Università di Roma, Piazzale Aldo Moro, 5, 00185 Roma, Italy
4   Centre for Neural Computation, Norwegian University of Science and Technology, 7491 Trondheim, Norway
*   Correspondence: ale@sissa.it; Tel.: +39-040-3787-623

Received: 2 August 2017; Accepted: 29 August 2017; Published: 3 September 2017

Abstract: We study latching dynamics in the adaptive Potts model network, through numerical simulations with randomly and also weakly correlated patterns, and we focus on comparing its slowly and fast adapting regimes. A measure, Q, is used to quantify the *quality of latching* in the phase space spanned by the number of Potts states S, the number of connections per Potts unit C and the number of stored memory patterns p. We find narrow regions, or *bands* in phase space, where distinct pattern retrieval and duration of latching combine to yield the highest values of Q. The bands are confined by the storage capacity curve, for large p, and by the onset of finite latching, for low p. Inside the band, in the slowly adapting regime, we observe complex structured dynamics, with transitions at high crossover between correlated memory patterns; while away from the band latching, transitions lose complexity in different ways: below, they are clear-cut but last such few steps as to span a transition matrix between states with few asymmetrical entries and limited entropy; while above, they tend to become random, with large entropy and bi-directional transition frequencies, but indistinguishable from noise. Extrapolating from the simulations, the band appears to scale almost quadratically in the p–S plane, and sublinearly in p–C. In the fast adapting regime, the band scales similarly, and it can be made even wider and more robust, but transitions between anti-correlated patterns dominate latching dynamics. This suggest that slow and fast adaptation have to be integrated in a scenario for viable latching in a cortical system. The results for the slowly adapting regime, obtained with randomly correlated patterns, remain valid also for the case with correlated patterns, with just a simple shift in phase space.

Keywords: neural network; Potts model; latching; recursion

---

1. Introduction

How can the human brain produce creative behaviour? Systems neuroscience has mainly focused on the states induced, in particular in the cortex, by external inputs, be these states simple distributions of neuronal activity or more complex dynamical trajectories. It has largely eschewed the question of how such states can be combined into novel sequences that express, rather than the reaction to an external drive, spontaneous cortical dynamics. However, the generation of novel sequences of states drawn from even a finite set has been characterized as the infinitely recursive process deemed to underlie language productivity, as well as other forms of creative cognition [1]. If the individual states, whether fixed points or stereotyped trajectories, are conceptualized as dynamical attractors [2], the cortex can be thought of as engaging in a kind of chaotic saltatory dynamics between such attractors [3]. Attractor dynamics has indeed fascinated theorists, and a major body of work has shown how to make relevant for neuroscience the concepts and analytical tools developed within statistical physics, but the focus has been on compact, homogeneous neural networks [4–7]. These have been regarded as simplified models of local cortical networks—as well as, e.g., of the CA3 hippocampal field—and have not been analysed in their potential saltatory dynamics, given that it would make no sense to consider local cortical networks as isolated systems. Even in the case of a ground-breaking investigation of putative spatial trajectory planning [8], the hippocampal activity that expressed it was thought not to be entirely endogeneous, but rather guided by external inputs, including those representing goals and path integration. Therefore, formal analyses of model networks endowed with attractor dynamics have been largely confined to the simple paradigm of cued retrieval from memory. Attempts have been made to explore methodologies to study mechanisms beyond simple cued retrieval [9,10], for example those involved in drawing, confabulation, thought processes in general, and language, which are all considered to be largely independent of external stimuli, at their core, and to combine generativity with recursion [11–16].

Potts neural networks, on the other hand, originally studied merely as a variant of mathematical or potentially applied interest [17–21], offer one approach to model spontaneous dynamics in extended cortical systems, in particular if simple mechanisms of temporal adaptation are taken into account [22]. They can be subject to rigorous analyses of e.g., their storage capacity [23], of the mechanics of saltatory transitions between states [24] and are amenable to a description in terms of distinct "thermodynamic" phases [25,26]. The dynamic modification of thresholds with timescales separate from that of retrieval, i.e., temporal adaptation, together with the correlation between cortical states, are key features characterizing cortical operations, and Potts network models may contribute to elucidate their roles. Adaptation and its role in semantic priming [27] have been linked to the instability manifested in schizophrenia [28].

The Potts description is admittedly an oversimplified effective model for an underlying two-level auto-associative memory network [29]. The even more drastically simplified model of latching dynamics considered by the Tsodyks group [30,31], however, has afforded spectacular success in explaining the scaling laws obtained for free recall in experiments performed 50 years ago. The Potts model may be relevant to a wide set of behaviours and to related experimental measures, once the correspondence between model parameters and the quantities characterizing the underlying two-level network are elucidated. On this correspondence, we elaborate in a separate study [32]. Here, we ask when does the Potts network latch?

2. The Model

We consider an attractor neural network model comprised of Potts units, as depicted in Figure 1. The rationale for the model is that each unit represents a local network of many neurons with its own attractor dynamics [4,6], but in a simplified/integrated manner, regardless of detailed local dynamics. Local attractor states are represented by $S + 1$ Potts states: S active ones and one quiescent state (intended to describe a situation of no retrieval in the local network), σ_i^k, $k = 0, 1, \cdots, S$, with the constraint that $\sum_{k=0}^{S} \sigma_i^k \equiv 1$. We call this autoassociative network of Potts units a Potts network, and refer to our earlier studies of some of its properties [22–25,33].

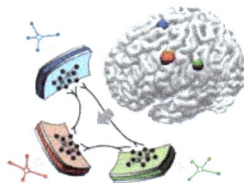

Figure 1. Global cortical model as a Potts neural network. Reprinted with permission from [25].

The "synaptic" connection between two Potts units is in fact a tensor summarizing the effect of very many actual connections between neurons in the two local networks, but still following the Hebbian learning rule [34], we write the connection weight between unit i in state k and unit j in state l as [23]

$$J_{ij}^{kl} = \frac{c_{ij}}{Ca(1 - a/S)} \sum_{\mu=1}^{p} \left(\delta_{\xi_i^\mu, k} - \frac{a}{S} \right) \left(\delta_{\xi_j^\mu, l} - \frac{a}{S} \right) (1 - \delta_{k0})(1 - \delta_{l0}), \tag{1}$$

where c_{ij} is 1 if two units i and j have a connection and 0 otherwise, C is the average number of connections per unit, a is the sparsity parameter, i.e., the fraction of active units in every stored global activity pattern ($\{\xi_i^\mu\}$, $\mu = 1, 2, \cdots, p$) and p is the number of stored patterns. The last two delta functions imply that the learned connection matrix does not affect the quiescent states. We will use the indices i, j for units, k, l for states and μ, ν for patterns. Units are updated in the following way:

$$\sigma_i^k = \frac{\exp (\beta r_i^k)}{\sum_{l=1}^{S} \exp (\beta r_i^l) + \exp [\beta(\theta_i^0 + U)]} \tag{2}$$

and

$$\sigma_i^0 = \frac{\exp [\beta(\theta_i^0 + U)]}{\sum_{l=1}^{S} \exp (\beta r_i^l) + \exp [\beta(\theta_i^0 + U)]}, \tag{3}$$

where r_i^k is the input to (active) state k of unit i integrated over a time scale τ_1, while U and θ_i^0 are, respectively, the constant and time-varying component of the effective overall threshold for unit i, which in practice act as inverse thresholds on its quiescent state. θ_i^0 varies with time constant τ_3, to describe local network adaptation and inhibitory effects. The stiffness of the local dynamics is parametrized by the inverse "temperature" β (or T^{-1}), which is then distinct from the standard notion of thermodynamic noise. The input-output relations (2) and (3) ensure that

$$\sum_{k=0}^{S} \sigma_i^k = 1.$$

In addition to the overall threshold, θ_i^k is the threshold for unit i specific to state k, and it varies with time constant τ_2, representing adaptation of the individual neurons active in that state, i.e., their neural or even synaptic fatigue. The time evolution of the network is then governed by equations that include three distinct time constants:

$$\tau_1 \frac{dr_i^k(t)}{dt} = h_i^k(t) - \theta_i^k(t) - r_i^k(t), \tag{4}$$

$$\tau_2 \frac{d\theta_i^k(t)}{dt} = \sigma_i^k(t) - \theta_i^k(t), \tag{5}$$

$$\tau_3 \frac{d\theta_i^0(t)}{dt} = \sum_{k=1}^{S} \sigma_i^k(t) - \theta_i^0(t), \tag{6}$$

where the field that the unit i in state k experiences reads

$$h_i^k = \sum_{j \neq i}^{N} \sum_{l=1}^{S} J_{ij}^{kl} \sigma_j^l + w \left(\sigma_i^k - \frac{1}{S} \sum_{l=1}^{S} \sigma_i^l \right). \tag{7}$$

The "local feedback term" w is a parameter, first introduced in [25], that modulates the inherent stability of Potts states, i.e., that of local attractors in the underlying network model. It helps the network converge to an attractor faster by giving positive feedback to the most active states and so it effectively deepens their basins of attraction. Note that, in this formulation, feedback is effectively spread over (at least) three time scales: w is positive feedback mediated by collective attractor effects at the neural activity time scale τ_1, θ_i^k is negative feedback mediated by fatigue at the slower time scale τ_2, while θ_i^0 is also negative, and it can be used to model both fast and slow inhibition; for analytical clarity, we consider the two options separately, as the "slowly adapting regime", with $\tau_3 > \tau_2$, and the "fast adapting regime", with $\tau_3 < \tau_1$. It would be easy, of course, to introduce additional time scales, for example by distinguishing a component of θ_i^0 that varies rapidly from one that varies slowly, but it would greatly complicate the observations presented in the following.

The overlap or correlation of the activity state of the network with the global memory pattern μ can be measured as

$$m_\mu = \frac{1}{Na\,(1 - a/S)} \sum_{j \neq i}^{N} \sum_{l \neq 0}^{S} \left(\delta_{\xi_j^\mu l} - \frac{a}{S} \right) \sigma_j^l. \tag{8}$$

Randomly correlated memory patterns are generated according to the following probability distribution

$$P(\xi_i^\mu = k) = \frac{a}{S}, \tag{9}$$

$$P(\xi_i^\mu = 0) = 1 - a,$$

while correlated patterns are generated by the multi-parent algorithm sketched in [22], which will be discussed in a separate study [35].

3. Results

When does robust latching, as a model of spontaneous sequence generation, occur? We address this question with extensive computer simulations, mostly focused on latching between randomly correlated patterns. We consider first the slowly adapting regime ($\tau_1 \ll \tau_2 \ll \tau_3$) in which active states (τ_2) adapt slower than activity propagation to other units (τ_1), while inhibitory feedback is restricted to an even slower timescale, τ_3. Next, we contrast with it the fast adapting regime ($\tau_3 \ll \tau_1 \ll \tau_2$) in which, instead, inhibitory feedback is immediate, relative to the other two time scales.

The critical parameters at play are the number of patterns, p, the number of active states, S, and the number of connections per unit, C, and we also look at the effect of the feedback term w. The other parameters, including T, τ_1, τ_2, and τ_3, are kept fixed during simulations, after having chosen a priori values that can lead to robust latching dynamics in the two regimes.

3.1. Slowly Adapting Regime

In the slowly adapting regime, over a (short) time of order τ_1 the network, if suitably cued, may reach one of the global attractors, and stay there for a while; whereupon, after an adaptation time of order τ_2, it may latch to another attractor, or else activity may die [25]. However, how distinct is the convergence to the new attractor? One may assess this as the difference between the two highest overlaps the network activity has, at time t, with any of the memory patterns, $m_1(t) - m_2(t)$: ideally, $m_1 \simeq 1$ and m_2 is small, so their difference approaches unity. A summary measure of memory pattern discrimination can be defined as $d_{12} \equiv \langle \int dt(m_1(t) - m_2(t)) \rangle_{\text{initial cue}}$, where, of course, the identity of patterns 1 and 2 changes over the sequence.

As discussed in [25], by looking at the latching length, how long a simulation runs before, if ever, the network falls into the global quiescent state, one can distinguish several "phases". Depending on the parameters, the dynamics exhibit finite or infinite latching behaviour, or no latching at all. Typically, when increasing the storage load p, the latching sequence is prolonged and eventually extends indefinitely, but, at the same time, its distinctiveness decreases, since memory patterns cannot be individually retrieved beyond the storage capacity; and, even before, each acquires neighbouring patterns, in the finite and more crowded pattern space, with which it is too correlated to be well discriminated.

In Figure 2, we see that, for each $S = (2, 3, 4)$, as p is increased beyond a certain value, latching dynamics rapidly picks up and extends eventually through the whole simulation, but, in parallel, its discriminative ability decreases and almost vanishes—the p-range where d_{12} is large is in fact when there is no latching, and d_{12} only measures the quality of the initial cued retrieval. For $S = 1$ no significant latching sequence is seen, whereas for higher values, at fixed p, its distinctiveness increases with S, but its length decreases from the peak value at $S = 2$.

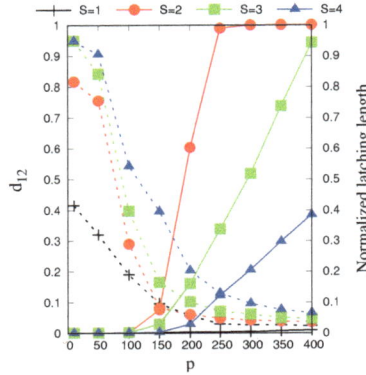

Figure 2. Trade-off between latching sequence length (solid lines) and retrieval discrimination (dashed lines). Different colors indicate different S values, while $C = 400$ throughout. The latching length l is in time steps (not in the number of transitions), normalized by the time of the simulation, $N_{update} = 6 \times 10^5$.

Since the latching length l is not itself sufficient to characterize latching and has to be complemented by discriminative ability, we find it convenient to quantify the overall *quality* of latching. With a new quantity Q defined as

$$Q = d_{12} \cdot l \cdot \eta, \tag{10}$$

where η is introduced to exclude cases in which the network gets stuck in the initial cued pattern, so that no latching occurs; however, high d_{12} and l are:

$$\eta = \begin{cases} 1: & \text{if at least one transition to a second memory pattern occurs,} \\ 0: & \text{otherwise.} \end{cases} \tag{11}$$

Q is therefore a positive real number between 0 and 1, and we report its color-coded value to delineate the relevant phases in phase space.

Thus, *low quality* latching with small Q may result from either small d_{12} or short l, or both. The parameters that determine Q which we focus on are S, C and p, after having suitably chosen all the other parameters, which are kept fixed. Their default values in the slowly adapting regime are $N = 1000$, $a = 0.25$, $U = 0.1$, $T = 0.09$, $w = 0.8$, $\tau_1 = 3.3$, $\tau_2 = 100.0$, $\tau_3 = 10^6$, unless explicitly noted otherwise. If activity does not die out before, simulations are terminated after $N_{update} = 6 \times 10^5$ steps, the total number of updates of the entire Potts network, and are repeated with different cued patterns. Re the values of S, C and p, we use the following notation, for simplicity:

$$Q = Q(S, C, p) = \begin{cases} Q(S, p): & C = 150 \quad \text{fixed,} \\ Q(C, p): & S = 5 \quad \text{fixed.} \end{cases} \tag{12}$$

Figure 3 shows that there are narrow regions in the S–p and C–p planes, which we call *bands*, where relatively *high quality* latching occurs. The values of p with the "best" latching scale almost quadratically in S, and sublinearly in C. Moreover, one notices that, below certain values of S and C, no latching is seen, i.e., the band effectively ends at $S \sim 2$, $p \sim 90$ in Figure 3a and at $C \sim 50$,

$p \sim 70$ in Figure 3b. Importantly, the band in Figure 3a is confined in the area delimited by the cyan solid and dashed curves above and below it. The dashed curve is for the onset of latching, i.e., the phase transition to finite latching [25], while the solid curve above is the storage capacity curve in a diluted network, given by the approximate relation beyond which retrieval fails [25]. It should also be noted that overall Q values are not large, in fact well below 0.5 throughout both S–p and C–p planes. The reason is, again, in the conflicting requirements of persistent latching, favoured by dense storage, high p, and good retrieval, allowed instead only at low storage loads (in practice, relatively low p/S^2 and p/C values):

$$p_c \simeq \frac{CS^2}{4a \ln \frac{2S}{a\sqrt{\ln \frac{S}{a}}}}. \tag{13}$$

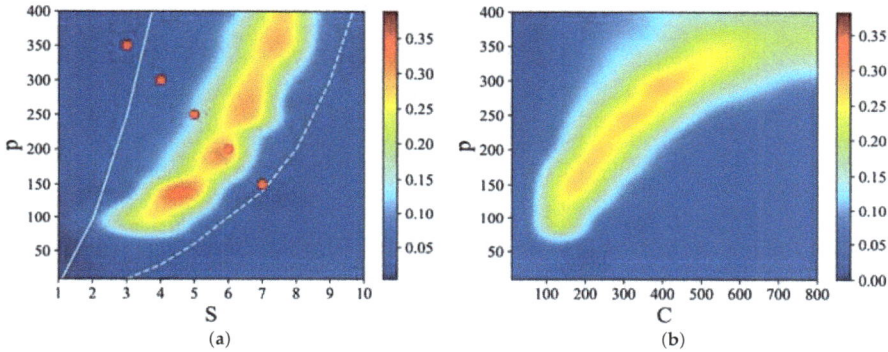

Figure 3. Phase space for $Q(S, p)$ in (**a**) and $Q(C, p)$ in (**b**) with randomly correlated patterns in the slowly adapting regime. The parameters are $C = 150$ and $S = 5$, if kept fixed, and $w = 0.8$. The red spots in (**a**) mark the parameter values used in the following analyses.

In Figure 4, we show representative latching dynamics at three selected points in the (S, p) plane, in terms of the time evolution of the overlap of the states with the stored activity patterns (see Equation (8)). The three points, marked in red, span across the band in Figure 3a, and we see that latching is indefinite but noisy in the example at (5, 250), which is apparently too close to storage capacity, while memory retrieval is good at (7, 150), but the sequence of states ends abruptly, as the network is in the phase of finite latching [25]. The two trends are representative of the two sides of the band, while in the middle, at (6, 200), one finds a reasonable trade-off, with relatively good retrieval combined with protracted latching.

We use two statistical measures, the *asymmetry* of the transition probability matrix and Shannon's *information entropy* [33,36,37] to characterize the essential features of the dynamics in different parameter regions. For that, we take all five red points from Figure 3a, such that they cut across the latching band in the S–p plane, and extend further upwards. We first compile a transition probability (or rather, frequency) matrix M from all distinct transitions observed along many latching sequences generated with the *same set* of stored patterns, as in [33]. The dimension of the matrix M is $(p + 1) \times (p + 1)$, as it includes all possible transitions between p patterns *plus* the global quiescent state. M is constructed from the transitions between states having both overlaps above a given threshold value, e.g., 0.5, in a data set of 1000 latching sequences, by accumulating their frequency between any two patterns into each element of the matrix and then normalizing to 1 row by row, so that $M_{\mu,\nu}$ reflects the probability of a transition from pattern μ to ν. A, the degree of asymmetry of M, is defined as

$$A = \frac{||M - M^T||}{||M||}, \tag{14}$$

where M^T is the transpose matrix of M and $||M|| = \sum_{\mu,\nu} |M_{\mu,\nu}|$. Note that A is small for unconstrained bi-directional dynamics and large for simpler stereotyped flows among global patterns, attaining its maximum value $A = 2$ for strictly uni-directional transitions. Note also that if the average had been taken over *different* realizations of the memory patterns, given sufficient statistics A would obviously vanish.

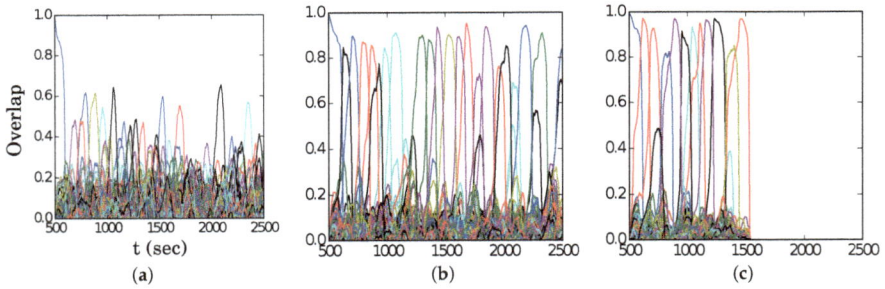

Figure 4. Latching behaviour for (S, p) equal to, respectively: (**a**) (5, 250); (**b**) (6, 200); and (**c**) (7, 150) in Figure 3a.

Another measure we apply to the transition matrix M is Shannon's information entropy, defined as

$$I_\mu = \left\langle \frac{1}{\log_2(p+1)} \sum_{\nu=1}^{p+1} M_{\mu,\nu} \log_2 \left(\frac{1}{M_{\mu,\nu}} \right) \right\rangle_\mu . \tag{15}$$

I_μ takes positive real values from 0 (deterministic, all transitions from one state are to a single other state) to 1 (completely random), since it is normalized by $\log_2(p+1)$, which corresponds to a completely random case.

We use these two measures, A and I_μ, on the points, marked red in Figure 3a.

$$(3, 350) - (4, 300) - (5, 250) - (6, 200) - (7, 150)$$

that lie on a segment going through the latching band observed in the slowly adapting regime. If we focus on transitions between states reaching at least a threshold overlap of 0.5, Figure 5 appears to show two complementary, almost opposite U-shaped curves as the two measures, asymmetry and entropy, are applied to the five points along the segment. One branch of each U shape extends over the range that includes the high-Q latching band: these are the right branches of the two curves, in which asymmetry decreases from a large value $A \simeq 1.6$ at (7, 150) to a smaller one $A \simeq 0.6$ at (5, 250), while concurrently the entropy increases from $I_\mu < 0.5$ at (7, 150) to $I_\mu > 0.8$ at (5, 250). As Figure 4 indicates, at (7, 150), latching sequences are distinct but very short, and few entries are filled in the transition matrix: generally either $M_{\mu\nu} = 0$ or $M_{\nu\mu} = 0$, so that asymmetry is high and entropy relatively low. This holds irrespective of the number of sequences that are averaged over. The opposite happens at (5, 250), where many transitions are observed, and in filling the transition matrix they approach the random limit. The point with the highest Q-value, (6, 200), is characterized by intermediate values of asymmetry and entropy which, we have previously observed, may be seen as a signature of complex dynamics [33]. Extending the range upwards, it seems as if the asymmetry, with threshold 0.5, were to eventually increase again, reaching its maximum $A = 2$ at (3, 350), with a decreasing entropy, vanishing at the same point (3, 350). These left branches are, however, dependent on the threshold values used, as Figure 5 shows, and do not imply that transitions become more deterministic because, in this region, there are simply fewer and fewer distinct transitions discernible above the noise (Figure 4). The left branches merely reflect the increasing arbitrariness with which one can identify significant correlations with memory states in the rambling dynamics observed at higher storage loads.

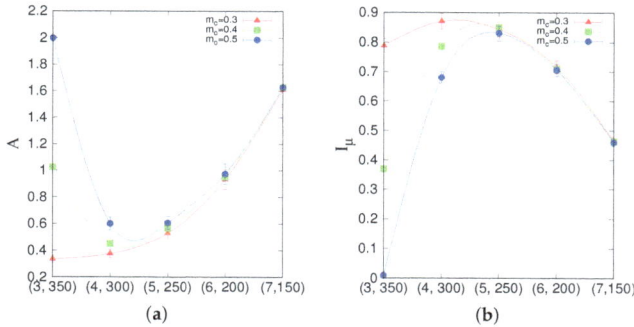

Figure 5. (**a**) asymmetry A of the transition matrix and (**b**) Shannon's information entropy, I_μ along the $(3, 350)$–$(4, 300)$–$(5, 250)$–$(6, 200)$–$(7, 150)$ parameter series from Figure 3. Different curves correspond to different thresholds for the overlap of the two states between which the network is defined to have a transition. The error bars report the standard deviation of either quantity for each of 1000 sequences.

In Figure 6, we see that the effect of the local feedback term, w, is first to enable latching sequences of reasonable quality, and then to also shift the latching band to higher values of S, effectively pushing this behaviour away from the storage capacity curve representing the retrieval capability of the Potts associative network. Hence, if one were to regard S as a structural parameter of the network, and w as a parameter that can be tuned, there is an optimal range of w values that allows good quality latching for higher storage. This argument has to be revised, however, by considering also the threshold U, since increasing w can be shown to be functionally equivalent, in terms of storage capacity, to decreasing U [32]. Also for U, in fact, one can find an optimal range for associative retrieval to occur, in the simple Potts network with no adaptation and with $w = 0$ [23]. This near equivalence between U and $-w$ does not hold anymore in the fast adapting regime, to which we turn next.

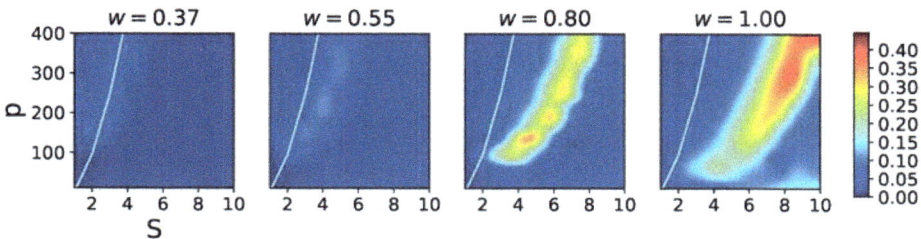

Figure 6. Latching quality $Q(S, p)$ with increasing local feedback, $w = 0.37$, 0.55, 0.8, and 1.0 in the slowly adapting regime. Randomly correlated patterns are used, with $C = 150$ as in Figure 3a.

3.2. Fast Adapting Regime

We characterize the fast adapting regime by the alternative ordering of time scales $\tau_3 < \tau_1 \ll \tau_2$, such that the mean activity in each Potts unit is rapidly regulated by fast inhibition, at the time scale τ_3. Equation (6) stipulates that $\sum_{k=1}^{S} \sigma_i^k(t)$, the total activity of each unit, is followed almost immediately, or more precisely at speed τ_3^{-1}, by the generic threshold $\theta_i^0(t)$. Extensive simulations, with the same parameters as for the slowly adapting regime, except for $w = 1.37$, $\tau_1 = 20$, $\tau_2 = 200$ and $\tau_3 = 10$, show that, similarly to the slowly adapting regime, there are latching bands in the $Q(S, p)$ and $Q(C, p)$

planes (see Figure 7). With these parameters, in particular, the larger value chosen for the feedback term w, the bands occupy a similar position as in the slowly adapting regime. Again, they appear to vanish below certain values of S and C, more precisely around $S \sim 3$, $p \sim 120$ in Figure 7a and around $C \sim 50$, $p \sim 90$ in Figure 7b, and to scale subquadratically in S and sublinearly in C. The band in the S–p plane is again confined by the storage capacity (solid cyan curve) and by the onset of (finite) latching (dashed curve). The storage capacity curve, which is independent of threshold adaptation, follows the same Equation (13).

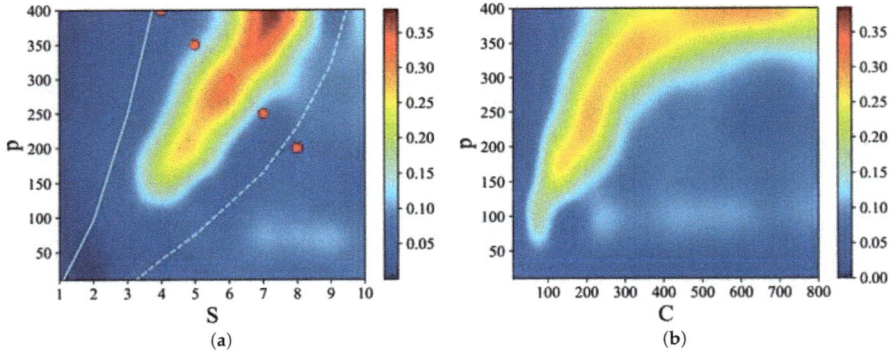

Figure 7. Phase space for $Q(S, p)$ in (**a**) and $Q(C, p)$ in (**b**) with randomly correlated patterns in the fast adapting regime. The parameters are identical to those in the slowly adapting regime, with the exception of $w = 1.37$, $\tau_1 = 20$, $\tau_2 = 200$, $\tau_3 = 10$. The red spots in (**a**) mark, again, the parameter values used in the Figures below.

Examples of latching behaviour outside and inside the band are presented in Figure 8, at the same values for S but shifted by $\Delta p = 100$, i.e., at the "red" points $(5, 350)$, $(6, 300)$, and $(7, 250)$ in the S–p plane. Again, we see from Figure 7a that $(5, 350)$ lies just above the band, while $(6, 300)$ is right on the centre. To the right of the band, e.g., at $(7, 250)$, the transitions are distinct but latching dies out very soon, while on the left, e.g., at $(5, 350)$, the progressively reduced overlaps are a manifestation of increasingly noisy retrieval dynamics. In all three examples, we observe that latching steps proceed slowly, even slower than the doubled time scale $\tau_2 = 200$ would have led to predict. This appears to be because often a significant time elapses between the decay of the overlap of the network with one pattern and the emergence of a new one.

Figure 9 shows the asymmetry and entropy measures, A and I_μ, along the points

$$(4, 400) - (5, 350) - (6, 300) - (7, 250) - (8, 200),$$

in Figure 7a, where, again, we have chosen a series shifted by $\Delta p = 100$ upwards in order to centre it better on the high quality latching band. Only an overlap threshold of 0.5 is considered. What one can see, in contrast with the slowly adapting regime, is that now the two measures are not quite complementary. The point $(6, 300)$ that lies inside the band, very much at its quality peak, shows again an intermediate value for the asymmetry, but the highest value, given the overlap threshold, for the entropy. The discrepancy may be ascribed to the different prevailing type of latching transition observed in the fast adapting regime, Figure 8. As discussed in [24], in a Potts network latching transitions with a high cross-over, which can only occur between memory patterns with a certain degree of correlation, can be distinguished from those with a vanishing cross-over, which are much more random. In the fast adapting regime, as indicated by the examples in Figure 8, all transitions tend to be of the latter type. A more careful analysis indicates, in fact, that they are quasi-random, in that they avoid a memory pattern in which largely the same Potts units are active as in the preceding pattern. In fact, the value of the entropy at (6, 300) implies that on average from each of the 300 memory patterns there are transitions to at least 190 other patterns (to 190 if they were equiprobable, in practice many more); therefore, only the few patterns that happen to be more (spatially) correlated are avoided.

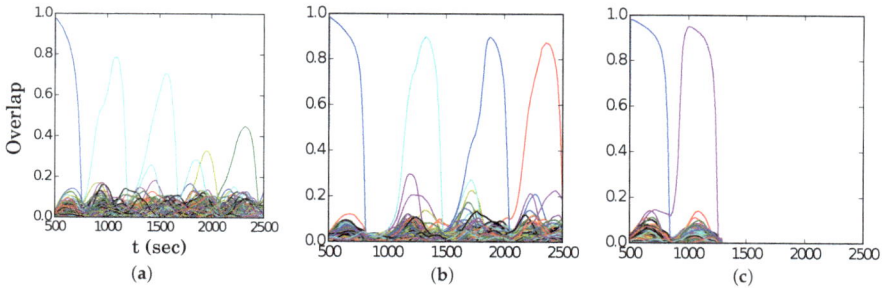

Figure 8. Latching behaviour for (S, p) equal to, respectively: (**a**) (5, 350); (**b**) (6, 300); and (**c**) (7, 250) in Figure 7a.

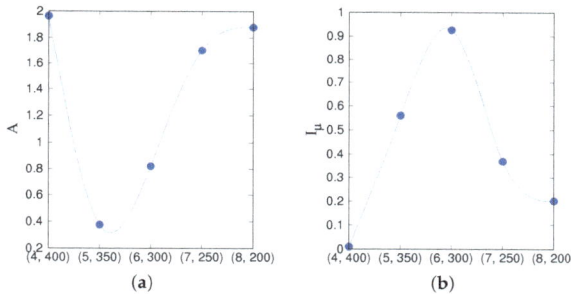

Figure 9. (**a**) asymmetry A of the transition matrix and (**b**) Shannon's information entropy, I_μ along the (4, 400)–(5, 350)–(6, 300)–(7, 250)–(8, 200) parameter series from Figure 7a, using only a threshold 0.5 for the overlaps before and after each transition.

Towards the left, the curves do not vary much depending on the threshold chosen for the overlaps, but the asymmetry eventually becomes maximal and the entropy vanishes simply because sequences of robustly retrieved patterns do not last long, so, in this particular case, it would take more than 1000 sequences to accumulate sufficient statistics.

The effects of increasing the w term in the fast adapting regime are shown in Figure 10, where one notices two main features. First, there is heightened sensitivity to the exact value of w, so that relatively close data points at $w = 1.33$, 1.37, 1.41, and 1.45 yield rather different pictures. Second, although again increasing w shifts the latching band rightward, by far the main effect is a widening of the band itself. This is because in the presence of rapid feedback inhibition a larger w term ceases to be functionally similar to a lower threshold, which in the slowly adapting regime was leading in turn to noisier dynamics and eventually indiscernible transitions. In the fast adapting regime, the increased positive feedback can be rapidly compensated by inhibitory feedback, so that in the high-storage region overlaps remain large, until they are suppressed by storage capacity constraints (the cyan curve, which remains at approximately the same distance from the larger and larger latching band).

We now turn to more explicit comparison of the transition dynamics in two regimes.

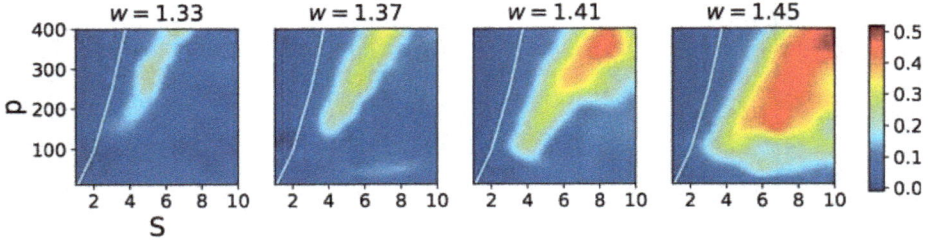

Figure 10. Latching quality $Q(S,p)$ with increasing local feedback, $w = 1.33, 1.37, 1.41$, and 1.45 in the fast adapting regime. Randomly correlated patterns are used, with $C = 150$ as in Figure 7a.

3.3. Comparison of Two Regimes

To look more closely at latching dynamics in the slowly and fast adapting regimes, we take the following points from Figures 3a and 7a, which allow us to cut through the bands at two different storage levels

$$\begin{cases} p = 200, & S = (4,5,6,7), \\ p = 400, & S = (6,7,8,9). \end{cases} \tag{16}$$

Figure 11 shows in different colors the overlaps of the state of the network with the global patterns, for sample sequences along the points (16), in the slowly adapting regime. For both $p = 200$ and 400, latching length is observed to decrease with S, unlike the discrimination between patterns, as measured by d_{12}, in agreement with Figure 2. Note that the two rows in the figure are similar, indicating that the shift $\Delta p = 200$ is approximately compensated by the rightward shift $\Delta S = 2$.

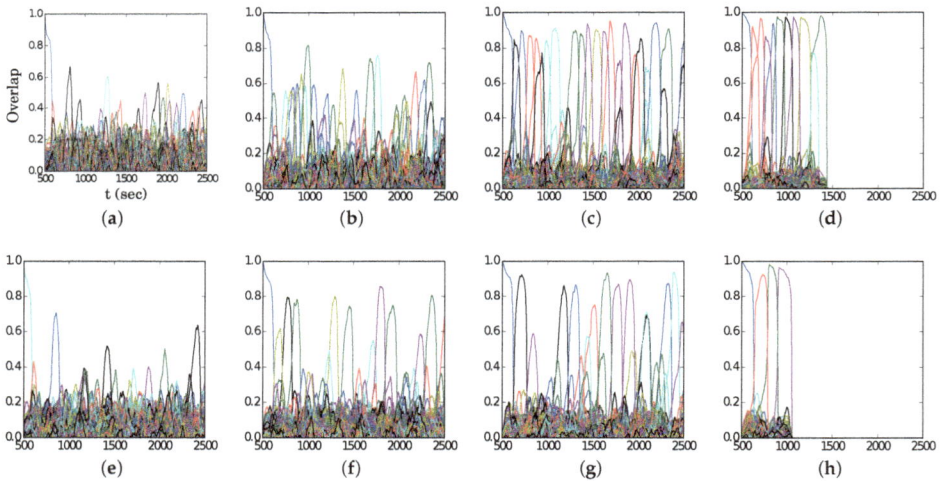

Figure 11. Latching behaviour in the slowly adapting regime. A sample of points (16) from Figure 3a. (**a**) $p = 200$, $S = 4$; (**b**) $p = 200$, $S = 5$; (**c**) $p = 200$, $S = 6$; (**d**) $p = 200$, $S = 7$; (**e**) $p = 400$, $S = 6$; (**f**) $p = 400$, $S = 7$; (**g**) $p = 400$, $S = 8$; (**h**) $p = 400$, $S = 9$.

The fast adapting regime shows the same trends, again one sees in Figure 12 the approximate compensation between the two shifts $\Delta p = 200$ and $\Delta S = 2$, but latching appears in general less noisy.

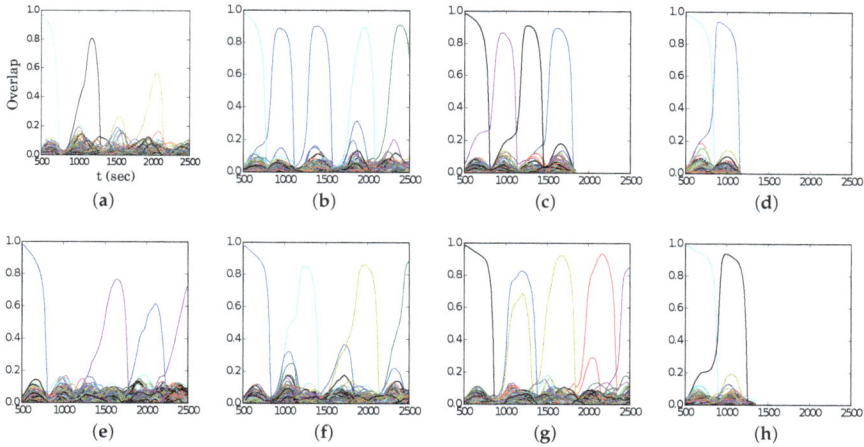

Figure 12. Latching behaviour in the fast adapting regime. A sample of points (16) from Figure 7a. (**a**) $p = 200$, $S = 4$; (**b**) $p = 200$, $S = 5$; (**c**) $p = 200$, $S = 6$; (**d**) $p = 200$, $S = 7$; (**e**) $p = 400$, $S = 6$; (**f**) $p = 400$, $S = 7$; (**g**) $p = 400$, $S = 8$; (**h**) $p = 400$, $S = 9$.

The main difference between the two regimes, however, is in the distribution of crossover values, those when the network has equal overlap with the preceding and the following pattern: their distribution (PDF, or probability density function) is shown in Figures 13 and 14.

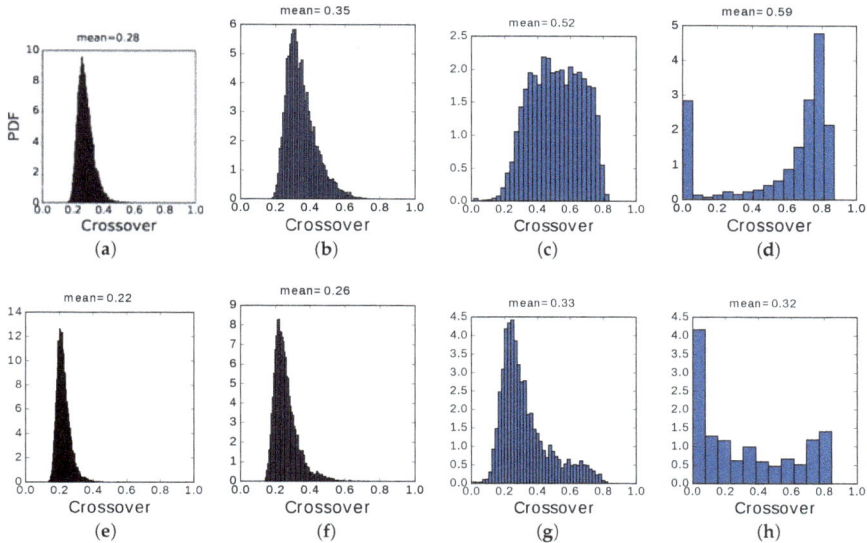

Figure 13. Probability density function (PDF) of crossover values in the slowly adapting regime. (**a**) $p = 200$, $S = 4$; (**b**) $p = 200$, $S = 5$; (**c**) $p = 200$, $S = 6$; (**d**) $p = 200$, $S = 7$; (**e**) $p = 400$, $S = 6$; (**f**) $p = 400$, $S = 7$; (**g**) $p = 400$, $S = 8$; (**h**) $p = 400$, $S = 9$.

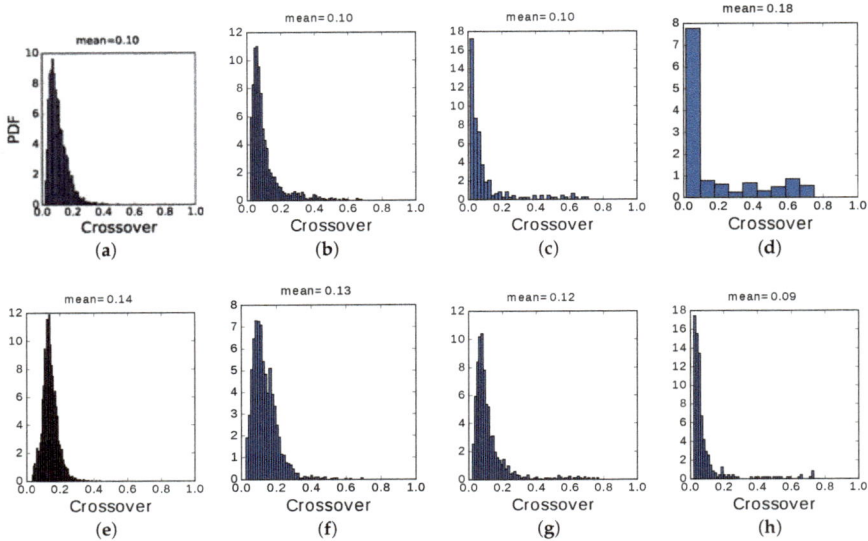

Figure 14. Probability density function (PDF) of crossover values in the fast adapting regime. (a) $p = 200$, $S = 4$; (b) $p = 200$, $S = 5$; (c) $p = 200$, $S = 6$; (d) $p = 200$, $S = 7$; (e) $p = 400$, $S = 6$; (f) $p = 400$, $S = 7$; (g) $p = 400$, $S = 8$; (h) $p = 400$, $S = 9$.

We see that, in the fast adapting regime, most transitions occur at very low crossover, i.e., the correlation with the preceding memory has to decay almost to zero before the next memory pattern can be activated. Only in regions of the (S, p) plane where latching sequences are very short, a few transitions only, we begin to see a small fraction of them with crossover values above 0.2. In most cases, the inhibitory feedback conveyed by the variable θ_i^0 is so fast as not to allow transitions to be carried through by positive correlations, i.e., by the subset of Potts units which are in the same active state in the preceding and successive pattern. The choice of the next pattern is not completely random, as indicated by the relative entropy values still below unity, but is determined essentially by negative selection, as mentioned above: the next pattern tends to have few active Potts units that coincide with those active in the preceding pattern.

In the slowly adapting regime, instead, due to the slow variation of the non-specific threshold, active Potts units can remain active, but they are *encouraged* by the variables θ_i^k to switch between active states if they have been in the same for too long. This can produce, particularly in the center of the latching band, sequences of patterns succeeding each other at high crossover, as shown by the distribution in Figure 13c. Even when latching is very noisy and approaches randomness, as in panels Figure 13a,e, crossover values are consistently above 0.2, indicating a preference for patterns insisting on the same set of active Potts units, unlike the fast adapting regime. Finally, when the number of states S is too large or, equivalently, that of patterns p too low, we observe some transitions with minimal crossover and a majority with very large crossover, as if occurring only with those patterns that were already partially retrieved when the network had still the largest overlap with the preceding pattern, but the main observation is that there are very few transitions at all, so that to plot a probability density distribution we need to used wide bins, in panels Figure 13d,h (and in Figure 14d).

This difference between the two regimes is confirmed by an analysis of the correlations between successive patterns in latching sequences. In the Potts network, at least two types of spatial correlation between patterns are relevant: how many active Potts units the two patterns share, and how many of these units are active and in the same state. We quantify them with C_1, the fraction of the units active in one pattern that are active also in the other, *and* in the same state; and with C_2, the fraction that are also active, but in a different state. In a large set of randomly determined patterns, the mean values are $\langle C_1 \rangle = a/S$ and $\langle C_2 \rangle = a(S-1)/S$. The full distribution, among all pairs, is scattered around these mean values. However, do transitions occur between any pair of patterns?

Figure 15 shows that relative to the full distribution, in blue, transitions tend to occur, in the slowly adapting regime on the left, only between patterns with C_1 *above* and C_2 *below* (or at most around) their average values. Thus, when the network has retrieved a memory representation, it looks for correlated ones, as it were, where to jump. In the fast adapting regime, this is not the case: transitions are almost random, except there appears to be a slight tendency to avoid those with C_1 well above its mean value. Note that the values of p and w are different in the two panels, and are chosen so as to be in roughly equivalent positions within the respective latching bands.

The analysis of the crossover points, therefore, affords insight into the rather different transition dynamics prevailing in the fast and slowly adapting regimes, in particular in the center of their latching bands, suggesting that in a more realistic cortical model, which combines both types of activity regulation, there should still be a significant component of "slow adaptation" for interesting sequences of correlated patterns to emerge. The preceding simulations, however, were all carried out with randomly correlated patterns, in which the occasional high or low correlation of a pair is merely the result of a statistical fluctuation. Does the insight carry over to a more stuctured model of the correlations among memory patterns? This is what we ask next.

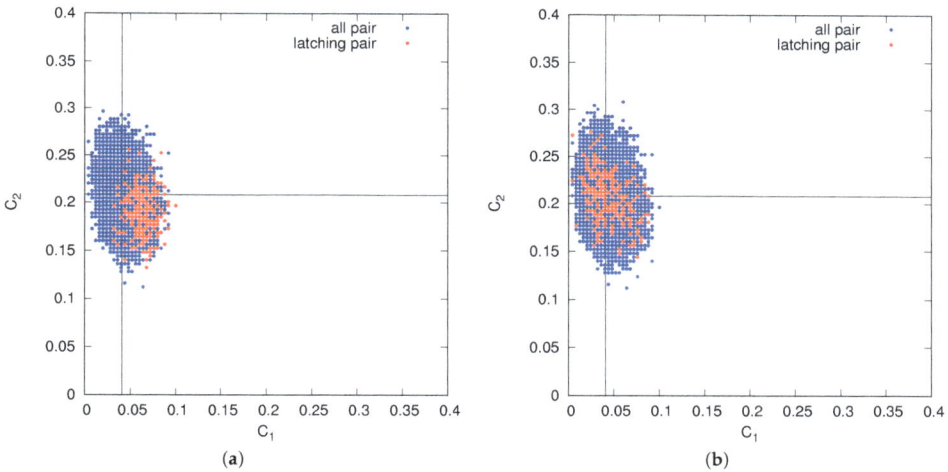

Figure 15. Scatterplots of the fractions C_1 and C_2 of Potts units active in one pattern that are active also in another, and in the same state or, respectively, in another active state. The panels show the full distribution between any pattern pair, in the slowly (**a**) and fast adapting (**b**) regimes, in blue; and the distribution between successive patterns in latching transitions, in red. The blue distribution for the fast adapting regime (for which $a = 0.25$, $S = 6$, $p = 300$ and $w = 1.32$) is similar to the one for the slowly adapting regime (for which again $a = 0.25$, $S = 6$, but $p = 200$ and $w = 0.65$), except that it is slightly wider, because of the higher storage load, while the red distributions are markedly different. Vertical lines indicate mean values: (**a**) slowly adapting regime; (**b**) fast adapting regime.

3.4. Analysis with Correlated Patterns

Correlated patterns were generated according to the algorithm mentioned by [22] and discussed in detail in [35]. The multi-parent pattern generation algorithm works in three stages. In the first step, a total set of Π random patterns are generated to act as parents. In the second step, each of the total set of parents are assigned to p_{par} randomly chosen children. Then, a "child" pattern is generated:

each pattern, receiving the influence of its parents with a probability a_p, aligns itself, unit by unit, in the direction of the largest field. In the third and final step, a fraction a of the units with the highest fields is set to become active. In this way, child patterns with a sparsity a are generated. In addition, another parameter ζ can be defined, according to which the field received by a child pattern is weighted with a factor $\exp(-\zeta k)$ where the index k runs through all parents. This is meant to express a non-homogeneous input from parents.

It is clear that such patterns, however, cannot be considered as independent and identically distributed, as in Equation (9), because their activity is drawn from a common pool of parents. In fact, they are correlated, in the sense that those children receiving congruent input from a larger number of common parents will tend to be more similar. All of these observations are studied in more detail in [35], and here we only focus on how correlations affect the phase diagrams. In the following simulations, the parameters pertaining to the patterns are $a_p = 0.4$, $\Pi = 100$, $\zeta = 0.1$ while p_{par}/p, the probability that a pattern be influenced by a parent is kept constant at 0.277.

Simulations with correlated patterns were carried out across the same S–p and C–p planes in phase space, in the slowly adapting regime, as shown in Figure 16. We focused on the slowly adapting regime based on the results of the crossover analysis. All other simulation parameters were kept at the values used with randomly correlated patterns.

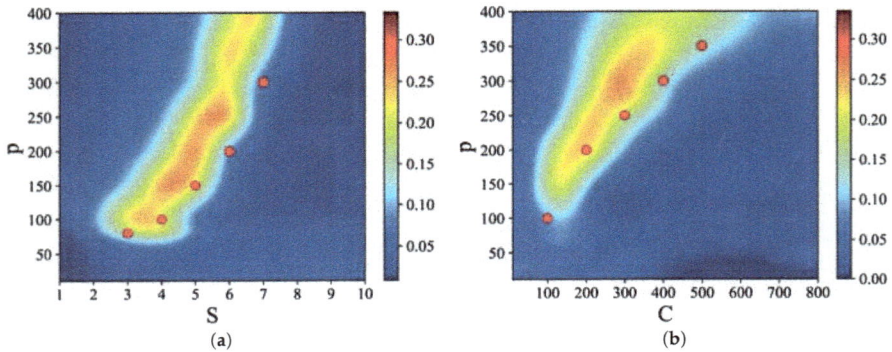

Figure 16. Phase space, cut across the $Q(S, p)$ plane in (**a**) and $Q(C, p)$ in (**b**), with correlated patterns in the slowly adapting regime. Red dots represent the quality peaks in the the same planes, with randomly correlated patterns. The parameters are $C = 150$ and $S = 5$, if kept fixed, and $w = 0.8$.

We see from the figure that the presence of non-random correlations among the memory patterns, albeit weak, shifts the bands to the left and upward in phase space, keeping approximately the dependence of the viable storage load p on S and C, but at somewhat higher values. It is as if more memories could "fit", if correlated, into the same latching dynamics.

Figure 17 shows the S–p plane cut along $p = 200$, to better compare the cases with correlated (blue) and random (red) patterns. It is apparent that there is a leftward shift, in the case of correlated patterns, from the red curve applying to the random case, but the dependence on S remains very similar.

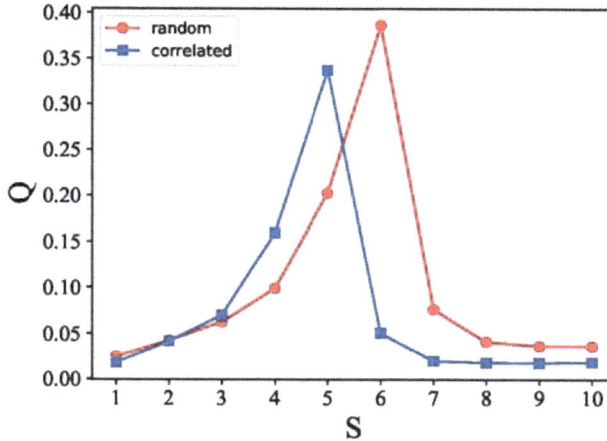

Figure 17. Comparison of S–p phase spaces along $p = 200$ with random (red dotted line) and correlated (blue dotted) patterns in the slow adapting regime.

4. Conclusions

In this paper, we have found the region in the Potts network phase space spanned by the number of Potts states S, the number of connections per unit C and the storage load p, where latching dynamics occur, and we have described their character, comparing and contrasting the slowly and fast adapting regimes. In relation to our earlier paper [22], where the possibility of such a latching region was pointed out on the basis of limited simulations, we have now a firmer basis to extrapolate to regions of parameter space of relevance to the human cortex, possibly a step toward quantitatively studying human specific capacities, including creative behaviour. A common hallmark in both regimes is that good quality latching occupies a band which scales almost quadratically in the p–S plane, while it is sublinear in the p–C plane. These bands are bounded by the storage capacity line, above, and by the boundary between no latching and finite latching, below. If, as discussed elsewhere [32], we were to take $C \approx 10^2$ and $S \approx 10^2$ as the orders of magnitude of interest for the human brain, we would conclude that the relevant storage load, or semantic depth, is in the region $p \approx 10^5$, in both regimes. At the center of the band in the slowly adapting regime, asymmetry and entropy take intermediate values, pointing at maximally complex and potentially useful dynamics, intermediate between the deterministic and the random extremes. High crossover values indicate that many transitions occur between highly correlated patterns. Using correlated patterns shifts the position of the band in phase space, but preserving the features observed with random patterns, still in the slowly adapting regime.

In the fast adapting regime, instead, in the center of the band, which can be made wider and more robust, the entropy is higher, and correspondingly only low crossover transitions are observed, indicating that the network latches most of the time from one pattern to any other among the many with which it is weakly or anti-correlated, avoiding only those few with which it is highly correlated.

Therefore, we can conclude that the fast adapting regime, modelling rapid inhibitory feedback, offers a robust framework for latching dynamics, but of an essentially random, not very useful nature; whereas in the slowly adapting regime, modelling slow inhibition or local fatigue, correlations can drive latching transitions, potentially enabling semantic content in a stream of thoughts or linguistic productions, but with fragile dynamics, living at the very edge between memory overload and sequence termination because of the inability of the network to jump forward. This suggests the opportunity of considering models that integrate both fast and slowly adapting dynamics in their non-specific thresholds, so as to combine the useful features of both regimes. It will be the object of future work.

We would like to note, in the end, the inherent limitation of considering a simple homogeneous Potts network, with no differentiation among its units and no internal structure. In order to make contact with cognitive processes, of any kind, this limitation has to be overcome, as perhaps attempted, with one first step among many possible ones, by arranging Potts units on a ring [38]. Nevertheless, even in its crudest form, the Potts network with its latching dynamics can be used to explore e.g., novel theories as to the evolutionary origin of complex cognition [39]. It establishes a quantitative framework to understand phase transitions [25], complementary to the perspective offered by other modelling approaches to sequence generation in cortical networks [40]. At the most abstract level, it can be considered an implementation of a fuzzy logic system [41,42], but with the critical advantage that its parameters can eventually be related to cortical parameters, as we begin to describe in a related study [32].

Acknowledgments: We are grateful to Leonardo Romor who optimized the Potts network code, and to the Human Frontier Science Program RGP0057/2016 collaboration, that has also supported open access publication.

Author Contributions: All authors conceived and designed the simulations, which were performed primarily by Chol Jun Kang, with major contributions by Michelangelo Naim and Vezha Boboeva. Alessandro Treves and Chol Jun Kang wrote the paper, with input from the other two authors. All authors have read and approved the final manuscript.

Conflicts of Interest: The authors declare no conflict of interest.

References

1. Hauser, M.D.; Chomsky, N.; Fitch, W.T. The Faculty of language: What is it, who has it, and how did it evolve? *Science* **2002**, *298*, doi:10.1126/science.298.5598.1569.
2. Amit, D.J. The Hebbian paradigm reintegrated: Local reverberations as internal representations. *Behav. Brain. Sci.* **1995**, *18*, doi:10.1017/S0140525X00040164.
3. Kaneko, K.; Tsuda, I. Dynamic link of memory—Chaotic memory map in nonequilibrium neural networks. *Chaos* **2003**, *13*, doi:10.1016/S0893-6080(05)80029-2.
4. Hopfield, J.J. Neural networks and physical systems with emergent collective computational abilities. *Proc. Natl. Acad. Sci. USA* **1982**, *79*, 2554–2558
5. Amit, D.J.; Gutfreund, H.; Sompolinsky, H. Statistical mechanics of neural networks near saturation. *Ann. Phys.* **1987**, *173*, doi:10.1016/0003-4916(87)90092-3.
6. Amit, D.J. *Modeling Brain Function*; Cambridge University Press: New York, NY, USA, 1992.
7. Rolls, E.T.; Treves, A. *Neural Networks and Brain Function*; Oxford University Press: New York, NY, USA, 1998.
8. Pfeiffer, B.E.; Foster, D.J. Hippocampal place cell sequences depict future paths to remembered goals. *Nature* **2013**, *497*, 74–79.
9. Abeles, M. *Local Cortical Circuits: An Electrophysiological Study*; Springer Science & Business Media: Berlin, Germany, 2012.
10. Chossat, P.; Krupa, M.; Lavigne, F. Latching dynamics in neural networks with synaptic depression. *arXiv* **2016**, arXiv:1611.03645v2 .
11. Burgess, P.W.; Shallice, T. Confabulation and the control of recollection. *Memory* **1996**, *4*, 359–411.
12. Epstein, R. The neural-cognitive basis of the Jamesian stream of thought. *Conscious. Cognit.* **2000**, *9*, 550–575.
13. Abeles, M. Time is precious. *Science* **2004**, *304*, 523–524.
14. Pulvermüller, F. Brain mechanisms linking language and action. *Nat. Rev. Neurosci.* **2005**, *6*, 576–582.
15. Shmiel, T.; Drori, R.; Shmiel, O.; Ben-Shaul, Y.; Nadasdy, Z.; Shemesh, M.; Teicher, M.; Abeles, M. Temporally precise cortical firing patterns are associated with distinct action segments. *J. Neurophysiol.* **2006**, *96*, 2645–2652.
16. Sosnik, R.; Shemesh, M.; Abeles, M. The point of no return in planar hand movements: An indication of the existence of high level motion primitives. *Cognit. Neurodyn.* **2007**, *1*, 341–358.
17. Kanter, I. Potts-glass models of neural networks. *Phys. Rev. A* **1988**, *37*, 2739–2742.

18. Bollé, D.; Dupont, P.; Mourik, J.V. Stability properties of potts neural networks with biased patterns and low loading. *J. Phys. A Math. Gen.* **1991**, *24*, doi:10.1088/0305-4470/24/5/021.
19. Bollé, D.; Dupont, P.; Huyghebaert, J. Thermodynamic properties of the Q-state potts-glass neural network. *Phys. Rev. A* **1992**, *45*, doi:10.1103/PhysRevA.45.4194.
20. Bollé, D.; Inck, B.; Zagrebnov, V.A. On the parallel dynamics of the Q-state potts and Q-ising neural networks. *J. Stat. Phys.* **1993**, *70*, 1099–1119.
21. Bollé, D.; Cools, R.; Dupont, P.; Huyghebaert, J. Mean-field theory for the Q-state potts-glass neural network with biased patterns. *J. Phys. A Math. Gen.* **1993**, *26*, doi:10.1088/0305-4470/26/3/017.
22. Treves, A. Frontal latching networks: A possible neural basis for infinite recursion. *Cogn. Neuropsychol.* **2005**, *22*, 276–291.
23. Kropff, E.; Treves, A. The storage capacity of potts models for semantic memory retrieval. *J. Stat. Mech. Theor. Exp.* **2005**, doi:10.1088/1742-5468/2005/08/P08010.
24. Russo, E.; Namboodiri, V.M.K.; Treves, A.; Kropff, E. Free association transitions in models of cortical latching dynamics. *New J. Phys.* **2008**, *10*, doi:10.1088/1367-2630/10/1/015008.
25. Russo, E.; Treves, A. Cortical free-association dynamics: Distinct phases of a latching network. *Phys. Rev. E* **2012**, *85*, doi:10.1103/PhysRevE.85.051920.
26. Abdollah-nia, M.F.; Saeedghalati, M.; Abbassian, A. Optimal region of latching activity in an adaptive Potts model for networks of neurons. *J. Stat. Mech. Theory Exp.* **2012**, doi:10.1088/1742-5468/2012/02/P02018.
27. Lerner, I.; Bentin, S.; Shriki, O. Spreading activation in an attractor network with latching dynamics: Automatic semantic priming revisited. *Cogn. Sci.* **2012**, *36*, 1339–1382.
28. Lerner, I.; Bentin, S.; Shriki, O. Excessive attractor instability accounts for semantic priming in schizophrenia. *PLoS ONE* **2012**, *7*, doi:10.1371/journal.pone.0040663.
29. O'Kane, D.; Treves, A. Short-and long-range connections in autoassociative memory. *J. Phys. A Math. Gen.* **1992**, *25*, doi:10.1088/0305-4470/25/19/018.
30. Romani, S.; Pinkoviezky, I.; Rubin, A.; Tsodyks, M. Scaling laws of associative memory retrieval. *Neural Comput.* **2013**, *25*, 2523–2544.
31. Recanatesi, S.; Katkov, M.; Romani, S.; Tsodyks, M. Neural network model of memory retrieval. *Front. Comput. Neurosci.* **2015**, *9*, doi:10.3389/fncom.2015.00149.
32. Naim, M.; Boboeva, V.; Kang, C.J.; Treves, A. From multi-modular Hopfield networks to the Potts network and its storage capacity. Unpublished work, 2017.
33. Kropff, E.; Treves, A. The complexity of latching transitions in large scale cortical networks. *Nat. Comput.* **2007**, *6*, 169–185.
34. Hebb, D.O. *The Organization of Behavior: A Neuropsychological Theory*; John Wiley & Sons: New York, NY, USA, 2005.
35. Boboeva, V.; Treves, A. The storage capacity of the Potts network with correlated patterns. Unpublished work, 2017.
36. Cover, T.M.; Thomas, J.A. *Elements of Information Theory*, 2nd ed.; John Wiley & Sons: New York, NY, USA, 2006.
37. Russo, E.; Pirmoradian, S.; Treves, A. Associative latching dynamics vs. syntax. In *Advances in Cognitive Neurodynamics (II)*; Wang, R., Gu, F., Eds.; Springer: Dordrecht, The Netherlands, 2011; pp. 111–115.
38. Song, S.; Yao, H.; Treves, A. A modular latching chain. *Cogn. Neurodyn.* **2014**, *8*, 37–46.
39. Amati, D.; Shallice, T. On the emergence of modern humans. *Cognition* **2007**, *103*, 358–385.
40. Rajan, K.; Harvey, C.D.; Tank, D.W. Recurrent network models of sequence generation and memory. *Neuron* **2016**, *9*, 128–142.
41. Jiang, Y.; Chung, F.L.; Deng, Z.; Wang, S. Multitask TSK fuzzy system modeling by mining intertask common hidden structure. *IEEE Trans. Cybern.* **2015**, *45*, 548–561.
42. Tu, C.C.; Juang, C.F. Recurrent type-2 fuzzy neural network using Haar wavelet energy and entropy features for speech detection in noisy environments. *Expert Syst. Appl.* **2012**, *39*, 2479–2488.

entropy

MDPI

Article

Lifespan Development of the Human Brain Revealed by Large-Scale Network Eigen-Entropy

Yiming Fan, Ling-Li Zeng, Hui Shen, Jian Qin, Fuquan Li and Dewen Hu *

College of Mechatronics and Automation, National University of Defense Technology, 109 Deya Road, Changsha 410073, China; fanyimingendeavor@126.com (Y.F.); lingl.zeng@gmail.com (L.-L.Z.); shenhui_nudt@126.com (H.S.); qinjian714@126.com (J.Q.); lifuquan16@nudt.edu.cn (F.L.)
* Correspondence: dwhu@nudt.edu.cn; Tel.: +86-731-8457-4992

Received: 3 August 2017; Accepted: 1 September 2017; Published: 4 September 2017

Abstract: Imaging connectomics based on graph theory has become an effective and unique methodological framework for studying functional connectivity patterns of the developing and aging brain. Normal brain development is characterized by continuous and significant network evolution through infancy, childhood, and adolescence, following specific maturational patterns. Normal aging is related to some resting state brain networks disruption, which are associated with certain cognitive decline. It is a big challenge to design an integral metric to track connectome evolution patterns across the lifespan, which is to understand the principles of network organization in the human brain. In this study, we first defined a brain network eigen-entropy (NEE) based on the energy probability (EP) of each brain node. Next, we used the NEE to characterize the lifespan orderness trajectory of the whole-brain functional connectivity of 173 healthy individuals ranging in age from 7 to 85 years. The results revealed that during the lifespan, the whole-brain NEE exhibited a significant non-linear decrease and that the EP distribution shifted from concentration to wide dispersion, implying orderness enhancement of functional connectome over age. Furthermore, brain regions with significant EP changes from the flourishing (7–20 years) to the youth period (23–38 years) were mainly located in the right prefrontal cortex and basal ganglia, and were involved in emotion regulation and executive function in coordination with the action of the sensory system, implying that self-awareness and voluntary control performance significantly changed during neurodevelopment. However, the changes from the youth period to middle age (40–59 years) were located in the mesial temporal lobe and caudate, which are associated with long-term memory, implying that the memory of the human brain begins to decline with age during this period. Overall, the findings suggested that the human connectome shifted from a relatively anatomical driven state to an orderly organized state with lower entropy.

Keywords: functional connectome; graph theoretical analysis; eigenvector centrality; orderness; network eigen-entropy

---

1. Introduction

The human connectome undergoes complex transformations across the lifespan, and it can be mathematically modeled as a complex network by graph theoretical analysis. Imaging connectomics based on graph theory has become an effective and unique methodological framework for studying functional connectivity patterns of the developing and aging brain [1]. Normal brain development is characterized by continuous and significant network evolution through infancy, childhood, and adolescence, following specific maturational patterns. Normal aging is related to some resting state brain networks disruption, which are associated with certain cognitive decline. Specifically, a connectome (i.e., a large-scale resting-state functional connectivity network, RSFC) can be mapped by taking each parcellation unit of the whole-brain as nodes and estimating inter-unit correlations (i.e., functional connectivity, FC) in spontaneous blood oxygenation level dependent (BOLD) fluctuations as edges. Combined with graph theoretical analysis, this study aimed to explore typical age-related architecture changes of the human brain, which is a prerequisite for studying neuropsychiatric diseases such as autism, depression, schizophrenia, Alzheimer's disease, and Parkinson's disease.

Many connectomics studies using network science and graph theory have shown age-related changes of FC patterns. FC patterns between/within resting-state subsystems (a set of brain regions that exhibit coherent activity in a task-free state [2] and exhibit consistent spatial topographic patterns across the cerebral cortex [3,4]) have been investigated with age [5], and FC patterns for brain regions had also been examined across the lifespan [6,7]. In addition, several whole-brain FC pattern studies on RSFC networks have focused either on early or late age-related changes [3,8–13], while other studies have focused on specific subsystems involving default networks [14–17], and cognitive-control [18–20]. Moreover, findings on large-scale RSFC network properties have revealed that modularity and local efficiency decreased over aging [8,21–23]. In contrast, another study reported that modularity was similar in older and younger participants [24]. Furthermore, studies of connectomics on global efficiency have not reached a consensus [8,21]. These studies focused on studying the FC pattern or some network properties in the developmental stage or across the lifespan. Little attention has been focused on the changes to whole-brain centralities with age.

Nodal connectivity or centralities account for a given region's relationship with the entire functional connectome and not just its relationship to specified regions (seed-based analysis), or to brain subsystems (independent component analysis). As such, centrality measures allow us to capture the given region's importance in the functional connectome. Prior studies of nodal connectivity or centralities have focused primarily on identifying "cortical hubs" [8,25–29] and finding rich-clubs [21,30,31] in healthy adulthood participants. A few researchers have reported that the degree distribution in the normal adult human brain conformed to an exponentially truncated power law [25,32,33]. Recent neuroscience studies have employed eigenvector centrality (EC) mapping for measuring FC in both diseases [34,35] and task states of brain activity [26]; however, these studies have not explored EC pattern changes of the whole-brain RSFC network across the lifespan in healthy individuals.

The network eigen-entropy (NEE) was defined based on EC and the Shannon entropy. In a complex system, it characterizes the orderness change of a system indicating the state or the development direction of a system. EC defined in this way assigned each node a centrality that depended on both the quantity and the quality of its connections. It integrated the centralities of a node's neighbors and each node was ranked by its EC scores in a network. Therefore, EC was extended to capture the global features of the graph [36,37]. The energy probability (EP) of each node was derived from EC scores in a graph network. The resulting set of energy probabilities generated a frequency distribution. The EP can be regarded as a random variable that follows this frequency distribution, and the distribution measured the functional connectome heterogeneity [36]. If the nodes' function in the network is not specialized, and the communication of information is not preferential, the EP distribution is uniform, and the network is homogenous. However, if the nodes' function is specialized, and the communication of information is preferential, the EP distribution is distributed, and the network is heterogeneous. Collectively, the more distributed the EP histogram of a system is, the lower the entropy is and vice versa. Thus, the NEE allows us to analyze the organizational mechanisms of the functional connectome development.

We initiated this study under the assumption that the orderness of the functional connectome was related to the age of the brain.

2. Materials and Methods

2.1. fMRI Data Acquisition

The analyses described in the study used two separate resting-state functional magnetic resonance imaging (rs-fMRI) datasets. During the scan, participants were instructed to relax and to fixate on a crosshair while remaining as still as possible. Unless specifically noted, these datasets were handled identically in processing and analysis.

Dataset I was selected from the Nathan Kline Institute-Rockland Sample (NKI-RS) (http://fcon_1000.projects.nitrc.org/indi/pro/nki.html). All approvals and procedures for the collection and sharing of data were approved by the NKI institutional review board. Dataset I consisted of 173 healthy participants in total (aged 7–85 years; mean age, 36.53 ± 20.28 years; 102 males). All participants were divided into four groups in terms of the standard age division recommended by the World Health Organization (WHO), as shown in Table 1. MRI data was acquired using a 3.0 T SIMENS Trio scanner. Structural images were acquired using a magnetization-prepared rapid gradient echo (MPRAGE) sequence (repetition time [TR] = 2500 ms, echo time [TE] = 3.5 ms, flip angle = 90°, thickness = 1.0 mm, slices = 192, matrix = 256 × 256, and FOV = 256 mm). Functional images were acquired using echo-planar imaging (EPI) sequence (TR/TE = 2500/30 ms, FA = 80°, FOV = 216 mm, matrix = 64 × 64, slices = 38, thickness = 3.0 mm). A 10-minute (260 volumes) R-fMRI scan was collected for each participant.

Table 1. Demographics of the four age groups.

Groups	Classification	Age Range (years)	Number of Participants	Gender (M [a]/F [b])
Group I	flourishing	7–20	40	23/17
Group II	youth period	23–38	35	17/18
Group III	middle age	40–59	37	27/10
Group IV	old age	61–85	32	15/17

M [a] = Male; F [b] = Female.

Dataset II consisted of 100 healthy participants randomly selected from the dataset of the Human Connectome Project (HCP) (http://www.humanconnectome.org). Each participant underwent two R-fMRI sessions on separate days. The ages of all participants ranged from 26 to 30 years. Functional images were acquired using multi-band gradient EPI sequences (time repetition/echo time = 720/33.1 ms, flip angle = 52°, field of view = 208 × 180 mm^2, matrix size = 104 × 90, thickness = 2.0 mm, slices = 72). A 14-minute and 24-second resting-state fMRI scan was collected for each participant (1200 volumes). For more detailed parameters, see [38].

2.2. Data Preprocessing

For the functional imaging data of each participant, the first five volumes were discarded to remove possible T1 stabilization effects. Next, imaging data were preprocessed using the statistical parametric mapping software package SPM8 (http://www.fil.ion.ucl.ac.uk/spm) following the pipeline in previous research described in [39,40]. Preprocessing steps mainly included the removal of sudden spikes caused by significant head motion, slicing time correction, spatial realignment correction, normalization, spatial smoothing, and temporal band-pass filtering. Slicing time correction was utilized to compensate for slice acquisition dependent time shifts of each volume. Six rigid body translation and rotation parameters were applied to correct head motion. Normalization included R-fMRI data written into the Montreal Neurological Institute (MNI) space at a 3 × 3 × 3 mm^3 resolution using the nonlinear transformation calculated on the corresponding anatomic images. The normalized functional volumes were spatially smoothed using a 6-mm full width at half maximum (FWHM) Gaussian filter kernel. Briefly, the linear trend over each scan was removed and a temporal band pass filter from 0.01 to 0.08 Hz was performed. Prior to spatial smoothing, to reduce hardware noise, the draining vessel effect, and motion artifacts on each voxel of gray matter, we further regressed the following nuisance variables: six rigid body motion parameters, the white matter (WM), cerebrospinal fluid (CSF), and whole-brain global signals. The residuals from the functional data were used for further analysis.

2.3. Construction of Resting-State Functional Connectivity Network

The whole brain was divided into many regions using a parcellating method, which offered multi-scale parcellations. A single divided brain region included a set number of voxels with spatially coherent brain activity [41]. Each voxel time series within the whole brain was extracted after data preprocessing. The time series of a region was obtained by averaging the time series of all the voxels in this region. The Pearson's correlation coefficients were then calculated between each pair of brain regions. To improve the normality of the correlation coefficients, Fisher's r-to-z transform was performed to convert the correlation coefficients to z-values. Only positive coefficients were preserved to avoid ambiguous biological explanations of negative coefficients. Next, we utilized Bonferroni correction to set the correlation coefficients with *p*-values greater than 0.05 as 0; otherwise, we set the correlation coefficients as 1. Finally, an adjacent matrix *A* was generated for each participant to represent an undirected binary RSFC network.

For estimating NEE change in each subsystem during a lifespan, we used a previously established functional parcellation of the human cerebral cortex [4], which was derived by clustering the whole-brain FC network of 500 subjects according to the similarity of regionals FC profiles. This procedure resulted in seven clusters, whose boundaries shared a close correspondence to the known topographic boundaries of visual and sensory motor areas, the limbic system and distributed association networks for executive control (frontoparietal), attention (dorsal, ventral), and internally-directed cognition (default). Thus, we divided the whole-brain RSFC network nodes into seven subsystems. Specifically, we first calculated the number of voxels where all voxels of each brain region were dropped in a certain subsystem. Second, we put the brain region into the subsystem owning the maximum number of voxels. If all voxels of the brain regions were not dropped into any of the subsystems, we put those regions into an unknown category. The divided results of the seven subsystems are shown in Figure 1. The unknown brain regions are not shown in Figure 1.

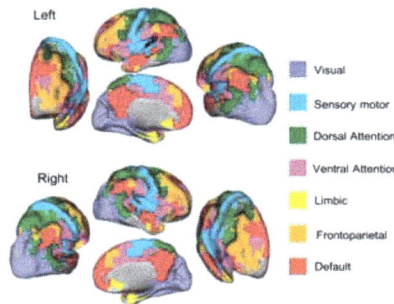

Figure 1. Surface rendering of all seven resting-state functional connectivity subsystems comprising 800 brain regions in total on both left hemisphere (**left**) and right hemisphere (**right**). The inflated surfaces are provided by Caret. The color map indicates colors picturing each of the seven components and their regions on the surfaces.

2.4. Calculate Network Eigen-Entropy

As the EC depends on both the quantity and the quality of its connections, the centrality score (EC_i) was proportional to the sum of the scores of the node's neighbors $\sum_{j=1}^{N} a_{ij} EC_j$.

$$EC_i = \frac{1}{\lambda} \sum_{j=1}^{N} a_{ij} EC_j \qquad (1)$$

where a_{ij} denotes the element of the adjacent matrix A, and N denotes the number of nodes of a binary RSFC network, and λ denotes unique largest real eigenvalue. Using the power iteration method, we found that the centrality vector (EC) was an eigenvector of the adjacency matrix associated with λ, which had strictly positive components using the Perron-Frobenius theorem [42]. The calculation process of EC was as follow. First, a starting positive amount of influence of every node was set with $1/N$. The iteration stopped after the maximum iterations had been reached. In the power iteration, every iteration vector was multiplied by the adjacency matrix of the network and normalized. The i-th component of the EC represented the relative centrality score of the node i in the network.

To calculate entropy, we first defined the intrinsic probability of each node in the binary RSFC network. We know that the unique positive centrality score of each node is the final steady state of the system through the power iteration processing. The eigenvector was only defined up to a common factor, so the ratios of the centralities of the nodes (I) were defined as an intrinsic state probability of the binary RSFC network. Thus, the intrinsic probability of each node was defined as

$$I_i = \frac{EC_i}{\sum_{j=1}^{N} EC_j} \tag{2}$$

We also called the intrinsic probability of each unit in the system the EP of each node in the RSFC network. Thus, I_i represents the EP of each node in a binary RSFC network. Quoting the Shannon entropy equation, NEE was constructed:

$$E = -\sum_{i=1}^{N} I_i \ln(I_i) \tag{3}$$

The entropy E is related to the eigenvector of a RSFC network, so we called it NEE. When the network is completely uniform $EC_i = \sqrt{\frac{1}{N}}$, the EP distribution is uniform and the NEE is maximum.

$$E_{max} = -\sum_{i=1}^{N} \frac{1}{N} \ln \frac{1}{N} = \ln N \tag{4}$$

When all nodes are not connected to each other, $EC_i = 0$, the NEE is minimum theoretically, but in the real RSFC networks, the network is connected, and the NEEs are greater than minimum.

$$E_{min} = 0 \tag{5}$$

2.5. Statistical Analyses

We used multiple linear regression to model NEE with *age* and *age²* as predictors, along with three other nuisance variables: *sex*, head-motion mean frame-wise displacement (*meanFD*) [43] and temporal signal-to-noise ratio (*tSNR*) [44]. The regression model was formulated as follows:

$$E = \beta_0 + \beta_1 \times age + \beta_2 \times age^2 + \beta_3 \times sex + \beta_4 \times meanFD + \beta_5 \times tSNR \tag{6}$$

In this study, the Mann–Whitney rank-sum test was used to test whether the whole-brain NEE or brain node EP values of pairwise groups were significantly different. This is useful for examining the significant differences between two independents, but in non-normal populations, which is a sample rank instead of the sample value of the test.

2.6. Test–Retest Reliability

Test–retest reliability is an important group level statistic, referring to the temporal or intra-individual stability of an index measured across multiple occasions in a group of participants. Intra-class correlation (ICC) [45] was used on dataset II to quantify NEE's test–retest reliability in this functional connectome analysis.

The ICC values were categorized into five common intervals [46]: $0 < ICC \leq 0.2$ (slight); $0.2 < ICC \leq 0.4$ (fair); $0.4 < ICC \leq 0.6$ (moderate); $0.6 < ICC \leq 0.8$ (substantial) and $0.8 < ICC < 1.0$ (almost perfect). However, several limitations of the ICC approach has been described in [47]. In addition to the absolute values of ICC, we calculated the *p*-values of ICC using the permutation test. The *p*-value allowed a precise statistical analysis to evaluate the significance of the extracted ICC.

3. Results

3.1. Age-Related Changes of Topologic Energy Probability Distribution and Energy Probability Histograms between Groups

The topologic EP maps and EP histograms for the four groups are shown in Figure 2. Each region's EP within each group was obtained by calculating the average EP in this region, in this group. Consequently, we obtained four group-level EP maps. The EP distribution of Group I was concentrated on $0.6 \times 10^{-3} - 2 \times 10^{-3}$ as shown in the bottom panel of Figure 2, whereas the EP distribution interval of Group II was amplified by $0.6 \times 10^{-3} - 3 \times 10^{-3}$, which sloped to the right and tended to have a tail. The EP distribution of Group III had a right long tail, and the distribution interval ranged from 0.2×10^{-3} to 3.5×10^{-3}. Corresponding to the EP histograms, the topographic EP map in Group I showed that the EP of each region was approximate to equal, as shown in the top panel of Figure 2. While the EP in other regions (except the sensory motor area and ventral attention network) began to decrease in the other three groups.

To quantitatively characterize the transformation of EP distribution in individuals with age, the whole-brain NEE was calculated for each participant.

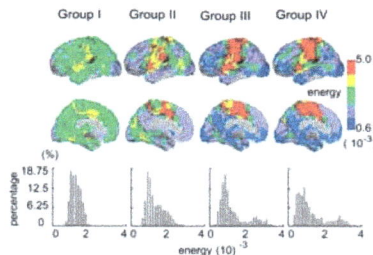

Figure 2. The topographic energy probability maps and energy probability histograms for the four groups.

3.2. Age-Related Changes of the Whole-Brain Network Eigen-Entropy across Lifespan

Model selection analyses were carried out using the Akaike information criterion (AIC). The probable models of functional connectome orderness are shown in Figure S1. These analyses showed that the functional connectome orderness levels during a lifespan, as measured by whole-brain NEE, are best fit by the polynomial fitting curve (Figure 3). The positive quadratic trajectory showed that the whole-brain NEE values decreased fast with continuous *age* before close to 60 years (Figure 3a). In a multiple linear regression model, no significant impact of *sex* ($t = 0.23$, $p = 0.81$), *meanFD* ($t = -0.67$, $p = 0.50$) or *tSNR* ($t = -0.49$, $p = 0.62$) on NEE was exhibited. In contrast, significant impacts of *age* ($t = -5.67$, $p < 0.0001$) and age^2 ($t = 3.43$, $p = 0.0007$) on NEE were exhibited. Pairwise comparison analysis through the four age groups using the Mann–Whitney rank-sum test showed that significant decreasing NEEs between groups were tested consistently (Figure 3b). In detail, the NEE of Group II significantly decreased relative to that of Group I ($p = 0.001$), and the NEE of Group III significantly decreased relative to that of Group II ($p < 0.001$); however, there was no significant difference between Groups III and IV for NEE ($p = 0.56$). Consistent with the results of a comparison of Groups III and IV, the NEE fitting curve maintained its trend during the normal aging stage. An important issue was whether the whole-brain NEE changes were driven by the seven subsystems.

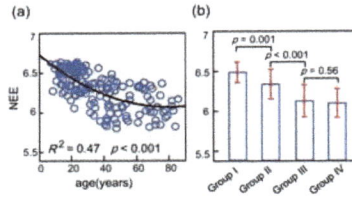

Figure 3. (a) The network eigen-entropy values reduced with age and the polynomial fitting curve was significant with an R^2 of 0.47 and p of 0.001; (b) Using the Mann–Whitney rank-sum test by pairwise comparison, significant network eigen-entropy differences between groups were exhibited.

3.3. Age-Related Changes of Subsystems' Network Eigen-Entropy across Lifespan

To assess how whole-brain NEEs in seven subsystems exhibited age-related changes, MLRs were fit to lifespan trajectories of the seven subsystems. The trajectories of the values of NEE in the seven subsystems showed linear or non-linear decreases with age (Figure 4). The overall trends were the same as the trend of the values of whole-brain NEE. Excluding the values of NEE of the limbic system and sensory motor area, the NEEs of the other five subsystems showed significant linear decreases with age. The age-related ($t = -8.47$, $p = 1.18 \times 10^{-14}$) NEE trajectory within the dorsal attention network slipped the fastest (F (1, 167) = 71.82, $p = 1.18 \times 10^{-14}$), and the age-related ($t = -4.24$, $p = 3.69 \times 10^{-5}$) NEE trajectory within the frontoparietal network slipped slightly (F (1, 167) = 17.98, $p = 3.69 \times 10^{-5}$). In contrast, the NEE trajectory within the limbic system exhibited positive quadratic changes (F (2, 166) = 20.85, $p = 8.34 \times 10^{-9}$) with age, and the NEE trajectory of the sensory motor area exhibited negative quadratic changes (F (2, 166) = 19.92, $p = 1.76 \times 10^{-8}$) with age. Significant linear age effects ($t = -3.69$, $p = 3.06 \times 10^{-4}$) and quadratic age effects ($t = 2.45$, $p = 0.016$) on limbic NEE were detected, and significant linear age effects ($t = 2.45$, $p = 0.015$) and quadratic age effects ($t = -3.66$, $p = 3.44 \times 10^{-4}$) on the sensory motor area NEE were also detected.

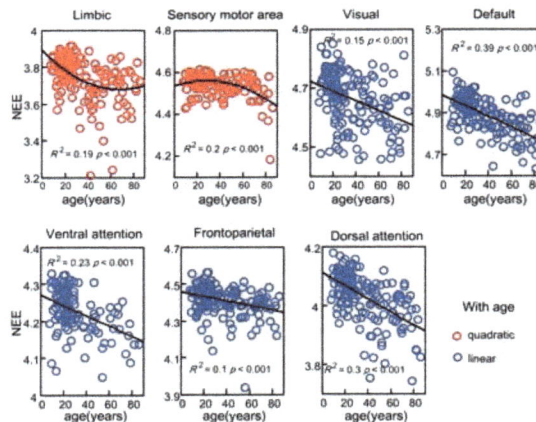

Figure 4. The network eigen-entropy trajectories of seven subsystems with *age*, the polynomial fitting curve of the limbic system and sensory motor area were significant with an R^2 of 0.19 and 0.20; the linear decreased fitting curve of the other five subsystems were significantly detected.

3.4. Brain Regions with Significant Energy Probablity Changes

There was a specific attempt to unearth specific brain regions which prominently drive NEE decline during different periods of the lifespan. The Mann–Whitney rank-sum test analysis for EP on whole-brain regions (800 nodes) between groups was undertaken though False Discovery Rate (FDR) correction. When comparing Group II with Group I, the results exhibited a total of 14 regions that showed statistically significant age-related change, representing 1.75% of whole-brain region comparisons ($q < 0.05$, FDR-corrected; corresponding to an uncorrected $p = 8.4 \times 10^{-4}$). These brain regions were involved with six of the seven subsystems, excluding a visual network. We mapped these significantly changed brain regions to the cerebral cortex and discovered that they were mainly comprised of the right inferior frontal gyrus, right supplementary motor area, and basal ganglia (containing the cingulate gyrus, right caudate nucleus, left globus pallidum, and left putamen). The topographic brain region distribution is shown in detail in Figure 5a. In addition to a few brain regions that showed a significant increase in EP, all the rest showed a significant decline. In addition, most of the brain regions were on the right hemisphere. When comparing Group III with Group II (Figure 5b), the results exhibited a total of 12 regions that showed statistically significant age-related change, representing 1.5% of whole-brain regions comparisons ($q < 0.05$, FDR-corrected; corresponding to an uncorrected $p = 8.5 \times 10^{-4}$). These brain regions were involved with two of the seven subsystems and when these brain regions were mapped to the cerebral cortex, it was discovered that they were primarily located within the right temporal pole, left lingual gyrus, left inferior occipital gyrus and parahippocampal gyrus, left hippocampus and right caudate nucleus, which are part of the visual subsystem and limbic system. In contrast to the results of the comparison between Group II and Group I, all significantly changed brain regions showed a reduction in EP.

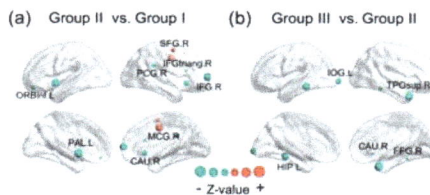

Figure 5. Statistically significant differences in brain regions in energy probability among three groups. The red represents the positive and the teal color represents the negative Z-value. The size of the node represents the absolute Z-value. The results were presented using the BrainNet Viewer package (http://www.nitrc.org/projects/bnv). (SFG.R = right superior frontal gyrus, IFG.R = right inferior frontal gyrus, MCG.R = right median cingulate gyrus, PCG.R = right posterior cingulate gyrus, IFGtriang.R = the triangular part of the right inferior frontal gyrus, CAU.R = right caudate, ORBinf.L = the orbital part of left inferior frontal gyrus, PAL.L = left globus pallidum, IOG.L = left inferior occipital gyrus, HIP.L = left hippocampus, TPOsup.R = the temporal pole of right superior temporal gurus, FFG.R = right fusiform gyrus).

3.5. Control Analyses

The scale of the brain networks and the scanning duration of the data may have influenced the calculation of the NEE. Therefore, we evaluated the potential confounding effects of these two factors on NEE.

To explore how the network scales influenced NEE, a series of cortical parcellations [41] ranging from 50 to 800 regions were chosen. The top panel of Figure 6a illustrates the increased NEE values of dataset II with the network scales. The NEE value of each network scale in the top panel curve was calculated by averaging the NEE values of all participants. The bottom panel of Figure 6a shows the increased growth rate of NEE with the network scales. The network scale (the number of regions

$N \geq 150$) was conducted using the elbow criterion [48] of the growth rate in the top panel of Figure 6a. We selected brain parcellation with 800 regions in this study for elaboration and stability.

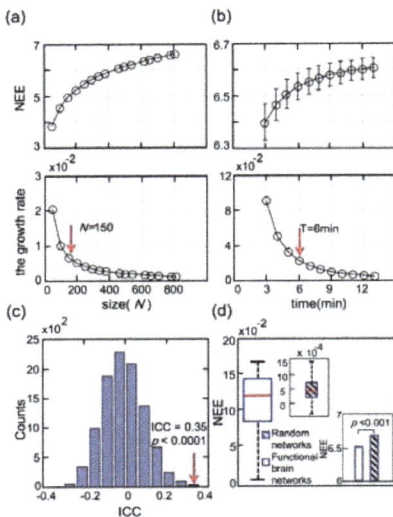

Figure 6. (**a**) The network eigen-entropy values changed with the network scales; (**b**) The values of network eigen-entropy changed with different scanning durations; (**c**) The permutation distribution of the estimate of Intra-class correlation when disrupting sessions; (**d**) Significant differences of network eigen-entropy values between the RSFC networks and random networks. The variance distribution of resting-state functional connectivity network NEE values was two orders of magnitude larger than that of the random networks' shown by a box diagram.

To investigate the effects of scanning duration on NEE, we calculated each participant's NEE with different continuous time points in dataset II, ranging from 3 to 13 min, which were extracted from the total scanning duration (1200-time points) by adjusting the starting time point randomly. The top panel of Figure 6b shows the NEE curve with different scanning durations and the bottom panel of Figure 6b shows the growth rate curve of NEE with different scanning durations. As shown in Figure 6a, the NEE value was calculated by averaging each participant's NEE for this scanning duration. Similarly, the scanning duration ($T \geq 6$ min) was assessed using the elbow criterion of the growth rate curve, in the top panel of Figure 6b. The NEE values slightly increased with scanning duration, but the growth rates of NEE decreased sharply. To obtain stable FC, the scanning duration should be longer than six minutes.

Additionally, the test–retest reliability measures the temporal or intra-individual stability across multiple occasions in a group of participants. In this study, Intra-class correlation was used to weigh the NEE's test–retest reliability.

The ICC value for dataset II was 0.35, which indicated fair test–retest reliability. However, several limitations of ICC approaches have been described in [47]. In addition to the absolute values of ICC, we calculated the *p*-values of ICC using a permutation test. The *p*-value allowed a precise statistical analysis to evaluate the significance of the extracted ICC. The distribution of the ICC was displayed in Figure 6c through 10,000 repeated permutation tests. According to the absolute value and *p*-value ($p < 0.0001$) of the ICC, the metric NEE was considerably reliable in temporal or intra-individual conditions.

Moreover, the significance of NEE should be established to compare it with the results of null-hypothesis networks. Null-hypothesis networks (random networks) shared the same size, density and binary degree distribution of the original network as dataset II, and the random network sample corresponded to each participant network. As shown in Figure 6d, the significant differences between the RSFC networks and random networks ($p < 0.001$, Mann–Whitney rank-sum test) were revealed.

4. Discussion

In this study, we calculated the whole-brain NEE from the EP of parcellation units of the cerebral cortex. By systematically investigating the development and aging of human connectome from 7 to 85 years, we found that the whole-brain NEE showed a non-linear decrease with age. The EP distribution within the whole-brain shifted from concentration to wide dispersion with age at group level. Moreover, the NEE values of subsystems showed non-linear or linear decreases with age. Interestingly, the NEE values of the sensory motor area showed negative quadratic changes with age, while the NEE of limbic system showed positive quadratic changes with age, and the other five subsystems showed linear decreases. Furthermore, brain regions with significant EP changing from flourishing (7–20 years) to the youth period (23–38 years) were mainly located in the right prefrontal cortex and basal ganglia, the areas involved in emotion regulation and executive function in coordination with the action of the sensory system, thus implying that self-awareness and voluntary control performance significantly changed during neurodevelopment. Those from the youth period through to middle (40–59 years) age were located in the mesial temporal lobe and caudate (which are associated with long-term memory), implying that the memory of the human brain begins to decline with age during this period. These results may be relevant for understanding normal developmental and aging changes in neural circuits that underlie age–related variation in cognition and behavior. These novel findings are discussed below in detail.

4.1. The Orderness of the Functional Connectome Measured by the Network Eigen-Entropy

Previous studies on structural and functional connectomics have revealed that the human brain has a small-world structure [33,49,50], which possesses a greater clustering coefficient than the random network and lower characteristic path length than the regular network. Unlike the properties of the small-world structure, whole-brain NEE characterizes the orderness of network structure from a spatial perspective based on the EC and statistical entropy. As EC conveys the global structural information of the network, the uniformity of the EC distribution reflects the degree of complex network heterogeneity. So, NEE can also capture the heterogeneity of the network. The entropy of a single discrete random variable is a measure of its average uncertainty, characterizing the degree of underlying randomness of the random variable. The orderness/certainty implied by NEE is dependent on the similarity of the intrinsic structure on multi-levels in a network. The more uniform energy distribution, the greater the entropy, the more uncertainty in the system; on the other hand, the more orderness/certainty in the system.

On the application of entropy in the human brain, there are several studies related to our study. Yao and his colleagues [51] regarded the correlation coefficients between brain regions as a random variable, and applied relative entropy to measure the variability in the distribution of correlation coefficients. The results demonstrated that the relative entropy increased with age. These results investigated whether the FC changed with age. Jia and his colleagues [51] regarded the dynamic correlation coefficients between brain regions as a random variable, and applied sample entropy to measure the complexity of dynamic FC, while the results demonstrated that the sample entropy reduced with age. Mcintosh and his colleagues [52] used multiscale entropy (MSE) to measure brain signal variability through a collection of the electroencephalography (EEG) and magnetoenphalography (MEG) signals in tasks. MSE characterized the temporal complexity of the brain signal, while NEE depicted the spatial complexity of the interaction between brain regions signal activities across the whole brain. However, these entropy analyses cannot capture the orderness of the brain network structure from a spatial perspective.

Through the analysis of the Mann–Whitney rank-sum test, we found that the mean NEE value of the RSFC networks was lower than that of random networks and their variance box distribution differed by two magnitudes (Figure 6d). As the connections between nodes in the random network were random, the EC distribution was uniform, and the intrinsic architecture of the functional brain network was regular relative to the random network. The degree of orderness of the functional brain network was higher than the random network, and the NEE could robustly distinguish functional brain networks from random networks.

4.2. The Orderness Variability of the Whole-Brain with Age

To specifically assess the association of the whole-brain NEE with age, we divided the fMRI dataset I into four age groups. In the analysis of different aged individuals in dataset I, we found that at the group level, the EP distribution of the RSFC networks shifted from concentration to dispersion (Figure 1). Through a statistical analysis of whole-brain NEE of different aged individuals, we found that the whole-brain NEE was robust when compared among the flourishing, youth period and middle age groups. Furthermore, the whole-brain NEE was significantly different among the entropies of the three sample groups, and further analysis over continuous aging showed that an approximate U-shaped trajectory of the whole-brain NEE fitted. The distribution of EP in the middle age group was more widely spread than that of the flourishing and youth periods. This led to the middle age group having a smaller NEE than that of the youth period group, which had an even smaller NEE than that of the flourishing. This revealed that there was a higher level of heterogeneity of functional connectome during middle age group than the youth period, and the youth group had a higher level of heterogeneity of functional connectome than that of the flourishing. This implied that the orderness/certainty of functional connectome enhanced rapidly until middle age. In contrast, the EP distribution was partially adjusted from middle age to old age. This led to no significant change of NEE between middle age and old age (61–85 years), and these findings may be associated with the degree of functional connectome adjustment with age. This finding verified the opinion in *What is Life* written by Schrodinger [53] that life continues to draw on negative entropy from the environment. The finding also was consistent with the evolution theory of the life system proposed by Darwin that the development of the organism and the highly complex brain is a process of diminishing entropy. These results may partly reveal the architectural transition of the human brain shift from a relatively anatomically driven configuration to a functional well-organized configuration with age. Furthermore, the current findings seem consistent with the notion of free energy minimization in the brain, as the entropy-like term (NEE) did indeed decrease with aging subjects [54–56]. Considering the notion of entropy in the context of free energy minimization may extend the value of this study to the broader community.

4.3. The Orderness Variability of Functional Subsystems with Age

In addition to determining the NEE of the whole brain, we determined the NEE of seven intra-subsystems of the brain across a lifespan. The subsystem NEE exhibited the FC pattern within this brain functional system, which indicated the orderness variability of the specific brain functional system. We observed distinct trajectories of these subsystem NEEs with age. Concretely, the NEE of limbic system decreased quickly over the early part of the lifespan (peak age of about 60 years); thereafter it showed a slightly increasing trend. This meant that the orderness of the limbic system was enhanced before 60 years, and weakened afterwards. The limbic system, as a subcortical cortex controlling human emotions and spiritual activities, showed a consistent trajectory with the whole-brain NEE. In contrast, the NEE of the sensory motor area remained at a high level before the age of 50, indicating that it maintained high homogeneity before the age of 50. After 50 years of age, its NEE decreased, which may be associated with the increases in the inter-connections and EP, making the heterogeneity of the sensory motor area and information exchange between the sensory motor area and other subsystems increase. The EP in the sensory motor area was slightly higher than other subsystems, which may be related to the earlier maturation of the sensory motor subsystem [10,57,58], while the heteromodal areas of the association cortex governing higher cognitive function were still developmental [12,18,29,59]. The finding implied that the orderness/certainty of all subsystems were the same as the orderness/certainty of whole-brain.

The energy probabilities of part brain regions of the ventral attention subnetwork and sensory motor area were apparently higher than other networks in the youth period (Group II), while the energy probabilities of the default network, frontoparietal network and limbic subsystem decreased. This finding may imply that youth is a prominent period of change to the subsystems' heterogeneity. The energy probabilities of the sensory motor network and part of the ventral attention network brain regions were highest in the whole-brain in the middle age (Group III). However, the energy probabilities of the visual network declined, while the energy probabilities of the default, frontoparietal, and limbic networks decreased further. This result may be consistent with the further enhanced heterogeneity of whole-brain functional subsystems [60–65]. The values of EP in the brain regions of the frontoparietal network increased slightly. And the energy probabilities of part brain regions of the default network decreased further in old age (Group IV). In summary, complicated changes to subsystems were not synchronous across the lifespan, which revealed the compensatory function among functional brain areas during normal aging [66–68], which may partly reflect the human brain plasticity.

4.4. Brain Regions with Significant Energy Probablity Changing

Through the Mann–Whitney rank-sum test on whole-brain EPs between Group II (youth period) and Group I (flourishing), we found that the significantly changed regions were mainly located in the right prefrontal cortex and basal ganglia, which are involved in emotion regulation and executive function in coordination with the action of the sensory system, implying that self-awareness and voluntary control performance significantly changed during neurodevelopment [69–71]. In addition, with the exception of the energy probabilities of the right median cingulate gyrus and right superior frontal gyrus which increased with age, other regions decreased with age. In contrast, a small set of significantly different regions was revealed between Group III (middle age) and Group II (youth period), which were less than that between Group II and Group I, and belonged to subcortical regions. These regions are mainly involved in long-term memory, implying that the memory of the human brain begins to decline with age during this period. Furthermore, we found that the brain regions where EP changes only comprised a small part of the whole brain and these changed brain regions were not synchronized. Between Groups III and IV, no brain region with significant EP changes was detected.

4.5. Limitations and Future Directions

It is worth noting that there were several issues to be considered when interpreting our findings. First, two strategies are commonly used for thresholding an FC matrix to format a graph adjacency matrix: (1) the correlation criterion using a fixed correlation value for all subjects, and (2) edges density criterion using a fixed density value for all subjects. In this study, we used a constant significance criterion. Although age and edges density were unrelated ($r = -0.08$, $p = 0.35$), more consistent results might be obtained if graph analyses could be conducted on density criterion. This issue should be studied carefully in future. Second, we regressed out the global signal to partly reduce physiological and other global noise. Third, it was a challenge to map the brain's parcellation-based functional connectome appropriately and precisely. In the study, we used a random-generated high-resolution template and explored the scale effect of a parcellation scheme on whole-brain NEE values. To eliminate the effect of templates on estimating brain system orderness, we will consider establishing voxel-wise RSFC networks in future work. Fourth, we attempted to explore the age-related differences of functional connectome over a continuous age range that covered both development and aging. However, the analyzed samples were not perfectly distributed across the entire lifespan as the number of young adults was greater than the number of older people. We utilized a quadratic model to explore the age changes across the human lifespan. The incomplete distribution of ages in our sample may have affected parametric curve fitting. Fifth, the RSFC network we constructed was binary and undirected; however, we know that an effective connection between two brain regions indicated a causal relationship, which contained more precise information. Furthermore, weighted connections indicated the strength of the information communication between two brain regions, so in future, we will also consider establishing weighted directed RSFC networks. Sixth, the present work discussed the orderness variability of functional connectome with normal developmental and aging, while structural and functional connectomics were of mutual interaction. Brain structural connectome supported and constrained functional connectome, meanwhile brain network organization refined brain structure gradually. We will consider discussing the orderness variability of structural connectomics with age in further works to reveal the consistency and discrepancy between them.

There were some disadvantages of NEE. It can describe the integral topological properties of the system, but it cannot locate local development of the system. In addition, the model used in this study is relatively simple and may impede its generalization and application to broader areas. In the future, we will expand to encompass more graph theoretic formalisms, and pursue more accurate and detailed models. The whole-brain NEE discussed in this study was estimated only for healthy individuals. We know that there is abnormal functional connectivity specific to some RSFC subsystems in brain diseases (e.g., depression, schizophrenia, autism, Alzheimer's disease) [72,73]. So, we would examine whether the NEE of the whole brain or specific subsystems could be used to predict the course of brain diseases and to evaluate treatment effects at a mesoscale level, shedding light on the pathophysiology of brain diseases in the future.

5. Conclusions

Using a public dataset (age range 7–85 years), the results demonstrated the heterogeneity changes of RSFC networks among four groups by exploring the EP distribution. Characterizing this property through network entropy, we found a trend towards the non-linear reduction of whole-brain NEE with continuous age. Going beyond detecting whole-brain NEE, we explored seven subsystems' NEE variability with age. The results demonstrated their consistence with trends, and the asynchronization of development and aging. In summary, our findings provided new insights into tracking the evolution of the human connectome.

Supplementary Materials: The following are available online at www.mdpi.com/1099-4300/19/9/471/s1, Figure S1: Individual NEE levels between the ages of 7 to 85 years, Table S1: Model selection and curve fitting.

Acknowledgments: We gratefully thank the NKI-RS and HCP for collecting and sharing data. This work was supported by the National Science Foundation of China (61420106001, 61503397 and 9142030002) and the National Basic Research Program of China (2013CB329401).

Author Contributions: Yiming Fan and Ling-Li Zeng conceived the algorithm and wrote the manuscript. Hui Shen and Dewen Hu revised the article. Jian Qin and Fuquan Li performed the data prepared. All authors have read and approved the final manuscript.

Conflicts of Interest: The authors declare no conflict of interest.

References

1. Wainwright, M.J.; Jordan, M.I. Graphical models, exponential families, and variational inference. *Found. Trends Mach. Learn.* **2008**, *1*, 1–305. [CrossRef]
2. Buckner, R.L.; Krienen, F.M.; Yeo, B.T.T. Opportunities and limitations of intrinsic functional connectivity MRI. *Nat. Neurosci.* **2013**, *16*, 832–837. [CrossRef] [PubMed]
3. Power, J.D.; Fair, D.A.; Schlaggar, B.L.; Petersen, S.E. The development of human functional brain networks. *Neuron* **2010**, *67*, 735–748. [CrossRef] [PubMed]
4. Yeo, B.T.T.; Krienen, F.M.; Sepulcre, J.; Sabuncu, M.R.; Lashkari, D.; Hollinshead, M.; Roffman, J.L.; Smoller, J.W.; Zöllei, L.; Polimeni, J.R.; et al. The organization of the human cerebral cortex estimated by intrinsic functional connectivity. *J. Neurophysiol.* **2011**, *106*, 1125–1165. [PubMed]

5. Betzel, R.F.; Byrge, L.; He, Y.; Goni, J.; Zuo, X.N.; Sporns, O. Changes in structural and functional connectivity among resting-state networks across the human lifespan. *Neuroimage* **2014**, *102*, 345–357. [CrossRef] [PubMed]

6. Zuo, X.-N.; Kelly, C.; Martino, A.D.; Mennes, M.; Margulies, D.S.; Bangaru, S.; Grzadzinski, R.; Evans, A.C.; Zang, Y.-F.; Castellanos, F.X.; et al. Growing together and growing apart-regional and sex differences in the lifespan developmental trajectories of functional homotopy. *J. Neurosci.* **2010**, *30*, 15034–15043. [CrossRef] [PubMed]

7. Wang, L.; Su, L.; Shen, H.; Hu, D. Decoding lifespan changes of the human brain using resting-state functional connectivity mri. *PLoS ONE* **2012**, *7*, e44530. [CrossRef] [PubMed]

8. Achard, S.; Bullmore, E. Efficiency and cost of economical brain functional networks. *PLoS Comput. Biol.* **2007**, *3*, e17. [CrossRef] [PubMed]

9. Fair, D.A.; Cohen, A.L.; Power, J.D.; Dosenbach, N.U.; Church, J.A.; Miezin, F.M.; Schlaggar, B.L.; Petersen, S.E. Functional brain networks develop from a "local to distributed" organization. *PLoS Comput. Biol.* **2009**, *5*, e1000381. [CrossRef] [PubMed]

10. Fransson, P.; Aden, U.; Blennow, M.; Lagercrantz, H. The functional architecture of the infant brain as revealed by resting-state fMRI. *Cereb. Cortex* **2011**, *21*, 145–154. [CrossRef] [PubMed]

11. Uddin, L.Q.; Supekar, K.S.; Ryali, S.; Menon, V. Dynamic reconfiguration of structural and functional connectivity across core neurocognitive brain networks with development. *J. Neurosci.* **2011**, *31*, 18578–18589. [CrossRef] [PubMed]

12. Supekar, K.; Musen, M.; Menon, V. Development of large-scale functional brain networks in children. *PLoS Biol.* **2009**, *7*, e1000157. [CrossRef] [PubMed]

13. Gu, S.; Satterthwaite, T.D.; Medaglia, J.D.; Yang, M.; Gur, R.E.; Gur, R.C.; Bassett, D.S. Emergence of system roles in normative neurodevelopment. *Proc. Natl. Acad. Sci. USA* **2015**, *112*, 13681–13686. [CrossRef] [PubMed]

14. Fair, D.A.; Cohen, A.L.; Dosenbach, N.U.F.; Church, J.A.; Miezin, F.M.; Barch, D.M.; Raichle, M.E.; Petersen, S.E.; Schlaggar, B.L. The maturing architecture of the brain's default network. *Proc. Natl. Acad. Sci. USA* **2008**, *105*, 4028–4032. [CrossRef] [PubMed]

15. Gao, W.; Zhu, H.; Giovanello, K.S.; Smith, J.K.; Shen, D.; Gilmore, J.H.; Lin, W. Evidence on the emergence of the brain's default network from 2-week-old to 2-year-old healthy pediatric subjects. *Proc. Natl. Acad. Sci. USA* **2009**, *106*, 6790–6795. [CrossRef] [PubMed]

16. Supekar, K.; Uddin, L.Q.; Prater, K.; Amin, H.; Greicius, M.D.; Menon, V. Development of functional and structural connectivity within the default mode network in young children. *Neuroimage* **2010**, *52*, 290–301. [CrossRef] [PubMed]

17. Yang, Z.; Chang, C.; Xu, T.; Jiang, L.; Handwerker, D.A.; Castellanos, F.X.; Milham, M.P.; Bandettini, P.A.; Zuo, X.-N. Connectivity trajectory across lifespan differentiates the precuneus from the default network. *Neuroimage* **2014**, *89*, 45–56. [CrossRef] [PubMed]

18. Fair, D.A.; Dosenbach, N.U.F.; Church, J.A.; Cohen, A.L.; Brahmbhatt, S.; Miezin, F.M.; Barch, D.M.; Raichle, M.E.; Petersen, S.E.; Schlaggar, B.L. Development of distinct control networks through segregation and integration. *Proc. Natl. Acad. Sci. USA* **2007**, *104*, 13507–13512. [CrossRef] [PubMed]

19. Kilford, E.J.; Garrett, E.; Blakemore, S.J. The development of social cognition in adolescence: An integrated perspective. *Neurosci. Biobehav. Rev.* **2016**, *70*, 106–120. [CrossRef] [PubMed]

20. Luna, B.; Padmanabhan, A.; O'Hearn, K. What has fmri told us about the development of cognitive control through adolescence? *Brain Cogn.* **2010**, *72*, 101–113. [CrossRef] [PubMed]

21. Cao, M.; Wang, J.H.; Dai, Z.J.; Cao, X.Y.; Jiang, L.L.; Fan, F.M.; Song, X.W.; Xia, M.R.; Shu, N.; Dong, Q.; et al. Topological organization of the human brain functional connectome across the lifespan. *Dev. Cogn. Neurosci.* **2014**, *7*, 76–93. [CrossRef] [PubMed]

22. Song, J.; Birn, R.M.; Boly, M.; Meier, T.B.; Nair, V.A.; Meyerand, M.E.; Prabhakaran, V. Age-related reorganizational changes in modularity and functional connectivity of human brain networks. *Brain Connect.* **2014**, *4*, 662–676. [CrossRef] [PubMed]

23. Geerligs, L.; Renken, R.J.; Saliasi, E.; Maurits, N.M.; Lorist, M.M. A brain-wide study of age-related changes in functional connectivity. *Cereb. Cortex* **2015**, *25*, 1987–1999. [CrossRef] [PubMed]

24. Meunier, D.; Achard, S.; Morcom, A.; Bullmore, E. Age-related changes in modular organization of human brain functional networks. *Neuroimage* **2009**, *44*, 715–723. [CrossRef] [PubMed]

25. Joyce, K.E.; Laurienti, P.J.; Burdette, J.H.; Hayasaka, S. A new measure of centrality for brain networks. *PLoS ONE* **2010**, *5*, e12200. [CrossRef] [PubMed]

26. Lohmann, G.; Margulies, D.S.; Horstmann, A.; Pleger, B.; Lepsien, J.; Goldhahn, D.; Schloegl, H.; Stumvoll, M.; Villringer, A.; Turner, R. Eigenvector centrality mapping for analyzing connectivity patterns in fmri data of the human brain. *PLoS ONE* **2010**, *5*, e10232. [CrossRef] [PubMed]

27. He, Y.; Wang, J.; Wang, L.; Chen, Z.J.; Yan, C.; Yang, H.; Tang, H.; Zhu, C.; Gong, Q.; Zang, Y.; et al. Uncovering intrinsic modular organization of spontaneous brain activity in humans. *PLoS ONE* **2009**, *4*, e5226. [CrossRef] [PubMed]

28. Power, J.D.; Schlaggar, B.L.; Lessov-Schlaggar, C.N.; Petersen, S.E. Evidence for hubs in human functional brain networks. *Neuron* **2013**, *79*, 798–813. [CrossRef] [PubMed]

29. Cao, M.; Huang, H.; Peng, Y.; Dong, Q.; He, Y. Toward developmental connectomics of the human brain. *Front. Neuroanat.* **2016**, *10*, 25. [CrossRef] [PubMed]

30. Van den Heuvel, M.P.; Sporns, O. Rich-club organization of the human connectome. *J. Neurosci.* **2011**, *31*, 15775–15786. [CrossRef] [PubMed]

31. Nigam, S.; Shimono, M.; Ito, S.; Yeh, F.C.; Timme, N.; Myroshnychenko, M.; Lapish, C.C.; Tosi, Z.; Hottowy, P.; Smith, W.C. Rich-club organization in effective connectivity among cortical neurons. *J. Neurosci.* **2016**, *36*, 670–684. [CrossRef] [PubMed]

32. Bullmore, E.; Sporns, O. Complex brain networks: Graph theoretical analysis of structural and functional systems. *Nat. Rev. Neurosci.* **2009**, *10*, 186–198. [CrossRef] [PubMed]

33. Achard, S.; Salvador, R.; Whitcher, B.; Suckling, J.; Bullmore, E. A resilient, low-frequency, small-world human brain functional network with highly connected association cortical hubs. *J. Neurosci.* **2006**, *26*, 63–72. [CrossRef] [PubMed]

34. Van, D.E.; Schoonheim, M.M.; Ijzerman, R.G.; Moll, A.C.; Landeirafernandez, J.; Klein, M.; Diamant, M.; Snoek, F.J.; Barkhof, F.; Wink, A.M. Altered eigenvector centrality is related to local resting-state network functional connectivity in patients with longstanding type 1 diabetes mellitus. *Hum. Brain Mapp.* **2017**, *38*, 3623–3636.

35. Binnewijzend, M.; Adriaanse, S.; Flier, W.M.; Teunissen, C.E.; Munck, J.C.; Stam, C.J.; Scheltens, P.; Berckel, B.N.M.; Barkhof, F.; Wink, A.M. Brain network alterations in alzheimer's disease measured by eigenvector centrality in fmri are related to cognition and csf biomarkers. *Hum. Brain Mapp.* **2014**, *35*, 2383–2393. [CrossRef] [PubMed]

36. Zuo, X.N.; Ehmke, R.; Mennes, M.; Imperati, D.; Castellanos, F.X.; Sporns, O.; Milham, M.P. Network centrality in the human functional connectome. *Cereb. Cortex* **2012**, *22*, 1862–1875. [CrossRef] [PubMed]

37. Wink, A.M.; de Munck, J.C.; Yd, V.D.W.; Oa, V.D.H.; Barkhof, F. Fast eigenvector centrality mapping of voxel-wise connectivity in functional magnetic resonance imaging: Implementation, validation, and interpretation. *Brain Connect.* **2012**, *2*, 265–274. [CrossRef] [PubMed]

38. Smith, S.M.; Beckmann, C.F.; Andersson, J.; Auerbach, E.J.; Bijsterbosch, J.; Douaud, G.; Duff, E.; Feinberg, D.A.; Griffanti, L.; Harms, M.P. Resting-state fmri in the human connectome project. *Neuroimage* **2013**, *80*, 144–168. [CrossRef] [PubMed]

39. Zeng, L.L.; Wang, D.; Fox, M.D.; Sabuncu, M.; Hu, D.; Ge, M.; Buckner, R.L.; Liu, H. Neurobiological basis of head motion in brain imaging. *Proc. Natl. Acad. Sci. USA* **2014**, *111*, 6058–6062. [CrossRef] [PubMed]

40. Zeng, L.L.; Shen, H.; Liu, L.; Hu, D. Unsupervised classification of major depression using functional connectivity mri. *Hum. Brain Mapp.* **2014**, *35*, 1630–1641. [CrossRef] [PubMed]

41. Craddock, R.C.; James, G.A.; Holtzheimer, P.E.; Hu, X.P.; Mayberg, H.S. A whole brain fmri atlas generated via spatially constrained spectral clustering. *Hum. Brain Mapp.* **2012**, *33*, 1914–1928. [CrossRef] [PubMed]

42. Strang, G. *Introduction to Linear Algebra*, 5th ed.; Wellesley-Cambridge Press: Wellesley, MA, USA, 2016.

43. Power, J.D.; Barnes, K.A.; Snyder, A.Z.; Schlaggar, B.L.; Petersen, S.E. Spurious but systematic correlations in functional connectivity mri networks arise from subject motion. *Neuroimage* **2012**, *59*, 2142–2154. [CrossRef] [PubMed]

44. Van Dijk, K.R.; Sabuncu, M.R.; Buckner, R.L. The influence of head motion on intrinsic functional connectivity mri. *Neuroimage* **2012**, *59*, 431–438. [CrossRef] [PubMed]

45. Zuo, X.N.; Xing, X.X. Test-retest reliabilities of resting-state fmri measurements in human brain functional connectomics: A systems neuroscience perspective. *Neurosci. Biobehav. Rev.* **2014**, *45*, 100–118. [CrossRef] [PubMed]

46. Landis, J.R.; Koch, G.G. The measurement of observer agreement for categorical data. *Biometrics* **1977**, *33*, 159–174. [CrossRef] [PubMed]

47. Vargha, P. A critical discussion of intraclass correlation coefficients. *Stat. Med.* **1997**, *16*, 821–823. [CrossRef]

48. Allen, E.A.; Damaraju, E.; Plis, S.M.; Erhardt, E.B.; Eichele, T.; Calhoun, V.D. Tracking whole-brain connectivity dynamics in the resting state. *Cereb. Cortex* **2014**, *24*, 663–676. [CrossRef] [PubMed]

49. Eguıluz, V.M.; Chialvo, D.R.; Cecchi, G.A.; Baliki, M.; Apkarian, A.V. Scale-free brain functional networks. *Phys. Rev. Lett.* **2005**, *94*, 018102. [CrossRef] [PubMed]

50. Liao, X.; Vasilakos, A.V.; Yong, H. Small-world human brain networks: Perspectives and challenges. *Neurosci. Biobehav. Rev.* **2017**, *77*, 286–300. [CrossRef] [PubMed]

51. Yao, Y.; Lu, W.L.; Xu, B.; Li, C.B.; Lin, C.P.; Waxman, D.; Feng, J.F. The increase of the functional entropy of the human brain with age. *Sci. Rep.* **2013**, *3*, 2853. [CrossRef] [PubMed]

52. Mcintosh, A.R.; Vakorin, V.; Kovacevic, N.; Wang, H.; Diaconescu, A.; Protzner, A.B. Spatiotemporal dependency of age-related changes in brain signal variability. *Cereb. Cortex* **2014**, *24*, 1806–1817. [CrossRef] [PubMed]

53. Schrödinger, E. *What is Life?* University Press: Cambridge, UK, 1944.

54. Tkacik, G.; Marre, O.; Mora, T.; Amodei, D.; Ii, M.J.B.; Bialek, W. The simplest maximum entropy model for collective behavior in a neural network. *J. Stat. Mech. Theor. Exp.* **2012**, *2013*, 829–837.

55. Maren, A.J. The cluster variation method: A primer for neuroscientists. *Brain Sci.* **2016**, *6*, E44. [CrossRef] [PubMed]

56. Tkačik's, G.; Marre, O.; Amodei, D.; Schneidman, E.; Bialek, W.; Nd, B.M. Searching for collective behavior in a large network of sensory neurons. *PLoS Comput. Biol.* **2014**, *10*, e1003408. [CrossRef] [PubMed]

57. Lin, W.; Zhu, Q.; Gao, W.; Chen, Y.; Toh, C.H.; Styner, M.; Gerig, G.; Smith, J.K.; Biswal, B.; Gilmore, J.H. Functional connectivity mr imaging reveals cortical functional connectivity in the developing brain. *AJNR Am. J. Neuroradiol.* **2008**, *29*, 1883–1889. [CrossRef] [PubMed]

58. Miao, C.; Yong, H.; Dai, Z.; Liao, X.; Jeon, T.; Ouyang, M.; Chalak, L.; Bi, Y.; Rollins, N.; Qi, D. Early development of functional network segregation revealed by connectomic analysis of the preterm human brain. *Cereb. Cortex* **2016**, *27*, 1949–1963.

59. Kaiser, M. Mechanisms of connectome development. *Trends Cogn. Sci.* **2017**, *9*, 703–717. [CrossRef] [PubMed]

60. Fjell, A.M.; Sneve, M.H.; Grydeland, H.; Storsve, A.B.; Walhovd, K.B. The disconnected brain and executive function decline in aging. *Cereb. Cortex* **2017**, *27*, 2303–2317. [CrossRef] [PubMed]

61. Sala-Llonch, R.; Bartres-Faz, D.; Junque, C. Reorganization of brain networks in aging: A review of functional connectivity studies. *Front. Psychol.* **2015**, *6*, 663. [CrossRef] [PubMed]

62. Andrewshanna, J.R.; Snyder, A.Z.; Vincent, J.L.; Lustig, C.; Head, D.; Raichle, M.E.; Buckner, R.L. Disruption of large-scale brain systems in advanced aging. *Neuron* **2007**, *56*, 924–935. [CrossRef] [PubMed]

63. Wu, T.; Zang, Y.; Wang, L.; Long, X.; Li, K.; Chan, P. Normal aging decreases regional homogeneity of the motor areas in the resting state. *Neurosci. Lett.* **2007**, *423*, 189–193. [CrossRef] [PubMed]

64. Onoda, K.; Ishihara, M.; Yamaguchi, S. Decreased functional connectivity by aging is associated with cognitive decline. *J. Cogn. Neurosci.* **2012**, *24*, 2186–2198. [CrossRef] [PubMed]

65. Tomasi, D.; Volkow, N.D. Aging and functional brain networks. *Mol. Psychiatry* **2012**, *17*, 471, 549–558. [CrossRef] [PubMed]

66. West, R.L. An application of prefrontal cortex function theory to cognitive aging. *Psychol. Bull.* **1996**, *120*, 272–292. [CrossRef] [PubMed]

67. Raz-Yaseef, N. Aging of the brain and its impact on cognitive performance: Integration of structural and functional findings. In *Handbook of Aging and Cognition*, 2nd ed.; Psychology Press: Hove, UK, 2000; pp. 1–90.

68. Woodruff-Pak, D.S. *The Neuropsychology of Aging*; Blackwell: Hoboken, NJ, USA, 1997; p. 119.

69. Albin, R.L.; Young, A.B.; Penney, J.B. The functional anatomy of basal ganglia disorders. *Trends Neurosci.* **1989**, *12*, 366–375. [CrossRef]

70. Alexander, G.E.; Crutcher, M.D. Functional architecture of basal ganglia circuits: Neural substrates of parallel processing. *Trends Neurosci.* **1990**, *13*, 266–271. [CrossRef]

71. Alexander, G.E.; DeLong, M.R.; Strick, P.L. Parallel organization of functionally segregated circuits linking basal ganglia and cortex. *Annu. Rev. Neurosci.* **1986**, *9*, 357–381. [CrossRef] [PubMed]

72. Wang, H.; Zeng, L.L.; Chen, Y.; Yin, H.; Tan, Q.; Hu, D. Evidence of a dissociation pattern in default mode subnetwork functional connectivity in schizophrenia. *Sci. Rep.* **2015**, *5*, 14655. [CrossRef] [PubMed]
73. Zeng, L.L.; Shen, H.; Liu, L.; Wang, L.; Li, B.; Fang, P.; Zhou, Z.; Li, Y.; Hu, D. Identifying major depression using whole-brain functional connectivity: A multivariate pattern analysis. *Brain* **2012**, *135*, 1498–1507. [CrossRef] [PubMed]

Article

Mutual Information and Information Gating in Synfire Chains

Zhuocheng Xiao [1,†], Binxu Wang [2,3,†], Andrew T. Sornborger [4,5,*] and Louis Tao [2,6,*]

[1]   Department of Mathematics, University of Arizona, Tucson, AZ 85721, USA; xiaoz@math.arizona.edu
[2]   Center for Bioinformatics, National Laboratory of Protein Engineering and Plant Genetic Engineering,
      School of Life Sciences, Peking University, Beijing 100871, China; lio50328@126.com
[3]   Yuanpei School, Peking University, Beijing 100871, China
[4]   Information Sciences, CCS-3, Los Alamos National Laboratory, Los Alamos, NM 87545, USA
[5]   Department of Mathematics, University of California, Davis, CA 95616, USA
[6]   Center for Quantitative Biology, Peking University, Beijing 100871, China
[*]   Correspondence: sornborg@lanl.gov (A.T.S.); taolt@mail.cbi.pku.edu.cn (L.T.)
[†]   These authors contributed equally to this work.

Received: 22 December 2017; Accepted: 30 January 2018; Published: 1 February 2018

Abstract: Coherent neuronal activity is believed to underlie the transfer and processing of information in the brain. Coherent activity in the form of synchronous firing and oscillations has been measured in many brain regions and has been correlated with enhanced feature processing and other sensory and cognitive functions. In the theoretical context, synfire chains and the transfer of transient activity packets in feedforward networks have been appealed to in order to describe coherent spiking and information transfer. Recently, it has been demonstrated that the classical synfire chain architecture, with the addition of suitably timed gating currents, can support the graded transfer of mean firing rates in feedforward networks (called synfire-gated synfire chains—SGSCs). Here we study information propagation in SGSCs by examining mutual information as a function of layer number in a feedforward network. We explore the effects of gating and noise on information transfer in synfire chains and demonstrate that asymptotically, two main regions exist in parameter space where information may be propagated and its propagation is controlled by pulse-gating: a large region where binary codes may be propagated, and a smaller region near a cusp in parameter space that supports graded propagation across many layers.

Keywords: pulse-gating; channel capacity; neural coding; feedforward networks; neural information propagation

1. Introduction

Faithful information transmission between neuronal populations is essential to computation in the brain. Correlated spiking activity has been measured experimentally between many brain areas [1–8]. Experimental and theoretical studies have shown that synchronized volleys of spikes can propagate within cortical networks and are thus capable of transmitting information between neuronal populations on millisecond timescales [9–17]. Many such mechanisms have been proposed for feedforward networks [14,15,17–23]. Commonly, mechanisms use transient synchronization to provide windows in time during which spikes may be transferred more easily from layer to layer [14,15,17–22].

For instance, the successful propagation of synchronous activity has been identified in "synfire chains" [18,19,21,24,25], wherein volleys of transiently synchronous spikes can be propagated through a predominantly excitatory feedforward architecture. Studies have shown that synchronous spike volleys can reliably drive responses through the visual cortex [13] with the temporal precision required for neural coding [26,27]. However, only sufficiently strong stimuli can elicit transient spike volleys that can successfully propagate through the network, and the waveform of spiking tends to an attractor with a single, fixed amplitude [21,28]. Thus, it is not possible to transfer graded information in the amplitudes of synchronously propagating spike volleys in this type of synfire chain.

Nonetheless, recent work has shown that synfire chains may be used as a pulse-gating mechanism coupled with a parallel "graded" chain ("synfire-gated synfire chain"—SGSC) to transfer arbitrary firing-rate amplitudes (graded information) through many layers in a feedforward neural circuit [14–17]. The addition of a companion gating circuit additionally provides a new mechanism for controlling information propagation in neural circuits [14–17]. Within a feedforward network, graded-rate transfer manifests as approximately time-translationally invariant spiking probabilities that propagate through many layers (guided by gating pulses) [17]. Using a Fokker–Planck (FP) approach, it has been demonstrated that this time-translational invariance arises near a cusp catastrophe in the parameter space of the gating current, synaptic strength and synaptic noise [17].

While many researchers have studied the dynamics of activity transmission in feedforward networks [21,23,25,29], few have examined these networks as information channels (however, see [30]). Shannon information [31,32] provides a natural framework with which to quantify the capacity of neural information transmission, by providing a measure of the correlation between input and output variables.

In particular, mutual information (MI), and measures based on MI, can be used to evaluate the expected reduction in entropy, for example, of the input, from the measurement of the output. Much work has focused on estimating probability distributions from spiking data (see, e.g., [33–38]). At the circuit level, studies have shown that MI is maximized in balanced networks (in cortical cultures, rats and macaques) that admit neuronal avalanches [39,40]. Furthermore, by examining the effects of gamma oscillations on MI in infralimbic and prelimbic cortex of mice, it has been shown that gamma rhythms enhance information transfer by reducing noise and signal amplification [7].

Theoretically, maximizing MI has been used to find nonlinear "infomax" networks [41] that can find statistically independent components capable of separating features in the visual scene [42]. MI has been used to assess the effectiveness and precision of population codes [43]. By combining decoding and MI, one can extract single-trial information from population activity and, at the same time, a quantitative estimate of how each neuron in the population contributes to the internal representation of the stimulus [44]. Furthermore, information-theoretic measures such as MI have been used on large-scale measurements of brain activity to estimate connectivity between different brain regions [45].

Here we consider information propagation properties in feedforward networks, and in particular, examine MI of mean firing-rate transfers across many layers of an SGSC neural circuit. We make use of an FP model to describe the dynamic evolution of membrane-potential probability densities in a pulse-gated, feedforward, integrate-and-fire (I&F) neuronal network [17]. We investigate the efficacy of information transfer in the parameter space near where graded mean firing-rate transfer is possible. We find that MI can be optimized by adjusting the strength of the gating (gating current), the feedforward synaptic strength, the level of synaptic noise, and the input distribution. Furthermore, our results reveal that via the coordination of pulse-gating and synaptic noise, a graded channel may be transformed into a binary channel. Our results demonstrate a wide range of possible information propagation choices in feedforward networks and the dynamic coding capacity of SGSCs.

2. Materials and Methods

We use a neuronal network model consisting of a set of $j = 1, \ldots, M$ populations, each with $i = 1, \ldots, N$ excitatory, current-based, I&F point neurons whose membrane potential, $V_{i,j}$, and (feedforward) synaptic current, $I_{i,j}^{\text{ff}}$, are described by

$$\frac{d}{dt} V_{i,j} = -g_L \left(V_{i,j} - V_R \right) + I_{g,j} + I_{i,j}^{\text{ff}} \tag{1a}$$

$$\tau \frac{d}{dt} I_{i,j}^{\text{ff}} = -I_{i,j}^{\text{ff}} + \begin{cases} \frac{S}{pN} \sum_k p_{jk} \sum_l \delta\left(t - t_{j-1,k}^l\right), & j > 1 \\ A\delta(t), & j = 1 \end{cases} \tag{1b}$$

124

where V_R is the reset voltage, τ is the synaptic timescale, S is the synaptic coupling strength, p_{jk} is a Bernoulli distributed random variable and $p = \langle p_{jk} \rangle$ is the mean synaptic coupling probability. The lth spike time of the kth neuron in layer $j-1$ is determined by $V(t^l_{j-1,k}) = V_{Th}$, that is, when the neuron reaches the threshold. The gating current, $I_{g,j}$, is a white-noise process with a square pulse envelope, $\theta(t-(j-1)T) - \theta(t-jT)$, where θ is a Heaviside theta function and T is the pulse length [14] of pulse height I_g and variance σ_0^2. We note that with the $j=1$ equation, an exponentially decaying current is injected into population 1, providing graded synchronized activity that subsequently propagates downstream through populations $j = 2, \ldots, M$.

Assuming the spike trains in Equation (1b) to be Poisson-distributed, the collective behavior of this feedforward network may be described by the FP equations:

$$\frac{\partial}{\partial t}\rho_j(V,t) = -\frac{\partial}{\partial V}J_j(V,t) \tag{2a}$$

$$\tau\frac{d}{dt}I_j^{ff} = -I_j^{ff} + \begin{cases} Sm_{j-1}, & j > 1 \\ A\delta(t), & j = 1 \end{cases} \tag{2b}$$

(While the output spike-train of a single neuron in general does not obey Poisson statistics, the spike train obtained from the summed output of all neurons in a single population does obey these statistics asymptotically for a large network size N. In the case of pulse-gating, the summed output spikes of a single population tend to a time-dependent Poisson process.) These equations describe the evolution of the probability density function (PDF), $\rho_j(V,t)$, in terms of the probability density flux, $J_j(V,t)$, the mean feedforward synaptic current, $I_j^{ff} \equiv \langle I_{i,j}^{ff} \rangle$, and the population firing rate, m_j. For each layer j, the probability density function gives the probability of finding a neuron with membrane potential $V \in (-\infty, V_{Th}]$ at time t.

The probability density flux is given by

$$J_j(V,t) = \left(\left[-g_L(V - V_R) + I_g + I_j^{ff} \right] - \sigma_j^2\frac{\partial}{\partial V} \right) \rho_j(V,t)$$

where I_g indicates the mean gating current. The effective diffusivity is

$$\sigma_j^2 = \sigma_0^2 + \frac{1}{2}\frac{S^2}{pN}m_{j-1}(t) \tag{3}$$

In the simulations reported below, we have taken $N \to \infty$, thus ignoring the second term in Equation (3) (i.e., ignoring diffusion due to finite size effects). The population firing rate is the flux of the PDF at threshold:

$$m_j(t) = J_j(V_{Th}, t) \tag{4}$$

The boundary conditions for the FP equations are

$$J_j(V_R^+, t) = J_j(V_{Th}, t) + J_j(V_R^-, t) \tag{5}$$
$$\rho_j(V_R^+, t) = \rho_j(V_{Th}, t) + \rho_j(V_R^-, t) \tag{6}$$

and

$$\rho_j(V = -\infty, t) = 0 \tag{7}$$

To efficiently obtain solutions to the FP equations [17], we have used an approximate Gaussian initial distribution, $\rho_j(V, t+jT) = (1/P)\exp\left(-(V-\mu(t))^2/2\sigma^2\right)$, with width σ and mean $\mu(t)$, where $P = \int_{-\infty}^{V_{Th}} \rho_0(V,0)$ is a normalization factor that accounts for the truncation of the Gaussian at threshold, V_{Th}. At the onset of gating, the distribution is advected toward the voltage threshold, V_{Th}, and the population starts to fire. The advection neglects a small amount of firing due to a diffusive flux across the firing threshold; thus the fold bifurcation occurs at a slightly larger value of synaptic coupling,

S, for this approximation, relative to numerical simulations [17]. Because the pulse is fast, neurons only have time to fire once (approximately). Thus, we neglect the re-emergent population at V_R, which does not have enough time to advect to V_{Th} and therefore does not contribute to firing during the transient pulse.

Using this approximation, Equation (2a) gives rise to $\dot{\mu} = -g_L(\mu - V_R) + I^g + I_u^{ff}$, where $\sigma^2 = \sigma_0^2/g_L$, with upstream current $I_u^{ff} = Ae^{-t/\tau}$. Setting $V_{Th} = 1$, this integrates to

$$\mu(t) = \mu_0 e^{-g_L t} + \frac{I^g}{g_L}(1 - e^{-g_L t}) + \frac{A}{\frac{1}{\tau} - g_L}(e^{-g_L t} - e^{-t/\tau}) \tag{8}$$

and from Equation (4), we have

$$m(t) = \left[(-g_L\mu(t) + I^g + Ae^{-t/\tau})\frac{1}{P}e^{-(1-\mu(t))^2/2\sigma^2}\right]^+ \tag{9}$$

which, from Equation (2b), results in a downstream synaptic current at $t = T$:

$$I_d^{ff} = Se^{-T/\tau}\int_0^T e^{t/\tau}m(t)\frac{dt}{\tau} \tag{10}$$

After the gating pulse terminates, the current decays exponentially. This decaying current feeds forward and is integrated by the next layer. For an exact transfer, $I_d^{ff}(S, I^g, A, T) = A$.

To compute MI, we generated a distribution of upstream current amplitudes $\{I_u\}$. These were typically within or near the range of fixed points of the map $I_d^{ff}(S, I^g, A, T) = A$. Using Equations (8)–(10), we generated a distribution of downstream current amplitudes $\{I_d\}$ with $\{I_u\}$ as initial values. The joint probability distribution $p(I_u, I_d)$ was estimated by forming a two-dimensional histogram with $\Delta I = 0.3/s$. MI, $I(I_u; I_d) = H(I_u) + H(I_d) - H(I_u, I_d)$, was computed (in bits) with $H(x) = -\sum_x p(x)\log_2(p(x))$ and $H(x, y) = -\sum_{x,y} p(x, y)\log_2(p(x, y))$, where the marginal probabilities $p(I_u)$ and $p(I_d)$ were computed from the histogram.

3. Results

We consider the transfer of spiking activity between two successive layers of a pulse-gated feedforward network. The spiking activity of feedforward propagation quickly converges to a stereotypical firing-rate waveform with arbitrary amplitudes (within a certain range) and with associated dynamics of the membrane potential PDF, $\rho(V, t)$, for each layer. This waveform essentially represents a volley of spikes that propagate downstream within the network. Furthermore, the temporal evolution of the first two moments (i.e., mean and variance) of $\rho(V, t)$ suffice to capture the dynamics during pulse-gating [17]. Therefore, as [17] showed, the population dynamics across a large region of parameter space can be mapped by using a Gaussian approximation to the membrane potential PDF (see Materials and Methods), revealing the bifurcation structure underlying a cusp catastrophe.

We first examined the cusp catastrophe in SGSC systems and its role in shaping the transfer of mean firing rates. Because of the existence of stable, attracting, translationally invariant firing-rate waveforms, we could capture and understand the dynamics in the SGSC system as an iterated map describing the firing-rate amplitude as it changed between successive network layers. Figure 1 shows two cases of the fixed points of this iterated map. Figure 1a plots the firing-rate amplitude, *A*, at the fixed points at the end of the pulse-gating period as a function of the strength of the gating current, I_g, and synaptic coupling strength, *S*, for fixed $\sigma = 1.05$. For small I_g, there existed a range in *S* where the system was bistable; however, as we increased the strength of the gating, the region of bistability (two stable attracting solutions with an unstable solution in between) disappeared at a cusp where the manifold of the fixed point was nearly vertical. It was shown in [17] that near this cusp, the slow dynamics along the unstable manifold of the unstable fixed point allow for the nearly graded transfer of firing rates between successive layers, giving rise to an approximate line attractor in the amplitude of the output firing rates. Figure 1b shows that this cusp catastrophe also existed at fixed gating current $I_g = 24$ as we varied both the feedforward synaptic coupling strength S and the variance of the initial PDF, $\rho(V, t = 0)$, just before pulse-gating. As the variance, σ^2, was increased, the bistable region in S also disappeared.

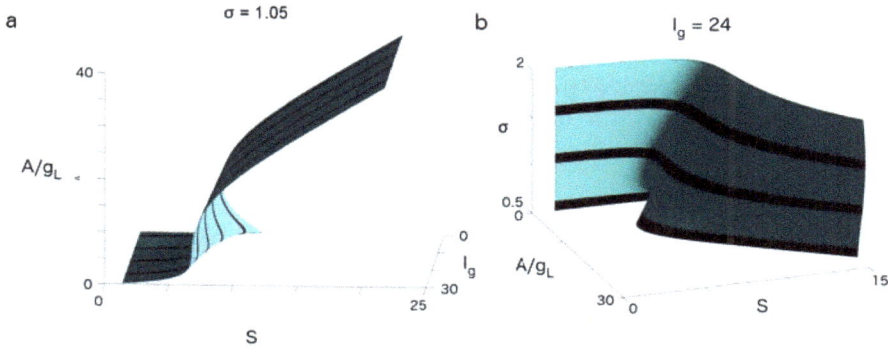

Figure 1. Cusp enabling graded and binary propagation in a synfire-gated synfire chain (SGSC). (a) View for fixed noise, $\sigma = 1.05$, as a function of firing-rate amplitude, A/g_L; synaptic strength, S; and gating current, I_g. (b) View for fixed gating current, $I_g = 24$, as a function of A/g_L, S, and σ. Plotted in both panels are zeros of the function $I(A, I_g, \sigma) - A$ (the difference between input and output firing-rate amplitudes in a given layer). This function becomes zero at a fixed point. At a fixed point, a firing-rate amplitude propagates exactly from one layer to the next. When this function is small but non-zero, the firing-rate amplitude only changes slowly as it propagates. We note that the cusp is high-dimensional and can be viewed in various projections. Graded propagation (approximately exact amplitude propagation) is enabled as a result of ghost (slow) dynamics near the cusp, such that the firing-rate amplitude A varies only slowly from layer to layer. At the cusp, an approximate line attractor exists where an input firing rate in a given layer changes only slowly as it propagates downstream. Away from the cusp, firing rates rapidly approach attractors. For many parameter regions, there are two attractors giving rise to the propagation of a binary code.

Next, we examine the evolution of MI through the network. Figure 2 shows the transmission of the mean firing rate, A/g_L, in a pulse-gated feedforward network, as well as the associated MI for three types of channels. Figure 2a plots the firing rate for an idealized exact transfer across many layers ($j = 1, \ldots, 100$), where we utilized a thresholded linear f–I curve to model each layer [14]. In this toy model, the transfer was exact and the mean firing rate propagated indefinitely without change. Figure 2b shows the propagation of firing rates for an SGSC situated near the cusp of Figure 1a. The long-term propagation of the initially uniform input firing rates revealed the dynamical structure near the cusp, namely, two stable, attracting firing rates (top and bottom) with an unstable saddle in between. Because of the slowness of the dynamics along the unstable manifold of the saddle, the transfer of mean firing rates through the network was approximately graded for many layers ($j \approx 1$–30). Furthermore, this transfer of the mean firing rates was order preserving (i.e., the relative ordering of amplitudes was maintained) across many more layers (j up to 100). Figure 2c demonstrates the propagation of firing rates for a binary transfer, where the unstable saddle was strongly repelling; for most initial conditions, the rates converged to one of two rates within 10–15 layers. To investigate the effects of the SGSC dynamics on information propagation, we computed the MI for each of the three cases. Figure 2d demonstrates the effect of the SGSC dynamics on information propagation. In the exact transfer case, the MI between the input and each layer remained constant and represented an upper bound on the information transfer. In the binary transfer case, the MI quickly decayed to 1 bit, but, as for the exact case, was stable over long timescales. Near the cusp, where the firing-rate transfer was approximately graded, the MI decayed slowly, so that even after $j = 100$ transfers, the channel retained almost 4 bits of information. In Figure 2e, the joint probabilities from which the MI was computed for representative layers are plotted. For the exact case, the distribution is always along the diagonal. The fast transition from diagonal to binary is evident for binary transfers, and a slow transition from diagonal to binary is seen for the graded case.

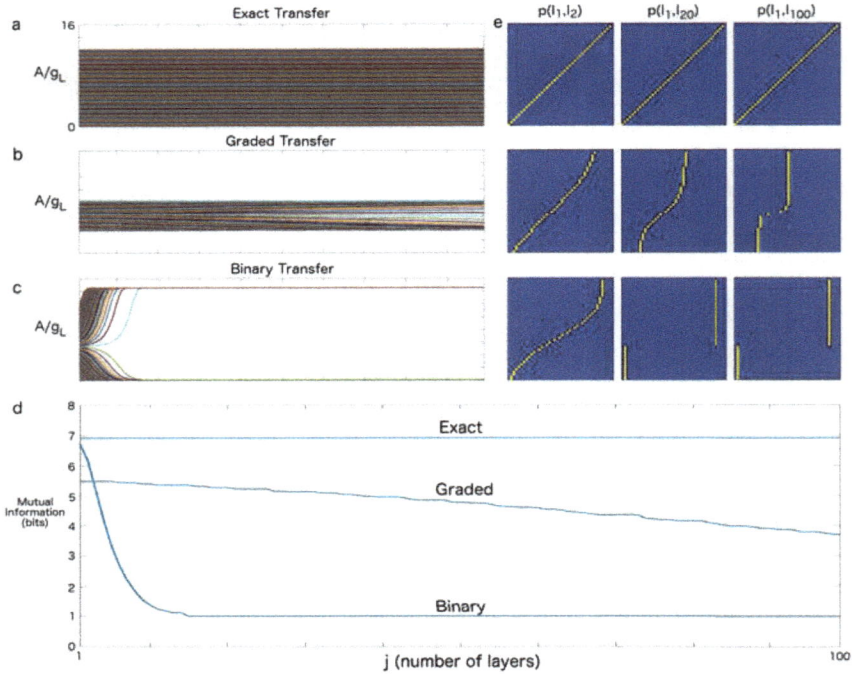

Figure 2. Mutual information (MI) for transfers across a 100 layer synfire-gated synfire chain (SGSC). (a) Exact transfer: Firing-rate amplitudes for theoretically perfect transfer. The range of A/g_L is the same for (**a–c**). Amplitudes transfer exactly from one layer to the next ($j = 1, \dots, 100$). This practically unattainable communication mode is shown for comparison and attains the maximum possible information transfer (MI; see (**d**)) as a function of the layer. (**b**) Graded transfer: Firing-rate amplitudes from approximate Fokker–Planck (FP) solutions in a multi-layer SGSC. Parameters: $I_g = 24.03$; $S = 7.001$; $\sigma = 1.0496$. We note that as the number of layers through which the initial amplitudes propagate increases, the amplitudes drift slowly away from an unstable attractor in the center of the amplitude distribution towards stable attractors at the sides of the distribution. (**c**) Binary transfer: Firing-rate amplitudes from approximate FP solutions in a multi-layer SGSC. Parameters: $I_g = 5.0$; $S = 7$; $\sigma = 0.6$. For these parameters, the unstable attractor rapidly repels the amplitudes toward stable attractors on either side, resulting in binary transfer. Binary transfer is extremely stable in the SGSC. (**d**) MI as a function of layer, j, for each of the above types of channel (**a–c**). (**e**) The joint probabilities, $p(I_1, I_2)$, $p(I_1, I_{20})$, and $p(I_1, I_{100})$, from which the MI was computed for exact, graded and binary transfer. Here, the exact case is as would be expected. The graded case shows a gradual deformation away from the diagonal that approaches binary transfer asymptotically. The binary case rapidly approaches the propagation of only two states. We note that in much of the parameter space, the approach to binary is much faster than that shown here. We used these parameters so that the transition was slower and therefore more evident.

An FP analysis of the SGSC system reveals four important system parameters: strength of the gating current, I_g; strength of the feedforward synaptic coupling, S; mean membrane potential distribution at the beginning of pulse-gating, $\rho_j(V, t = jT)$; and variance of the synaptic current, σ^2 (see Materials and Methods). Figure 3 shows where graded and binary codes are supported. Figure 3a,b plots the MI as a function of S and the network layer (which is equivalent to propagation time) for fixed I_g, σ and $\rho_j(V, t = jT)$ (which equals 0 for all panels). As we can see, for both cases, there exists an optimal S for which MI (and approximate graded activity) can be maintained for many layers. As we move away from this optimal S, we quickly go into a binary coding regime (MI = 1). Figure 3c demonstrates that similar qualitative behavior can be obtained by varying σ.

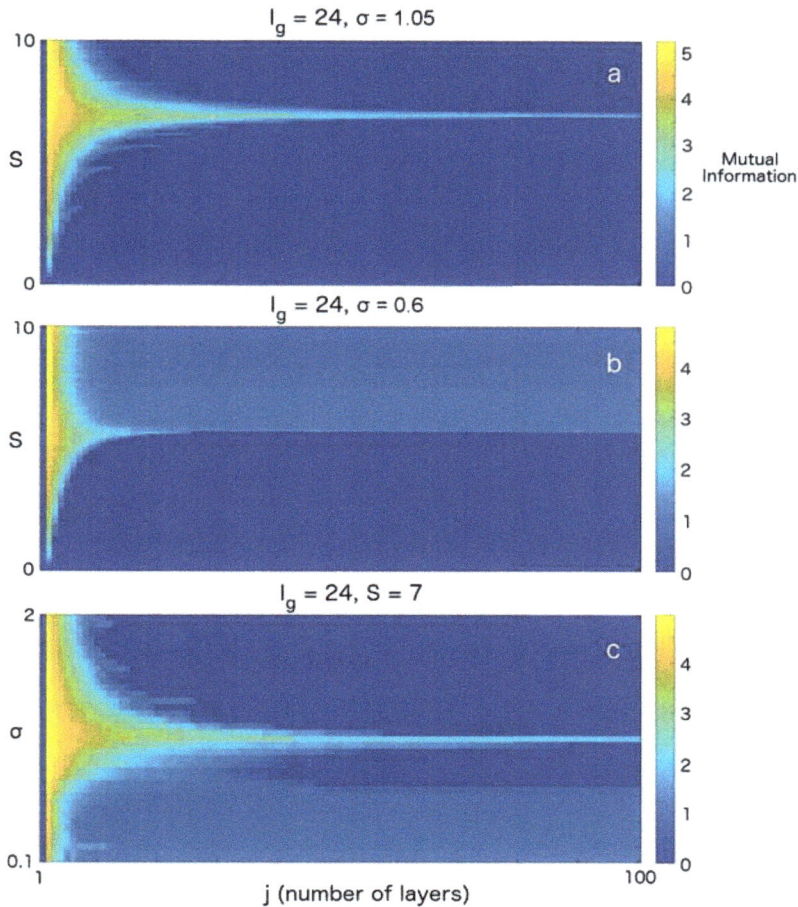

Figure 3. Mutual information (MI) for parameters supporting graded and binary codes. (**a**) MI propagation across 100 layers for $I_g = 24$, $\sigma = 1.05$ and S ranging from 0 to 10. We note that graded transfer allows high values of MI to propagate across 100 layers at $S \approx 7$. (**b**) MI propagation across 100 layers for $I_g = 24$, $\sigma = 0.6$ and S from 0 to 10. We note that above $S \approx 5$, there is a large range of S for which binary propagation (MI of 1 bit; see colorbar) is supported. (**c**) MI propagation for $I_g = 24$, $S = 7$ and σ from 0.1 to 2. Here, both graded and binary information propagate depending on the value of σ.

Out of the four system parameters, it is easiest to manipulate either the gating current or the variance of the synaptic current. Both can be viewed as controls independent of the SGSC system. Therefore, in Figure 4, we examine the evolution of MI as a function of I_g and σ through the network. As we expect from our results thus far, large regions of parameter space support binary coding; however, there is a thin line that materializes towards the bottom of the binary region that corresponds to the location of the cusp, where MI can be maintained at high levels (MI > 3) through many layers. We note that asymptotically, as $j \to \infty$, MI approaches 1 (a binary channel) because of the existence of two attractors, even near the cusp.

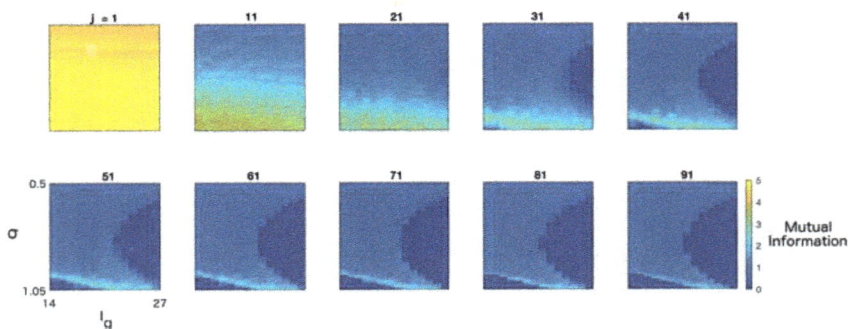

Figure 4. Regions of parameter space supporting graded and binary information propagation. Visualization of mutual information (MI) for σ between 0.5 and 1.05 and I_g between 14 and 27 at successive layers j; σ and I_g axes as denoted in the lower left panel are the same for all panels. Colorbar for lower right panel is the same for all panels. We note that across a few layers, MI remains high for a large region of parameter space, but by $j = 21$, only binary and graded codes persist. By $j = 91$, it is seen that a large region of parameter space supports binary propagation. Along a thin diagonal line at the bottom of the binary region, corresponding to the location of the cusp, graded information may be transferred.

4. Discussion

Although coherent activity has now been measured in many regions of the mammalian brain, the precise mechanism and the extent to which the brain can make use of synchronous spiking activity to transfer information have remained unclear. Many mechanisms have been proposed for information transfer via transiently synchronous spiking that relies on oscillations and gating [18,20,22,46–52]. These mechanisms make use of the fact that coherent input can provide temporal windows during which spiking activity may be more easily transferred between sending and receiving populations. However, from the theoretical perspective, how MI and other measures of communication capacity can be related to the underlying neuronal network architecture and the emergent network dynamics have remained unexplored.

Here we study the capacity for information transfer of feedforward networks by examining the evolution of mutual information through many layers of an SGSC system. Previous work has showed that by introducing suitably timed pulses, graded information could be transferred and controlled in a feedforward excitatory neuronal network [14–17]. In this context, pulse-gating has allowed us to understand information propagation as an iterated map. FP analysis of this map has enabled the identification of a cusp catastrophe in the relevant parameter space. Our results here demonstrate that the dynamics of the SGSC system naturally give rise to two different types of channel (as measured by MI). Large regions of parameter space support binary coding by using transiently synchronous propagation of high and low firing rates. (In the classical synfire chain case, the low firing-rate state is a silent state. In the SGSC case, as a result of the gating current, it is small, but non-zero.) In the binary regime, the distance between low and high firing rates is much larger than the variance of the distributions; therefore propagation is stable. Furthermore, by systematically varying the relevant SGSC system parameters, we were able to optimize graded-rate transfer near the fold of a cusp catastrophe, which enabled us to maintain a relatively high MI through many layers.

Evidence exists for graded information coding in visual and other cortices [53], and there is some evidence for binary coding in the auditory cortex [54]; sparse coding mechanisms for its use have been put forward [55,56]. Luczak et al. [51] have argued that spike packets and stereotypical and repeating sequences (similar to sequences observed in implementations of information processing algorithms in graded-transfer SGSCs [15]) underlie neural coding. More recently, Piet et al. [57] have used attractor networks to model frontal orienting fields in rat cortices to argue that bistable attractor dynamics can account for the memory observed in a perceptual decision-making task. Indeed, in many decision-making tasks, cortical activity appears to be holding graded information in *working memory*, before a *decision* forces the activity into a binary code [58]. Therefore, it is of interest that both graded and binary pulse-gated channels are supported by the SGSC mechanism and that it is fairly simple and rapid to convert a graded code to a binary code.

Examining the structure of the cusp catastrophe in the parameter space, it appears that in general, weaker I_g and higher S are correlated with bistability, while stronger I_g can support graded information transfer at lower S (see Figure 1a); at the same time, for fixed I_g, a lower σ tends to bistable states (see Figure 1b).

In previous theoretical studies of graded propagation and line attractors, it has been shown that some fine tuning of system parameters is required [59–61]. However, graded propagation and line attractors have been observed in many areas of the brain (see, e.g., [62–65]). Our results here demonstrate that an important consideration for understanding graded propagation is the depth of the circuit being used to propagate the information. That is, graded information propagation circuits with a depth of 20 to 30 layers are not particularly fine-tuned. In this depth range, about 1/10th of the parameter space (e.g., S or σ; see Figure 3, top and bottom plots) is capable of propagating graded information (with relatively high MI). As the depth of the circuit grows, $j \to \infty$, MI approaches 1 bit, and hence graded propagation in deep circuits is not possible (with an SGSC with parameters near the cusp).

A clear advantage of a graded information channel is that a vector of high-resolution graded information (resolution 2^n, where n is the number of bits of MI—up to 32 levels of resolution in the graded channel shown in Figure 2) could be rapidly processed in a network with linear synaptic connectivity. Thus, synaptic processing such as Gabor transforms, seen in the visual cortex, would most naturally operate on graded information, rapidly reducing the dimension of and orthogonalizing input data. However, the stable processing of information through deep and complicated neural logic circuits would take better advantage of binary channels. Here, essentially exact pulse-gated binary transformations and decisions [15] can make use of the attractor structure of the channel to reduce noise and maintain discrete states. In order to make use of high-bandwidth graded processing, at some point in a neural circuit, graded information would need to be transformed to binary information. Mechanisms to do this are beyond the scope of this paper, but they could make use of dimension-reducing transforms on the input (e.g., Gabor transform) and subsequent digitization of the data subspace.

Finally, we note that the symmetric nature of the binary SGSC channel may be an advantage for logic circuit gating, as either of the parallel synfire chains can operate as information or the gate [15]. Finally, it should be remembered that with a stable binary code, binary digit coding can be constructed, effectively increasing the information resolution propagating in a binary circuit.

5. Conclusions

We have performed an investigation of the communication capabilities of SGSCs using the metric of MI. The main conclusion of this investigation is that SGSCs sustain two types of channel: the first is binary transfer (MI = 1, in bits), which is supported for a wide range of parameters. For circuit depths of up to 30–40 layers, in a narrower range of parameters, graded transfer is also supported. A secondary conclusion is that, because of the dependence of MI on the depth of a circuit in some parameter regimes, circuit depth should be taken into account when considering the communication capacity of a neural circuit.

Acknowledgments: This work was supported by the Natural Science Foundation of China, Grant Nos. 31771147 and 91232715 (B.W. and L.T.); by the Open Research Fund of the State Key Laboratory of Cognitive Neuroscience and Learning, Grant No. CNLZD1404 (B.W. and L.T.); by the Beijing Municipal Science and Technology Commission under contract Z151100000915070 (B.W. and L.T.); by the SLS-Qidong Innovation Fund (B.W. and L.T.); and by the LANL ASC Beyond Moore's Law project (A.T.S.).

Author Contributions: All authors conceived and designed the numerical simulations. A.T.S. and L.T. wrote the paper, with major input from Z.X. and B.W. All authors have read and approved the final manuscript.

Conflicts of Interest: The authors declare no conflict of interest.

References

1. Gray, C.; König, P.; Engel, A.; Singer, W. Oscillatory responses in cat visual cortex exhibit inter-columnar synchronization which reflects global stimulus properties. *Nature* **1989**, *338*, 334–337.
2. Livingstone, M. Oscillatory firing and interneuronal correlations in squirrel monkey striate cortex. *J. Neurophysiol.* **1996**, *66*, 2467–2485.
3. Bragin, A.; Jandó, G.; Nádasdy, Z.; Hetke, J.; Wise, K.; Buzsáki, G. Gamma (40–100 Hz) oscillation in the hippocampus of the behaving rat. *J. Neurosci.* **1995**, *15*, 47–60.
4. Brosch, M.; Budinger, E.; Scheich, H. Stimulus-related gamma oscillations in primate auditory cortex. *J. Neurophysiol.* **2002**, *87*, 2715–2725.
5. Bauer, M.; Oostenveld, R.; Peeters, M.; Fries, P. Tactile spatial attention enhances gamma-band activity in somatosensory cortex and reduces low-frequency activity in parieto-occipital areas. *J. Neurosci.* **2006**, *26*, 490–501.
6. Pesaran, B.; Pezaris, J.; Sahani, M.; Mitra, P.; Andersen, R. Temporal structure in neuronal activity during working memory in macaque parietal cortex. *Nat. Neurosci.* **2002**, *5*, 805–811.
7. Sohal, V.S.; Zhang, F.; Yizhar, O.; Deisseroth, K. Parvalbumin neurons and gamma rhythms enhance cortical circuit performance. *Nature* **2009**, *459*, 698–702.
8. Popescu, A.; Popa, D.; Paré, D. Coherent gamma oscillations couple the amygdala and striatum during learning. *Nat. Neurosci.* **2009**, *12*, 801–807.
9. Bair, W.; Koch, C. Temporal precision of spike trains in extrastriate cortex of the behaving macaque monkey. *Neural Comput.* **1996**, *8*, 1185–1202.
10. Butts, D.A.; Weng, C.; Jin, J.; Yeh, C.I.; Lesica, N.A.; Alonso, J.M.; Stanley, G.B. Temporal precision in the neural code and the timescales of natural vision. *Nature* **2007**, *449*, 92–95.
11. Varga, C.; Golshani, P.; Soltesz, I. Frequency-invariant temporal ordering of interneuronal discharges during hippocampal oscillations in awake mice. *Proc. Natl. Acad. Sci. USA* **2012**, *109*, E2726–E2734.
12. Reyes, A.D. Synchrony-dependent propagation of firing rate in iteratively constructed networks in vitro. *Nat. Neurosci.* **2003**, *6*, 593–599.
13. Wang, H.P.; Spencer, D.; Fellous, J.M.; Sejnowski, T.J. Synchrony of thalamocortical inputs maximizes cortical reliability. *Science* **2010**, *328*, 106–109.
14. Sornborger, A.T.; Wang, Z.; Tao, L. A mechanism for graded, dynamically routable current propagation in pulse-gated synfire chains and implications for information coding. *J. Comput. Neurosci.* **2015**, *39*, 181–195.
15. Wang, Z.; Sornborger, A.; Tao, L. Graded, dynamically routable information processing with synfire-gated synfire chains. *PLoS Comput. Biol.* **2016**, *12*, e1004979.
16. Shao, Y.; Sornborger, A.; Tao, L. A pulse-gated, predictive neural circuit. In Proceedings of the 50th Asilomar Conference on Signals, Systems and Computers, Pacific Grove, CA, USA, 6–9 November 2016.
17. Xiao, Z.; Zhang, J.; Sornborger, A.; Tao, L. Cusps enable line attractors for neural computation. *Phys. Rev. E* **2017**, *96*, 052308.
18. Abeles, M. Role of the cortical neuron: Integrator or coincidence detector? *Isr. J. Med. Sci.* **1982**, *18*, 83–92.

19. König, P.; Engel, A.; Singer, W. Integrator or coincidence detector? The role of the cortical neuron revisited. *Trends Neurosci.* **1996**, *19*, 130–137.

20. Fries, P. A mechanism for cognitive dynamics: Neuronal communication through neuronal coherence. *Trends Cogn. Sci.* **2005**, *9*, 474–480.

21. Diesmann, M.; Gewaltig, M.O.; Aertsen, A. Stable propagation of synchronous spiking in cortical neural networks. *Nature* **1999**, *402*, 529–533.

22. Rubin, J.; Terman, D. High frequency stimulation of the subthalamic nucleus eliminates pathological thalamic rhythmicity in a computational model. *J. Comput. Neurosci.* **2004**, *16*, 211–235.

23. Van Rossum, M.C.; Turrigiano, G.G.; Nelson, S.B. Fast propagation of firing rates through layered networks of noisy neurons. *J. Neurosci.* **2002**, *22*, 1956–1966.

24. Kistler, W.; Gerstner, W. Stable propagation of activity pulses in populations of spiking neurons. *Neural Comput.* **2002**, *14*, 987–997.

25. Litvak, V.; Sompolinsky, H.; Segev, I.; Abeles, M. On the transmission of rate code in long feedforward networks with excitatory-inhibitory balance. *J. Neurosci.* **2003**, *23*, 3006–3015.

26. Kumar, A.; Rotter, S.; Aertsen, A. Conditions for propagating synchronous spiking and asynchronous firing rates in a cortical network model. *J. Neurosci.* **2008**, *28*, 5268–5280.

27. Nemenman, I.; Lewen, G.D.; Bialek, W.; de Ruyter van Steveninck, R.R. Neural coding of natural stimuli: Information at sub-millisecond resolution. *PLoS Comput. Biol.* **2008**, *4*, e1000025.

28. Moldakarimov, S.; Bazhenov, M.; Sejnowski, T.J. Feedback stabilizes propagation of synchronous spiking in cortical neural networks. *Proc. Natl. Acad. Sci. USA* **2015**, *112*, 2545–2550.

29. Kumar, A.; Rotter, S.; Aertsen, A. Spiking activity propagation in neuronal networks: Reconciling different perspectives on neural coding. *Nat. Rev. Neurosci.* **2010**, *11*, 615–627.

30. Cannon, J. Analytical calculation of mutual information between weakly coupled poisson-spiking neurons in models of dynamically gated communication. *Neural Comput.* **2017**, *29*, 118–145.

31. Shannon, C. A mathematical theory of communication. *Bell Syst. Tech. J.* **1948**, *27*, 379–423.

32. Cover, T.; Thomas, J. *Elements of Information Theory*, 2nd ed.; Wiley: Hoboken, NJ, USA, 2006.

33. Rieke, F.; Warland, D.; de Ruyter van Steveninck, R.; Bialek, W. *Spikes: Exploring the Neural Code*; The MIT Press: Cambridge, MA, USA, 1997.

34. Paninski, L. Estimation of entropy and mutual information. *Neural Comput.* **2003**, *15*, 1191–1253.

35. Victor, J.D. Binless strategies for estimation of information from neural data. *Phys. Rev. E Stat. Nonlinear Soft Matter Phys.* **2002**, *66*, 051903.

36. Nemenman, I.; Bialek, W.; de Ruyter van Steveninck, R. Entropy and information in neural spike trains: progress on the sampling problem. *Phys. Rev. E Stat. Nonlinear Soft Matter Phys.* **2004**, *69*, 056111.

37. Shapira, A.H.; Nelken, I. Binless Estimation of Mutual Information in Metric Spaces. In *Spike Timing: Mechanisms and Function*; DiLorenzo, P., Victor, J., Eds.; CRC Press: Boca Raton, FL, USA, 2013; Chapter 5, pp. 121–135.

38. Lopes-dos Santos, V.; Panzeri, S.; Kayser, C.; Diamond, M.E.; Quian Quiroga, R. Extracting information in spike time patterns with wavelets and information theory. *J. Neurophysiol.* **2015**, *113*, 1015–1033.

39. Shew, W.L.; Plenz, D. The functional benefits of criticality in the cortex. *Neuroscientist* **2013**, *19*, 88–100.

40. Shew, W.L.; Yang, H.; Yu, S.; Roy, R.; Plenz, D. Information capacity and transmission are maximized in balanced cortical networks with neuronal avalanches. *J. Neurosci.* **2011**, *31*, 55–63.

41. Bell, A.J.; Sejnowski, T.J. An information-maximization approach to blind separation and blind deconvolution. *Neural Comput.* **1995**, *7*, 1129–1159.

42. Bell, A.J.; Sejnowski, T.J. The "independent components" of natural scenes are edge filters. *Vis. Res.* **1997**, *37*, 3327–3338.

43. Yarrow, S.; Challis, E.; Series, P. Fisher and Shannon information in finite neural populations. *Neural Comput.* **2012**, *24*, 1740–1780.

44. Quiroga, R.Q.; Panzeri, S. Extracting information from neuronal populations: Information theory and decoding approaches. *Nat. Rev. Neurosci.* **2009**, *10*, 173–185.

45. Bullmore, E.; Sporns, O. Complex brain networks: Graph theoretical analysis of structural and functional systems. *Nat. Rev. Neurosci.* **2009**, *10*, 186–198.

46. Fries, P.; Reynolds, J.; Rorie, A.; Desimone, R. Modulation of oscillatory neuronal synchronization by selective visual attention. *Science* **2001**, *291*, 1560–1563.

47. Salinas, E.; Sejnowski, T. Correlated neuronal activity and the flow of neural information. *Nat. Rev. Neurosci.* **2001**, *2*, 539–550.

48. Csicsvari, J.; Jamieson, B.; Wise, K.; Buzsáki, G. Mechanisms of gamma oscillations in the hippocampus of the behaving rat. *Neuron* **2003**, *37*, 311–322.

49. Womelsdorf, T.; Schoffelen, J.; Oostenveld, R.; Singer, W.; Desimone, R.; Engel, A.; Fries, P. Modulation of neuronal interactions through neuronal synchronization. *Science* **2007**, *316*, 1609–1612.

50. Colgin, L.; Denninger, T.; Fyhn, M.; Hafting, T.; Bonnevie, T.; Jensen, O.; Moser, M.; Moser, E. Frequency of gamma oscillations routes flow of information in the hippocampus. *Nature* **2009**, *462*, 75–78.

51. Luczak, A.; McNaughton, B.L.; Harris, K.D. Packet-based communication in the cortex. *Nat. Rev. Neurosci.* **2015**, *16*, 745–755.

52. Yuste, R.; MacLean, J.; Smith, J.; Lansner, A. The cortex as a central pattern generator. *Nat. Rev. Neurosci.* **2005**, *6*, 477–483.

53. Hubel, D.; Wiesel, T. Receptive fields and functional architecture in two non striate visual areas (18 and 19) of the cat. *J. Neurophysiol.* **1965**, *28*, 229–289.

54. DeWeese, M.; Wehr, M.; Zador, A. Binary coding in auditory cortex. *J. Neurosci.* **2003**, *23*, 7940–7949.

55. Olshausen, B.; Field, D. Sparse coding of sensory inputs. *Curr. Opin. Neurobiol.* **2004**, *14*, 481–487.

56. Hromádka, T.; DeWeese, M.; Zador, A. Sparse representation of sounds in the unanesthetized auditory cortex. *PLoS Biol.* **2008**, *6*, e16.

57. Piet, A.T.; Erlich, J.C.; Kopec, C.D.; Brody, C.D. Rat Prefrontal Cortex Inactivations during Decision Making Are Explained by Bistable Attractor Dynamics. *Neural Comput.* **2017**, *29*, 2861–2886.

58. Brody, C.D.; Hernandez, A.; Zainos, A.; Romo, R. Timing and neural encoding of somatosensory parametric working memory in macaque prefrontal cortex. *Cereb. Cortex* **2003**, *13*, 1196–1207.

59. Seung, H.S. How the brain keeps the eyes still. *Proc. Natl. Acad. Sci. USA* **1996**, *93*, 13339–13344.

60. Koulakov, A.A.; Raghavachari, S.; Kepecs, A.; Lisman, J.E. Model for a robust neural integrator. *Nat. Neurosci.* **2002**, *5*, 775–782.

61. Goldman, M. Memory without feedback in a neural network. *Neuron* **2008**, *61*, 621–634.

62. Brody, C.D.; Romo, R.; Kepecs, A. Basic mechanisms for graded persistent activity: Discrete attractors, continuous attractors, and dynamic representations. *Curr. Opin. Neurobiol.* **2003**, *13*, 204–211.

63. Major, G.; Tank, D. Persistent neural activity: Prevalence and mechanisms. *Curr. Opin. Neurobiol.* **2004**, *14*, 675–684.

64. Aksay, E.; Olasagasti, I.; Mensh, B.D.; Baker, R.; Goldman, M.S.; Tank, D.W. Functional dissection of circuitry in a neural integrator. *Nat. Neurosci.* **2007**, *10*, 494–504.

65. Mante, V.; Sussillo, D.; Shenoy, K.V.; Newsome, W.T. Context-dependent computation by recurrent dynamics in prefrontal cortex. *Nature* **2013**, *503*, 78–84.

entropy

MDPI

Article

A Measure of Information Available for Inference

Takuya Isomura (ORCID)

Laboratory for Neural Computation and Adaptation, RIKEN Center for Brain Science, 2-1 Hirosawa, Wako, Saitama 351-0198, Japan; takuya.isomura@riken.jp; Tel.: +81-48-467-9644

Received: 11 May 2018; Accepted: 6 July 2018; Published: 7 July 2018

check for updates

Abstract: The mutual information between the state of a neural network and the state of the external world represents the amount of information stored in the neural network that is associated with the external world. In contrast, the surprise of the sensory input indicates the unpredictability of the current input. In other words, this is a measure of inference ability, and an upper bound of the surprise is known as the variational free energy. According to the free-energy principle (FEP), a neural network continuously minimizes the free energy to perceive the external world. For the survival of animals, inference ability is considered to be more important than simply memorized information. In this study, the free energy is shown to represent the gap between the amount of information stored in the neural network and that available for inference. This concept involves both the FEP and the infomax principle, and will be a useful measure for quantifying the amount of information available for inference.

Keywords: free-energy principle; internal model hypothesis; unconscious inference; infomax principle; independent component analysis; principal component analysis

1. Introduction

Sensory perception comprises complex responses of the brain to sensory inputs. For example, the visual cortex can distinguish objects from their background [1], while the auditory cortex can recognize a certain sound in a noisy place with high sensitivity, a phenomenon known as the cocktail party effect [2–7]. The brain (i.e., a neural network) has acquired these perceptual abilities without supervision, which is referred to as unsupervised learning [8–10]. Unsupervised learning, or implicit learning, is defined as the learning that happens in the absence of a teacher or supervisor; it is achieved through adaptation to past environments, which is necessary for higher brain functions. An understanding of the physiological mechanisms that mediate unsupervised learning is fundamental to augmenting our knowledge of information processing in the brain.

One of the consequent benefits of unsupervised learning is inference, which is the action of guessing unknown matters based on known facts or certain observations, i.e., the process of drawing conclusions through reasoning and estimation. While inference is thought to be an act of the conscious mind in the ordinary sense of the word, it can occur even in the unconscious mind. Hermann von Helmholtz, a 19th-century physicist/physiologist, realized that perception often requires inference by the unconscious mind and coined the word *unconscious inference* [11]. According to Helmholtz, conscious inference and unconscious inference can be distinguished based on whether conscious knowledge is involved in the process. For example, when an astronomer computes the positions or distances of stars in space based on images taken at various times from different parts of the orbit of the Earth, he or she performs conscious inference because the process is "based on a conscious knowledge of the laws of optics"; by contrast, "in the ordinary acts of vision, this knowledge of optics is lacking" [11]. Thus, the latter process is performed by the unconscious mind. Unconscious inference is crucial for estimating the overall picture from partial observations.

In the field of theoretical and computational neuroscience, unconscious inference has been translated as the successive inference of the generative process of the external world (in terms of Bayesian inference) that animals perform in order to achieve perception. One hypothesis, the so-called internal model hypothesis [12–19], states that animals reconstruct a model of the external world in their brain through past experiences. This internal model helps animals infer hidden causes and predict future inputs automatically; in other words, this inference process happens unconsciously. This is also known as the predictive coding hypothesis [20,21]. In the past decade, a mathematical foundation for unconscious inference, called the free-energy principle (FEP), has been proposed [13–17], and is a candidate unified theory of higher brain functions. Briefly, this principle hypothesizes that parameters of the generative model are learned through unsupervised learning, while hidden variables are inferred in the subsequent inference step. The FEP provides a unified framework for higher brain functions including perceptual learning [14], reinforcement learning [22], motor learning [23,24], communication [25,26], emotion, mental disorders [27,28], and evolution. However, the difference between the FEP and a related theory, namely the information maximization (infomax) principle, which states that a neural network maximizes the amount of sensory information preserved in the network [29–32], is still not fully understood.

In this study, the relationship between the FEP and the infomax principle is investigated. As one of the most simple and important examples, the study focuses on blind source separation (BSS), which is the task of separating sensory inputs into hidden sources (or causes) [33–36]. BSS is shown to be a subset of the inference problem considered in the FEP, and variational free energy is demonstrated to represent the difference between the information stored in the neural network (which is the measure of the infomax principle [29]) and the information available for inferring current sensory inputs.

2. Methods

2.1. Definition of a System

Let us suppose $s \equiv (s_1, \ldots, s_N)^T$ as hidden sources that follow $p(s|\lambda) \equiv \prod_i p(s_i|\lambda)$ parameterized by a hyper-parameter set λ; $x \equiv (x_1, \ldots, x_M)^T$ as sensory inputs; $u \equiv (u_1, \ldots, u_N)^T$ as neural outputs; $z \equiv (z_1, \ldots, z_M)^T$ as background noises that follow $p(z|\lambda)$ parameterized by λ; $\epsilon \equiv (\epsilon_1, \ldots, \epsilon_M)^T$ as reconstruction errors; and $f \in \mathbb{R}^M$, $g \in \mathbb{R}^N$, and $h \in \mathbb{R}^M$ as nonlinear functions (see also Table 1). The generative process of the external world (or the environment) is described by a stochastic equation as:

$$\text{Generative process}: \ x = f(s) + z. \tag{1}$$

Recognition and generative models of the neural network are defined as follows:

$$\text{Recognition model}: \ u = g(x), \tag{2}$$

$$\text{Generative model}: \ x = h(u) + \epsilon. \tag{3}$$

Figure 1 illustrates the structure of the system under consideration. For the generative model, the prior distribution of u is defined as $p^*(u|\gamma) = \prod_i p^*(u_i|\gamma)$ with a hyper-parameter set γ and the likelihood function as $p^*(x|h(u), \gamma) = \mathcal{N}[x; h(u), \Sigma_\epsilon(\gamma)]$, where p^* indicates a statistical model and $\mathcal{N}$ is a Gaussian distribution characterized by the mean $h(u)$ and covariance $\Sigma_\epsilon(\gamma)$. Moreover, suppose θ, $W \in \mathbb{R}^{N \times M}$, and $V \in \mathbb{R}^{M \times N}$ as parameter sets for f, g, and h, respectively, λ as a hyper-parameter set for $p(s|\lambda)$ and $p(z|\lambda)$, and γ as a hyper-parameter set for $p^*(u|\gamma)$ and $p^*(x|h(u), \gamma)$. Here, hyper-parameters are defined as parameters that determine the shape of distributions (e.g., the covariance matrix). Note that W and V are assumed as synaptic strength matrices for feedforward and backward paths, respectively, while γ is assumed as a state of neuromodulators similarly to [13–15]. In this study, unless specifically mentioned, parameters and hyper-parameters refer to slowly changing variables, so that W, V, and γ can change their values. Equations (1)–(3) are transformed into probabilistic representations.

Table 1. Glossary of expressions.

Expression	Description
Generative process	A set of stochastic equations that generate the external world dynamics
Recognition model	A model in the neural network that imitates the inverse of the generative process
Generative model	A model in the neural network that imitates the generative process
$s \in \mathbb{R}^N$	Hidden sources
$x \in \mathbb{R}^M$	Sensory inputs
θ	A set of parameters
λ	A set of hyper-parameters
$\vartheta \equiv \{s, \theta, \lambda\}$	A set of hidden states of the external world
$u \in \mathbb{R}^N$	Neural outputs
$W \in \mathbb{R}^{N \times M}, V \in \mathbb{R}^{M \times N}$	Synaptic strength matrices
γ	State of neuromodulators
$\varphi \equiv \{u, W, V, \gamma\}$	A set of the internal states of the neural network
$z \in \mathbb{R}^M$	Background noises
$\epsilon \in \mathbb{R}^M$	Reconstruction errors
$p(x)$	The actual probability density of x
$p(\varphi|x), p(x, \varphi), p(\varphi)$	Actual probability densities (posterior densities)
$p^*(u|\gamma), p^*(\varphi) \equiv p^*(u|\gamma)p^*(W, V, \gamma)$	Prior densities
$p^*(x|\varphi) \equiv p^*(x|u, V, \gamma)$	Likelihood function
$p^*(x), p^*(\varphi|x), p^*(x, \varphi)$	Statistical models
$\Delta_x \equiv \prod_i \Delta_{x_i}$	Finite spatial resolution of x, $\Delta_{x_i} > 0$
$\langle \cdot \rangle_{p(x)} \equiv \int \cdot\, p(x)dx$	Expectation of $\cdot$ over $p(x)$
$H[x] \equiv \langle -\log(p(x)\Delta_x) \rangle_{p(x)}$	Shannon entropy of $p(x)\Delta_x$
$\langle -\log(p^*(x)\Delta_x) \rangle_{p(x)}$	Cross entropy of $p^*(x)\Delta_x$ over $p(x)$
$\mathcal{D}_{KL}[p(\cdot)||p^*(\cdot)] \equiv \left\langle \log \frac{p(\cdot)}{p^*(\cdot)} \right\rangle_{p(\cdot)}$	KLD between $p(\cdot)$ and $p^*(\cdot)$
$I[x; \varphi] \equiv \mathcal{D}_{KL}[p(x, \varphi)||p(x)p(\varphi)]$	Mutual information between x and φ
$S(x) \equiv \log \frac{p(x)}{p^*(x)}$	Surprise
$\overline{S} \equiv \langle S(x) \rangle_{p(x)}$	Surprise expectation
$F(x) \equiv S(x) + \mathcal{D}_{KL}[p(\varphi|x)||p^*(\varphi|x)]$	Free energy
$\overline{F} \equiv \langle F(x) \rangle_{p(x)}$	Free energy expectation
$X[x; \varphi] \equiv \left\langle \log \frac{p^*(x, \varphi)}{p(x)p(\varphi)} \right\rangle_{p(x, \varphi)}$	Utilizable information between x and φ

$$\text{Generative process}: \ p(s, x|\theta, \lambda) = p(x|s, \theta, \lambda)p(s|\lambda)$$
$$= \int \delta(x - f(s; \theta) - z)p(z|\lambda)p(s|\lambda)dz \tag{4}$$
$$= p(z = x - f|s, \theta, \lambda)p(s|\lambda),$$

$$\text{Recognition model}: \ p(x, u|W) = p(x|u, W)p(u|W)$$
$$= p(u|x, W)p(x) \tag{5}$$
$$= p(u - g(x; W)|x, W)p(x),$$

$$\text{Generative model}: \ p^*(x, u|V, \gamma) = p^*(x|u, V, \gamma)p^*(u|\gamma)$$
$$= \int \delta(x - h(u; V) - \epsilon)p^*(\epsilon|V, \gamma)p^*(u|\gamma)d\epsilon \tag{6}$$
$$= p^*(\epsilon = x - h|u, V, \gamma)p^*(u|\gamma).$$

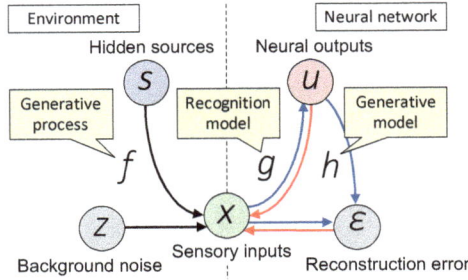

Figure 1. Schematic images of a generative process of the environment (left) and recognition and generative models of the neural network (right). Note that the neural network can access only the states in the right side of the dashed line, including x (see text in Section 2.2). Black arrows are causal relationships in the external world. Blue arrows are information flows of the neural network (i.e., actual causal relationships in the neural network), while red arrows are hypothesized causal relationships (to imitate the external world) when the generative model is considered. See main text and Table 1 for meanings of variables and functions.

Note that $\delta(\cdot)$ is Dirac's delta function and $p^*(x|u, V, \gamma) \equiv p(x|u, V, \gamma, m)$ is a statistical model given a model structure m. For simplification, let $\vartheta \equiv \{s, \theta, \lambda\}$ be a set of hidden states of the external world and $\varphi \equiv \{u, W, V, \gamma\}$ be a set of internal states of the neural network. By multiplying $p(\theta, \lambda)$ by Equation (4), $p(W, V, \gamma)$ by Equation (5), and $p^*(W, V, \gamma) = p^*(W)p^*(V)p^*(\gamma)$ by Equation (6), Equations (4)–(6) become

$$\text{Generative process}: \ p(x, \vartheta) = p(x|\vartheta)p(\vartheta) = p(z = x - f|\vartheta)p(\vartheta), \tag{7}$$

$$\text{Recognition model}: \ p(x, \varphi) = p(x|\varphi)p(\varphi) = p(\epsilon = x - h|\varphi)p(\varphi), \tag{8}$$

$$\text{Generative model}: \ p^*(x, \varphi) = p^*(x|\varphi)p^*(\varphi) = p^*(\epsilon = x - h|\varphi)p^*(\varphi), \tag{9}$$

where $p^*(\varphi) = p^*(u|\gamma)p^*(W, V, \gamma)$ is the prior distribution for φ and $p^*(x, \varphi) \equiv p(x, \varphi|m)$ is a statistical model given a model structure m, which is determined by the shapes of $p^*(\varphi)$ and $p^*(x|\varphi) \equiv p^*(x|u, V, \gamma)$. The expression of $p^*(x, \varphi)$ is used instead of $p(x, \varphi|m)$ to emphasize the difference between $p(x, \varphi)$ and $p^*(x, \varphi)$. While $p(x, \varphi) \equiv p(u|x, W)p(W, V, \gamma|x)p(x)$ is the actual joint probability of (x, φ) (which corresponds to the posterior distribution), $p^*(x, \varphi)$, i.e., the product of the likelihood function and the prior distribution, represents the generative model that the neural network expects (x, φ) to follow. Typically, elements of $p^*(W, V, \gamma)$ are supposed to be independent of each other, $p^*(W, V, \gamma) = \prod_{ii'} p^*(W_{ii'}) \prod_{jj'} p^*(V_{jj'}) \prod_k p^*(\gamma_k)$. For example, sparse priors about parameters are sometimes used to prevent the over-learning [37], while a generative model with sparse priors for outputs is known as a sparse coding model [38,39]. As shown later, the inference and learning are achieved by minimizing the difference between $p(x, \varphi)$ and $p^*(x, \varphi)$. At that time, minimizing the difference between $p(V, W, \gamma)$ and $p^*(V, W, \gamma)$ acts as a constraint or a regularizer that prevents over-learning (see Section 2.3 for details).

2.2. Information Stored in the Neural Network

Information is defined as the negative log of probability [40]. When $\text{Prob}(x)$ is the probability of given sensory inputs x, its information is given by $-\log \text{Prob}(x)$ [nat], where 1 nat = 1.4427 bits. When x takes continuous values, by coarse graining, $-\log \text{Prob}(x)$ is replaced with $-\log(p(x)\Delta_x)$, where $p(x)$ is the probability density of x and $\Delta_x \equiv \prod_i \Delta_{x_i}$ is the product of the finite spatial resolutions of x's elements ($\Delta_{x_i} > 0$). The expectation of $-\log(p(x)\Delta_x)$ over $p(x)$ gives the Shannon entropy (or average information), which is defined by

$$H[x] \equiv \langle -\log(p(x)\Delta_x)\rangle_{p(x)} \ [\text{nat}], \tag{10}$$

where $\langle \cdot \rangle_{p(x)} \equiv \int \cdot p(x) dx$ represents the expectation of $\cdot$ over $p(x)$. Note that the use of $-\log(p(x)\Delta_x)$ instead of $-\log p(x)$ is useful because this $H[x]$ is non-negative ($d\text{Prob}(x) = p(x)\Delta_x$ takes a value between 0 and 1). This is a coarse binning of x and the spatial resolution Δ_x takes a small but nonzero value so that the addition of constant $-\log \Delta_x$ has no effect except for sliding the offset value. If and only if $p(x)$ is Dirac's delta function (strictly, $p(x) = 1/\Delta_x$ at one bin and 0 otherwise), $H[x] = 0$ is realized. For the system under consideration (Equations (7)–(9)), the information shared between the external world states (x, ϑ) and the internal states of the neural network φ is defined by mutual information [41]

$$I[(x, \vartheta); \varphi] \equiv \left\langle \log \frac{p(x, \vartheta, \varphi)}{p(x, \vartheta)p(\varphi)} \right\rangle_{p(x,\vartheta,\varphi)} \ [\text{nat}]. \tag{11}$$

Note that $p(x, \vartheta, \varphi)$ is the joint probability of (x, ϑ) and φ. Moreover $p(x, \vartheta)$ and $p(\varphi)$ are their marginal distributions, respectively. This mutual information takes a non-negative value and quantifies how much (x, ϑ) and φ are related with each other. High mutual information indicates the internal states are informative for explaining the external world states, while zero mutual information means they are independent of each other.

However, the only information that the neural network can directly access is the sensory input. This is the case because the system under consideration can be described as a Bayesian network (see [42,43] for details on the Markov blanket). Hence, the entropy of the external world states under a fixed sensory input gives information that the neural network cannot infer. Moreover, there is no feedback control from the neural network to the external world in this setup. Thus, under a fixed x, ϑ and φ are conditionally independent of each other. From $p(\vartheta, \varphi|x) = p(\vartheta|x)p(\varphi|x)$, we can obtain

$$I[(x, \vartheta); \varphi] = \left\langle \log \frac{p(\vartheta|x)p(\varphi|x)p(x)}{p(\vartheta|x)p(x)p(\varphi)} \right\rangle_{p(\vartheta|x)p(\varphi|x)p(x)} = \left\langle \log \frac{p(\varphi|x)}{p(\varphi)} \right\rangle_{p(\varphi,x)} = I[x; \varphi]. \tag{12}$$

Using Shannon entropy, $I[x; \varphi]$ becomes

$$I[x; \varphi] = H[x] - H[x|\varphi] \ [\text{nat}], \tag{13}$$

where

$$H[x|\varphi] \equiv \left\langle -\log\left(p(x|\varphi)\Delta_x\right)\right\rangle_{p(x,\varphi)} \tag{14}$$

is the conditional entropy of x given φ. Thus, maximization of $I[(x, \vartheta); \varphi]$ is the same as maximization of $I[x; \varphi]$ for this system. As $I[x; \varphi]$, $H[x]$, and $H[x|\varphi]$ are non-negative, $I[x; \varphi]$ has the range $0 \leq I[x; \varphi] \leq H[x]$. Zero mutual information occurs if and only if x and φ are independent, while $I[x; \varphi] = H[x]$ occurs if and only if x is fully explained by φ. In this manner, $I[x; \varphi]$ describes the information about the external world stored in the neural network. Note that this $I[x; \varphi]$ can be expressed using the Kullback–Leibler divergence (KLD) [44] as $I[x; \varphi] \equiv \mathcal{D}_{KL}\left[p(x, \varphi)||p(x)p(\varphi)\right]$. The KLD takes a non-negative value and indicates the divergence between two distributions.

The infomax principle states that "the network connections develop in such a way as to maximize the amount of information that is preserved when signals are transformed at each processing stage, subject to certain constraints" [29], see also [30–32]. According to the infomax principle, the neural network is hypothesized to maximize $I[x; \varphi]$ to perceive the external world. However, $I[x; \varphi]$ does not fully explain the inference capability of a neural network. For example, if neural outputs just express the sensory input itself ($u = x$), $I[x; \varphi] = H[x]$ is easily achieved, but this does not mean that the neural network can predict or reconstruct input statistics. This is considered in the next section.

2.3. Free-Energy Principle

If one has a statistical model determined by model structure m, the information calculated based on m is given by the negative log likelihood $-\log p(x|m)$, which is termed as the surprise (or the marginal likelihood) of the sensory input and expresses the unpredictability of the sensory input for the individual. The neural network is considered to minimize the surprise in the sensory input using the knowledge about the external world, to perceive the external world [13]. To infer if an event is likely to happen based on the past observation, a statistical (i.e., generative) model is necessary; otherwise it is difficult to generalize sensory inputs [45]. Note that the surprise is the marginal over the generative model; hence, the neural network can reduce the surprise by optimizing its internal states, while Shannon entropy of the input is determined by the environment. When the actual probability density and a generative model are given by $p(x)$ and $p^*(x) \equiv p(x|m)$, respectively, the cross entropy $\langle -\log(p^*(x)\Delta_x) \rangle_{p(x)}$ is always larger than or equal to Shannon entropy $H[x]$ because of the non-negativity of KLD. Hence, in this study, the input surprise is defined by

$$S(x) \equiv -\log p^*(x) + \log p(x) \quad [\text{nat}] \tag{15}$$

and its expectation over $p(x)$ by

$$\overline{S} \equiv \langle S(x) \rangle_{p(x)} = \mathcal{D}_{KL}[p(x)||p^*(x)] = \langle -\log(p^*(x)\Delta_x) \rangle_{p(x)} - H[x] \quad [\text{nat}]. \tag{16}$$

This definition of $S(x)$ is to ensure $\overline{S}$ is non-negative and $\overline{S} = 0$ if and only if $p^*(x) = p(x)$. Since $H[x]$ is determined by the environment and constant for the neural network, minimization of this $\overline{S}$ is the same meaning as minimization of $\langle -\log(p^*(x)\Delta_x) \rangle_{p(x)}$.

As the sensory input is generated by the external world generative process, consideration of the structure and dynamics placed in the background of the sensory input can provide accurate inference. According to the internal model hypothesis, animals develop the internal model in their brain to increase the accuracy and efficiency of inference [12–15,17–19]; thus, internal states of the neural network φ are hypothesized to imitate the hidden states of the external world ϑ. A problem is that $-\log p^*(x) = -\log(\int p^*(x, \varphi)d\varphi)$ is intractable for the neural network, because the integral of $p^*(x, \varphi)$ placed in the logarithm function. The FEP hypothesizes that the neural network calculates an upper bound of $-\log p^*(x)$ instead of the exact value as a proxy, which is more tractable [13] (because $-\log p(x)$ is fixed, the free energy is sometimes defined including or excluding this term). This upper bound is termed as variational free energy:

$$F(x) \equiv S(x) + \mathcal{D}_{KL}[p(\varphi|x)||p^*(\varphi|x)] = \langle -\log p^*(x, \varphi) + \log p(x, \varphi) \rangle_{p(\varphi|x)} \, [\text{nat}]. \tag{17}$$

Note that $p(\varphi|x) \equiv p(u|x, W)p(W, V, \gamma|x)$ expresses the belief about hidden states of the external world encoded by internal states of the neural network, termed as the recognition density. Due to the non-negativity of KLD, $F(x)$ is guaranteed to be an upper bound of $S(x)$ and $F(x) = S(x)$ holds if and only if $p^*(\varphi|x) = p(\varphi|x)$. Furthermore, the expectation of $F(x)$ over $p(x)$ is defined by

$$\begin{aligned} \overline{F} &\equiv \langle F(x) \rangle_{p(x)} = \mathcal{D}_{KL}[p(x, \varphi)||p^*(x, \varphi)] \\ &= \langle -\log(p^*(x|\varphi)\Delta_x) \rangle_{p(x,\varphi)} + \langle -\log(p^*(\varphi)\Delta_\varphi) \rangle_{p(\varphi)} - H[\varphi|x] - H[x] \quad [\text{nat}], \end{aligned} \tag{18}$$

where $\langle -\log(p^*(x|\varphi)\Delta_x) \rangle_{p(x,\varphi)}$ is the negative log likelihood and called the accuracy [15]. The second and third terms are the cross entropy of φ and the conditional entropy of φ given x, $H[\varphi|x] \equiv \langle -\log(p(\varphi|x)\Delta_\varphi) \rangle_{p(x,\varphi)}$, where the difference between them is called the complexity [15]. The last term $H[x]$ is a constant. $\overline{F}$ indicates the difference between the actual probability $p(x, \varphi)$ and the generative model $p^*(x, \varphi)$. Given the non-negativity of KLD, $\overline{F}$ is always larger than or equal to non-negative value $\overline{S}$, and $\overline{F} = \overline{S} = 0$ holds if and only if $p^*(x, \varphi) = p(x, \varphi)$. The FEP hypothesized that $\overline{F}$ is minimized by optimizing neural activities (u), synaptic strengths (W and V; i.e., synaptic plasticity), and activities of neuromodulators (γ).

The accuracy $\langle -\log(p^*(x|\varphi)\Delta_x)\rangle_{p(x,\varphi)}$ quantifies the amplitude of the reconstruction error. Minimization of the accuracy is the maximum likelihood estimation [10] and provides a solution that (at least locally) minimizes the reconstruction error. Whereas, minimization of the complexity $\langle -\log(p^*(\varphi)\Delta_\varphi)\rangle_{p(\varphi)} - H[\varphi|x]$ makes $p(\varphi)$ closer to $p^*(\varphi)$. As $p^*(\varphi) = p^*(u|\gamma)p^*(W, V, \gamma)$ usually supposes the elements of φ are mutually independent, this acts as the maximization of the entropy under a constraint. Hence, this leads to the increase of the independence between internal states, which helps neurons to establish an efficient representation, as pointed out by Jaynes' max entropy principle [46,47]. This is essential for BSS [33–36] because the optimal parameters that minimize the accuracy are not always uniquely determined. Due to this, the maximum likelihood estimation alone does not always identify the generative process behind the sensory inputs. As $\overline{F}$ is the sum of costs for the maximum likelihood estimation and BSS, free-energy minimization is the rule to simultaneously minimize the reconstruction error and maximize the independence of the internal states. It is recognized that animals perform BSS [2–7]. Interestingly, even *in vitro* neural networks perform BSS, which is accompanied by significant reduction of free energy in accordance with the FEP and Jaynes' max entropy principle [48].

2.4. Information Available for Inference

We now consider how free energy expectation $\overline{F}$ relates to mutual information $I[x; \varphi]$. According to unconscious inference and the internal model hypothesis, the aim of a neural network is to predict x, and for this purpose, it infers hidden states of the external world. While the neural network is conventionally hypothesized to express sufficient statistics of the hidden states of the external world [14], here it is hypothesized that internal states of the neural network are random variables and the probability distribution of them imitates the probability distribution of the hidden states of the external world. The neural network hence attempts to match the joint probability of the sensory inputs and the internal states with that of the sensory inputs and the hidden states of the external world. To do so, the neural network shifts the actual probability of internal states $p(x, \varphi) = p(x|\varphi)p(\varphi)$ closer to those of the generative model $p^*(x, \varphi) = p^*(x|\varphi)p^*(\varphi)$ that the neural network expects (x, φ) to follow (note that here, $p(x|\varphi) = p(x|u, W)$ and $p^*(x|\varphi) = p^*(x|u, V, \gamma)$). This means that the shape or structure of $p^*(x, \varphi)$ is pre-defined, but the argument (x, φ) can still change. From this viewpoint, the difference between these two distributions is associated with the loss of information.

The amount of information available for inference can be calculated using the following three values related to information loss: (i) because $H[x]$ is information of the sensory input and $I[x; \varphi]$ is information stored in the neural network, $H[x] - I[x; \varphi] = H[x|\varphi]$ indicates the information loss in the recognition model (Figure 2); (ii) the difference between actual and desired (prior) distributions of internal states $\mathcal{D}_{KL}[p(\varphi)||p^*(\varphi)]$ quantifies the information loss for inferring internal states using the prior (i.e., blind state separation). This is a common approach used in BSS methods [33–36]; and (iii) the difference between distributions of the actual reconstruction error and the reconstruction error under the given model $\langle\mathcal{D}_{KL}[p(x|\varphi)||p^*(x|\varphi)]\rangle_{p(\varphi)}$ quantifies the information loss for representing inputs using internal states. Therefore, by subtracting these three values from $H[x]$, a mutual-information-like measure representing the inference capability is obtained:

$$
\begin{aligned}
X[x; \varphi] &\equiv H[x] - H[x|\varphi] - \mathcal{D}_{KL}[p(\varphi)||p^*(\varphi)] - \langle\mathcal{D}_{KL}[p(x|\varphi)||p^*(x|\varphi)]\rangle_{p(\varphi)} \\
&= \left\langle \log \frac{p^*(x, \varphi)}{p(x)p(\varphi)} \right\rangle_{p(x,\varphi)} \quad [\text{nat}],
\end{aligned}
\tag{19}
$$

which is called utilizable information in this study. This utilizable information $X[x; \varphi]$ is defined by replacing $p(x, \varphi)$ in $I[x; \varphi]$ with $p^*(x, \varphi)$, immediately yielding

$$\overline{F} = I[x; \varphi] - X[x; \varphi] \quad [\text{nat}]. \tag{20}$$

Hence, $\overline{F}$ represents the gap between the amount of information stored in the neural network and the amount that is available for inference, which is equivalent to the information loss in the generative model. Note that the sum of losses in the recognition and generative models $H[x] - X[x; \varphi] = \overline{F} + H[x|\varphi]$ is an upper bound of $\overline{F}$ because of the non-negativity of $H[x|\varphi]$ (Figure 2). As $H[x|\varphi]$ is generally nonzero, $F(x) + H[x|\varphi]$ does not usually reach zero, even when $p(x, \varphi) = p^*(x, \varphi)$.

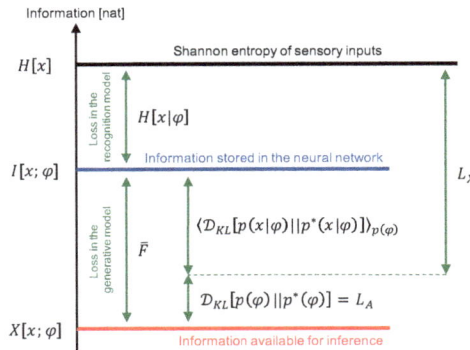

Figure 2. Relationship between information measures. The mutual information between the inputs and internal states of the neural network ($I[x; \varphi]$) is less than or equal to the Shannon entropy of the inputs ($H[x]$) because of the information loss in the recognition model. The utilizable information ($X[x; \varphi]$) is less than or equal to the mutual information, and the gap between them gives the expectation of the variational free energy ($\overline{F}$), which quantifies the loss in the generative model. The sum of the principal component analysis (PCA) and independent component analysis (ICA) costs ($L_X + L_A$) is equal to the gap between the Shannon entropy and the utilizable information, expressing the sum of losses in the recognition and generative models.

Furthermore, $X[x; \varphi]$ is transformed into

$$X[x; \varphi] = H[x] - L_X - L_A, \tag{21}$$

where

$$L_X \equiv \langle -\log(p^*(x|\varphi)\Delta_x) \rangle_{p(x, \varphi)} \tag{22}$$

is the so-called reconstruction error, which is similar to the reconstruction error for principal component analysis (PCA) [49], while

$$L_A \equiv \mathcal{D}_{KL}[p(\varphi)||p^*(\varphi)] \tag{23}$$

is a generalization of Amari's cost function for independent component analysis (ICA) [50].

PCA is one of the most popular dimensionality reduction methods. It is used to remove background noise and extract important features from sensory inputs [49,51]. In contrast, ICA is a BSS method used to decompose a mixture set of sensory inputs into independent hidden sources [34,36,50,52,53]. Theoreticians hypothesize that the PCA- and ICA-like learning underlies BSS in the brain [3]. This kind of extraction of the hidden representation is also an important problem in machine learning [54,55]. Equation (21) indicates that $X[x; \varphi]$ consists of PCA- and ICA-like parts, i.e., maximization of $X[x; \varphi]$ can perform both dimensionality reduction and BSS (Figure 2). Their relationship is discussed in the next section.

3. Comparison between the Free-Energy Principle and Related Theories

In this section, the FEP is compared with other theories. As described in the Methods, the aim of the infomax principle is to maximize mutual information $I[x; \varphi]$ (Equation (13)), while the aim of the FEP is to minimize free energy expectation $\bar{F}$ (Equation (18)), while maximization of utilizable information $X[x; \varphi]$ (Equation (19)) means to do both of them simultaneously.

3.1. Infomax Principle

The generative process and the recognition and generative models defined in Equations (1)–(3) are assumed. For the sake of simplicity, let us suppose W, V, and γ follow Dirac's delta functions; then, the goal of the infomax principle is simplified to maximization of the mutual information between the sensory inputs x and the neural outputs u:

$$I[x; u|W] = \left\langle \log \frac{p(x, u|W)}{p(x)p(u|W)} \right\rangle_{p(x,u,W)} = H[x] - H[x|u, W] = H[u|W] - H[u|x, W]. \quad (24)$$

Here W, V, and γ are still variables, and W is optimized according to the learning while V and γ do not directly contribute to minimization of $I[x; u|W]$. For the sake of simplicity, let us suppose $\dim(x) \geq \dim(u)$ and a linear recognition model $u = g(x) = Wx$, with full-rank matrix W. As $H[u|x, W] = \text{const.}$ is usually assumed and u has an infinite range, $I[x; u|W] = H[u|W] + \text{const.}$ monotonically increases as the variance of u increases. Thus, $I[x; u|W]$ without any constraint is insufficient for deriving learning algorithms for PCA or ICA. To perform PCA and ICA based on the infomax principle, one may consider mutual information between the sensory inputs and the nonlinearly transformed neural outputs $\psi(u) = (\psi(u_1), \ldots, \psi(u_N))^T$ with an injective nonlinear function $\psi(\cdot)$. This mutual information is given by:

$$I[x; \psi(u)|W] = \left\langle \log \frac{p(x, \psi(u)|W)}{p(x)p(\psi(u)|W)} \right\rangle_{p(x,\psi(u),W)} = H[\psi(u)|W] - H[\psi(u)|x, W]. \quad (25)$$

When nonlinear neural outputs have a finite range (e.g., between 0 and 1), the variance of u should be maintained in the appropriate range. The infomax-based ICA [52,53] is formulated based on this constraint. From $p(\psi(u)|W) = |\partial u/\partial \psi(u)| p(u|W) = (\prod_i \psi'(u_i))^{-1} p(u|W)$, $H[\psi(u)|W]$ becomes $H[\psi(u)|W] = \langle -\log\{(\prod_i \psi'(u_i))^{-1} p(u|W) \Delta_u\}\rangle_{p(u,W)} = H[u|W] + \langle \sum_i \log \psi'(u_i)\rangle_{p(u,W)}$. Since $H[\psi(u)|x, W] = \text{const.}$ holds, Equation (25) becomes:

$$I[x; \psi(u)|W] = H[u|W] + \left\langle \sum_i \log \psi'(u_i) \right\rangle_{p(u,W)} + \text{const.} \quad (26)$$

This is the cost function that is usually considered in the studies on the infomax-based ICA [52,53]. The following section shows that PCA and ICA are performed by the maximization of Equation (26) as well as the FEP.

3.2. Principal Component Analysis

Both the infomax principle and FEP yield a cost function of PCA. One of the most popular data compression methods, PCA is defined by minimization of the error when the inputs are reconstructed from the compressed representation (i.e., u in this study) [49]. It is known that PCA is derived from the infomax principle under a constraint on the internal states. Although maximization of the mutual information between x and u under the orthonormal constraint on W is usually considered [29], here let us consider another solution. Suppose $\dim(x) > \dim(u)$, $V = W^T$, and $\log \psi'(u_i) = u_i^2/2 + \text{const.}$

From Equation (24), $H[u|W] = H[x] - H[x|u, W] + \text{const.}$ holds. Since the reconstruction error is given by $\epsilon = x - W^T u = (I - W^T W)x$ for the linear system under consideration, we obtain $H[x|u, W] = \langle -\log\{p(x)|\partial x/\partial \epsilon|\Delta_x\}\rangle_{p(x,\varphi)} = H[x] + \langle \log|I - W^T W|\rangle_{p(\varphi)}$. Thus, Equation (26) becomes:

$$I[x; \psi(u)|W] = \left\langle -\log|I - W^T W| + \frac{1}{2}|u|^2 \right\rangle_{p(x,\varphi)} + \text{const.} \qquad (27)$$

The first term of Equation (27) is maximized if $WW^T = I$ holds (i.e., if W is an orthogonal matrix; here, a coarse graining with a finite resolution of W is supposed). To maximize the second term, outputs u need to be involved in a subspace spanned by the first to the N-th major principal components of x. Therefore, maximization of Equation (27) performs PCA.

Further, PCA is also derived by minimization of L_X (Equation (22)), under the assumption that the reconstruction error follows a Gaussian distribution $p^*(x|\varphi) = p^*(x|u, W, V, \gamma) = \mathcal{N}[x; W^T u, \gamma^{-1} I]$. Here, $\gamma > 0$ is a scalar hyper-parameter that scales the precision of the reconstruction error. Hence, the cost function is given by:

$$L_X = \left\langle \frac{\gamma}{2}\epsilon^T \epsilon - \frac{1}{2}\log|\gamma| \right\rangle_{p(\varphi)} + \text{const.} \qquad (28)$$

When γ is fixed, the derivative of Equation (28) with respect to W gives the update rule for the least mean square error PCA [49]. As this cost function quantifies the magnitude of the reconstruction error, the algorithm that minimizes Equation (28) yields the low-dimensional compressed representation that minimizes the loss incurred in reconstructing the sensory inputs. This algorithm is the same as Oja's subspace rule [51], up to an additional term that does not essentially change its behavior (see, e.g., [56] for a comparison between them). The L_X here is also in the same form as the cost function for an auto-encoder [54].

Moreover, when the priors of u, W, V, and γ are flat, $\langle -\log p^*(u|W)\rangle_{p(u,W)}$ and $\mathcal{D}_{KL}[p(W, V, \gamma)||p^*(W, V, \gamma)]$ are constants with respect to u, W, V, and γ, because $p(W, V, \gamma)$ is supposed to be a delta function. Hence, the free energy expectation (Equation (18)) becomes $\overline{F} = L_X - H[x|\varphi] - H[u|W] = L_X + \text{const.}$, where const. is a constant with respect to u, W, and V. In this case, the optimization of W gives the minimum of $\overline{F}$ because u and V are determined by W while γ is fixed. Thus, under this condition, $\overline{F}$ is equivalent to the cost function of the least mean square error PCA.

3.3. Independent Component Analysis

It is known that ICA yields independent representation of input data by maximizing the independence between the outputs [52,53]. Thus, ICA reduces the redundancy and yields an efficient representation. When sensory inputs are generated from hidden sources, representing the hidden sources is usually the most efficient representation. Both the infomax principle and FEP yield a cost function of ICA. Let us suppose that sources $s_1, \ldots, s_N$ independently follow an identical distribution $p_0(s_i|\lambda)$. The infomax-based ICA is derived from Equation (26) [52,53]. If $\psi(u_i)$ is defined to satisfy $\psi'(u_i) = p_0(u_i|\gamma)$, negative mutual information $-I[x; \psi(u)|W]$ becomes the KLD between the actual and prior distributions up to a constant term,

$$-I[x; \psi(u)|W] + \text{const.} = \left\langle \log p(u|W) - \log p_0(u|\gamma) \right\rangle_{p(\varphi)} = \langle \mathcal{D}_{KL}[p(u|W)||p_0(u|\gamma)]\rangle_{p(W,V,\gamma)} = L_A. \qquad (29)$$

The L_A here is known as Amari's ICA cost function [50], which is a reduction of (23). While both $-I[x; \psi(u)|W]$ and L_A provide the same gradient descent rule, formulating $I[x; \psi(u)|W]$ requires nonlinearly transformed neural outputs $\psi(u)$. By contrast, L_A straightforwardly represents that ICA is performed by minimization of the KLD between $p(u|W)$ and $p^*(u|\gamma) = p_0(u|\gamma)$. Indeed, if $\dim(u) = \dim(x) = N$, the background noise is small, and the priors of W, V, and γ are flat, we obtain $\overline{F} = \langle \mathcal{D}_{KL}[p(u|W)||p^*(u|\gamma)]\rangle_{p(W,V,\gamma)} = L_A$. Therefore, ICA is a subset of the inference problem considered in the FEP, and the derivation from the FEP is simpler, although both the infomax principle and FEP yield the same ICA algorithm.

Furthermore, when $\dim(x) > \dim(u)$, minimization of $\overline{F}$ can perform both dimensionality reduction and BSS. When the priors of $W, V,$ and γ are flat, free energy expectation (Equation (18)) approximately becomes $\overline{F} \approx L_X + L_A + \text{const.} = -X[x; u|W, V, \gamma] + \text{const.}$ Here, γ is fixed so that const. is a constant with respect to x, u, W and V. Conditional entropy $H[x|u, W]$ is ignored in the calculation because it is typically of a smaller order than L_X when $\Sigma(\gamma)$ is not fine-tuned. As γ parameterizes the precision of the reconstruction error, it controls the ratio of PCA to ICA. Hence, as γ decreases to zero, the solution shifts from a PCA-like to an ICA-like solution.

Unlike the case with the scalar γ described above, if $\Sigma_\epsilon(\gamma)$ is fine-tuned by high-dimensional γ to minimize $\overline{F}$, $\Sigma_\epsilon = \langle \epsilon\epsilon^T \rangle_{p(x,\varphi)}$ is obtained. Under this condition, L_X is equal to $H[x|u, W]$ up to a constant term, and thereby, $\overline{F} = L_A + \text{const.}$ is obtained. This indicates that $\overline{F}$ consists only of the ICA part. These comparisons suggest that low-dimensional γ is better for performing noise reduction than high-dimensional γ.

4. Simulation and Results

The difference between the infomax principle and the FEP is illustrated by a simple simulation using a linear generative process and a linear neural network (Figure 3). For simplification, it is assumed that u quickly converge to $u = Wx$ compared to the change of s (adiabatic approximation).

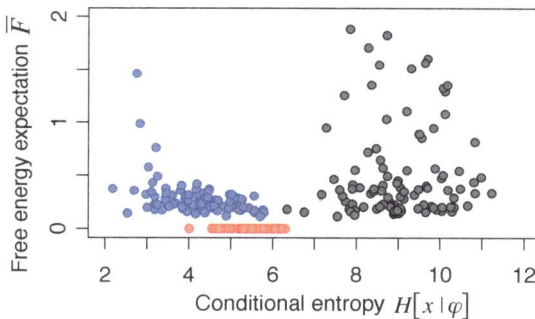

Figure 3. Difference between the infomax principle and free-energy principle (FEP) when sources follow a non-Gaussian distribution. Black, blue, and red circles indicate the results when W is a random matrix, optimized for the infomax principle (i.e., PCA), and optimized for the FEP, respectively.

For the results shown in Figure 3, s denotes two-dimensional hidden sources following an identical Laplace distribution with zero mean and unit variance; x denotes four-dimensional sensory inputs; u denotes two-dimensional neural outputs; z denotes four-dimensional background Gaussian noises following $\mathcal{N}[z; 0, \Sigma_z]$; θ denotes a 4×2-dimensional mixing matrix; W is a 2×4-dimensional synaptic strength matrix for the bottom-up path; and V is a 4×2-dimensional synaptic strength matrix for the top-down path. The priors of $W, V,$ and γ are supposed to be flat as in Section 3. Sensory inputs are determined by $x = \theta s + z$, while neural outputs are determined by $u = Wx$. The reconstruction error is given by $\epsilon = x - Vu$ and used to calculate $H[x|\varphi]$ and L_A. Horizontal and vertical axes in the figure are conditional entropy $H[x|\varphi]$ (Equation (14)) and free energy expectation $\overline{F}$ (Equation (18)), respectively. Simulations were conducted 100 times with randomly selected θ and Σ_z for each condition. For each simulation, 10^8 random sample points were generated and probability distributions were calculated using the histogram method.

First, when W is randomly chosen and V is defined by $V = W^T$, both $H[x|\varphi]$ and $\overline{F}$ are scattered (black circles in Figure 3) because neural outputs represent random mixtures of sources and noises.

Next, when W is optimized according to either Equation (27) or (28) under the constraint of $V = W^T$, the neural outputs express the major principal components of the inputs, i.e., the network performs PCA (blue circles in Figure 3). This is the case when $H[x|\varphi]$ is minimized. In contrast, when W, V, and $\Sigma_\epsilon(\gamma)$ are optimized according to the FEP (see Equation (18)), the neural outputs represent the independent components that match the prior source distribution; i.e., the network performs BSS or ICA while reducing the reconstruction error (red circles in Figure 3). For linear generative processes, the minimization of $\overline{F}$ can reliably and accurately perform both dimensionality reduction and BSS because the outputs become independent of each other and match the prior belief if and only if the outputs represent true sources up to permutation and sign-flip. As the utilizable information consists of PCA and ICA cost functions (see Equation (21)), the maximization of $X[x; \varphi]$ leads to a solution that is a compromise between the solutions for the infomax principle and the FEP. Interestingly, the infomax optimization (i.e., PCA) provides a W that makes $\overline{F}$ closer to zero than random states, which indicates that the infomax optimization contributes to the free energy minimization. Note that, for nonlinear systems, there are many different transformations that make the outputs independent of each other [57]. Hence, there is no guarantee that minimization of $\overline{F}$ can identify the true sources of nonlinear generative models.

In summary, the aims of the FEP and infomax principle are similar to each other. In particular, when both the sources and noises follow Gaussian distributions, their aims become the same. Conversely, the optimal synaptic weights under the FEP are different from those under the infomax principle when sources follow non-Gaussian distributions. Under this condition, the maximization of the utilizable information leads to a compromise solution between those for the FEP and the infomax principle.

5. Discussion

In this study, the FEP is linked with the infomax principle, PCA, and ICA. It is more likely that the purpose of a neural network in a biological system is to minimize the surprise of sensory inputs to realize better inference rather than maximize the amount of stored information. For example, the visual input captured by a video camera contributes to the stored information, but this amount of information is not equal to the amount of information available for inference. The surprise expectation represents the difference between actual and inferred observations; the free energy expectation provides the difference between recognition and generative models. Utilizable information is introduced to quantify the inference and generalization capability of sensory inputs. Using this approach, the free energy expectation can be explained as the gap between the information stored in the neural network and that available for inference.

To perform ICA based on the infomax principle, one needs to tune the nonlinearity of the neural outputs to ensure the derivative of the nonlinear I/O function matches the prior distribution. Conversely, under the FEP, ICA is straightforwardly derived from the KLD between the actual probability distribution and the prior distribution of u. Especially, in the absence of background noise and prior knowledge of the parameters and hyper-parameters, the free energy expectation is equivalent to the surprise expectation as well as Amari's ICA cost function, which indicates that ICA is a subproblem of the FEP.

The variational free energy quantifies the gap between the actual probability and the generative model and is a straightforward extension of the cost functions for BSS in the sense that it comprises the cost function for PCA [49] and ICA [50] in some special cases. Apart from that, there are studies that use the gap between the actual probability and the product of the marginal distributions to perform BSS [58] or to evaluate the information loss [59,60]. While the relationship between the product of the marginal distributions and the generative model is non-trivial, the comparison would lead to a deeper understanding about how the information of the external world is encoded by the neural network. In the subsequent work, we would like to see how the FEP and the infomax principle are related to those approaches.

Entropy **2018**, *20*, 512

The FEP is a rigorous and promising theory from theoretical and engineering viewpoints because various learning rules are derived from the FEP [14,15]. However, to be a physiologically plausible theory of the brain, the FEP needs to satisfy certain physiological requirements. There are two major requirements: first, physiological evidence that shows the existence of learning or self-organizing processes under the FEP is required. The model structure under the FEP is consistent with the structure of cortical microcircuits [19]. Moreover, *in vitro* neural networks performing BSS reduce free energy [48]. It is known that the spontaneous prior activity of a visual area enables it to learn the properties of natural pictures [61]. These results suggest the physiological plausibility of the FEP. Nevertheless, further experiments and consideration of information-theoretical optimization under physiological constraints [62] are required to prove the existence of the FEP in the biological brain. Second, the update rule must be a biologically plausible local learning rule, i.e., synaptic strengths must be changed by signals from connected cells or widespread liquid factors. While the synaptic update rule for a discrete system is local [17], the current rule for a continuous system [14] is a non-local rule. Recently developed biologically-plausible three-factor learning models in which Hebbian learning is mediated by a third modulatory factor [56,63–65] may help reveal the neuronal mechanism underlying unconscious inference. Therefore, it is necessary to investigate how actual neural networks infer the dynamics placed in the background of the sensory input and whether this is consistent with the FEP (see also [66] for the relationship between the FEP and spike-timing dependent plasticity [67,68]). This may help develop a biologically plausible learning algorithm through which an actual neural network might develop its internal model. Characterization of information from a physical viewpoint may also help understand how the brain physically embodies the information [69,70]. In the subsequent work, we would like to investigate this relationship.

In summary, this study investigated the differences between two types of information: information stored in the neural network and information available for inference. It was demonstrated that free energy represents the gap between these two types of information. This result clarifies the difference between the FEP and related theories and can be utilized for understanding unconscious inference from a theoretical viewpoint.

Acknowledgments: This work was supported by RIKEN Center for Brain Science.

Conflicts of Interest: The author declares no competing financial interests. The founding sponsor had no role in the design of the study; in the collection, analyses, or interpretation of data; in the writing of the manuscript, and in the decision to publish the results.

References

1. DiCarlo, J.J.; Zoccolan, D.; Rust, N.C. How does the brain solve visual object recognition? *Neuron* **2012**, *73*, 415–434. [CrossRef] [PubMed]
2. Bronkhorst, A.W. The cocktail party phenomenon: A review of research on speech intelligibility in multiple-talker conditions. *Acta Acust. United Acust.* **2000**, *86*, 117–128.
3. Brown, G.D.; Yamada, S.; Sejnowski, T.J. Independent component analysis at the neural cocktail party. *Trends Neurosci.* **2001**, *24*, 54–63. [CrossRef]
4. Haykin, S.; Chen, Z. The cocktail party problem. *Neural Comput.* **2005**, *17*, 1875–1902. [CrossRef] [PubMed]
5. Narayan, R.; Best, V.; Ozmeral, E.; McClaine, E.; Dent, M.; Shinn-Cunningham, B.; Sen, K. Cortical interference effects in the cocktail party problem. *Nat. Neurosci.* **2007**, *10*, 1601–1607. [CrossRef] [PubMed]
6. Mesgarani, N.; Chang, E.F. Selective cortical representation of attended speaker in multi-talker speech perception. *Nature* **2012**, *485*, 233–236. [CrossRef] [PubMed]
7. Golumbic, E.M.Z.; Ding, N.; Bickel, S.; Lakatos, P.; Schevon, C.A.; McKhann, G.M.; Schroeder, C.E. Mechanisms underlying selective neuronal tracking of attended speech at a "cocktail party". *Neuron* **2013**, *77*, 980–991.
8. Dayan, P.; Abbott, L.F. *Theoretical Neuroscience: Computational and Mathematical Modeling of Neural Systems*; MIT Press: London, UK, 2001.

9.   Gerstner, W.; Kistler, W. *Spiking Neuron Models: Single Neurons, Populations, Plasticity*; Cambridge University Press: Cambridge, UK, 2002.

10.   Bishop, C.M. *Pattern Recognition and Machine Learning*; Springer: New York, NY, USA, 2006.

11.   Von Helmholtz, H. Concerning the perceptions in general. In *Treatise on Physiological Optics*, 3rd ed.; Dover Publications: New York, NY, USA, 1962.

12.   Dayan, P.; Hinton, G.E.; Neal, R.M.; Zemel, R.S. The helmholtz machine. *Neural Comput.* **1995**, *7*, 889–904. [CrossRef] [PubMed]

13.   Friston, K.; Kilner, J.; Harrison, L. A free energy principle for the brain. *J. Physiol. Paris* **2006**, *100*, 70–87. [CrossRef] [PubMed]

14.   Friston, K.J. Hierarchical model in the brain. *PLoS Comput. Biol.* **2008**, *4*, e1000211. [CrossRef] [PubMed]

15.   Friston, K. The free-energy principle: A unified brain theory? *Nat. Rev. Neurosci.* **2010**, *11*, 127–138. [CrossRef] [PubMed]

16.   Friston, K. A free energy principle for biological systems. *Entropy* **2012**, *14*, 2100–2121.

17.   Friston, K.; FitzGerald, T.; Rigoli, F.; Schwartenbeck, P.; Pezzulo, G. Active inference: A process theory. *Neural Comput.* **2017**, *29*, 1–49. [CrossRef] [PubMed]

18.   George, D.; Hawkins, J. Towards a mathematical theory of cortical micro-circuits. *PLoS Comput. Biol.* **2009**, *5*, e1000532. [CrossRef] [PubMed]

19.   Bastos, A.M.; Usrey, W.M.; Adams, R.A.; Mangun, G.R.; Fries, P.; Friston, K.J. Canonical microcircuits for predictive coding. *Neuron* **2012**, *76*, 695–711. [CrossRef] [PubMed]

20.   Rao, R.P.; Ballard, D.H. Predictive coding in the visual cortex: A functional interpretation of some extra-classical receptive-field effects. *Nat. Neurosci.* **1999**, *2*, 79–87. [CrossRef] [PubMed]

21.   Friston, K. A theory of cortical responses. *Philos. Trans. R. Soc. Lond. B Biol. Sci.* **2005**, *360*, 815–836. [CrossRef] [PubMed]

22.   Friston, K.J.; Daunizeau, J.; Kiebel, S.J. Reinforcement learning or active inference? *PLoS ONE* **2009**, *4*, e6421. [CrossRef] [PubMed]

23.   Kilner, J.M.; Friston, K.J.; Frith, C.D. Predictive coding: An account of the mirror neuron system. *Cognit. Process.* **2007**, *8*, 159–166. [CrossRef] [PubMed]

24.   Friston, K.; Mattout, J.; Kilner, J. Action understanding and active inference. *Biol. Cybern.* **2011**, *104*, 137–160. [CrossRef] [PubMed]

25.   Friston, K.J.; Frith, C.D. Active inference, communication and hermeneutics. *Cortex* **2015**, *68*, 129–143. [CrossRef] [PubMed]

26.   Friston, K.; Frith, C. A duet for one. *Conscious. Cognit.* **2015**, *36*, 390–405. [CrossRef] [PubMed]

27.   Fletcher, P.C.; Frith, C.D. Perceiving is believing: A Bayesian approach to explaining the positive symptoms of schizophrenia. *Nat. Rev. Neurosci.* **2009**, *10*, 48–58. [CrossRef] [PubMed]

28.   Friston, K.J.; Stephan, K.E.; Montague, R.; Dolan, R.J. Computational psychiatry: The brain as a phantastic organ. *Lancet Psychiatry* **2014**, *1*, 148–158. [CrossRef]

29.   Linsker, R. Self-organization in a perceptual network. *Computer* **1988**, *21*, 105–117. [CrossRef]

30.   Linsker, R. Local synaptic learning rules suffice to maximize mutual information in a linear network. *Neural Comput.* **1992**, *4*, 691–702. [CrossRef]

31.   Lee, T.W.; Girolami, M.; Bell, A.J.; Sejnowski, T.J. A unifying information-theoretic framework for independent component analysis. *Comput. Math. Appl.* **2000**, *39*, 1–21. [CrossRef]

32.   Simoncelli, E.P.; Olshausen, B.A. Natural image statistics and neural representation. *Ann. Rev. Neurosci.* **2001**, *24*, 1193–1216. [CrossRef] [PubMed]

33.   Belouchrani, A.; Abed-Meraim, K.; Cardoso, J.F.; Moulines, E. A blind source separation technique using second-order statistics. *Signal Process. IEEE Trans.* **1997**, *45*, 434–444. [CrossRef]

34.   Choi, S.; Cichocki, A.; Park, H.M.; Lee, S.Y. Blind source separation and independent component analysis: A review. *Neural Inf. Process. Lett. Rev.* **2005**, *6*, 1–57.

35.   Cichocki, A.; Zdunek, R.; Phan, A.H.; Amari, S.I. *Nonnegative Matrix and Tensor Factorizations: Applications to Exploratory Multi-Way Data Analysis and Blind Source Separation*; John Wiley & Sons: Hoboken, NJ, USA, 2009.

36.   Comon, P.; Jutten, C. *Handbook of Blind Source Separation: Independent Component Analysis and Applications*; Academic Press: Oxford, UK, 2010.

37.   Palmer, J.; Rao, B.D.; Wipf, D.P. Perspectives on sparse Bayesian learning. *Adv. Neural Inf. Proc. Syst.* **2004**, *27*, 249–256.

38. Olshausen, B.A.; Field, D.J. Emergence of simple-cell receptive field properties by learning a sparse code for natural images. *Nature* **1996**, *381*, 607–609. [CrossRef] [PubMed]
39. Olshausen, B.A.; Field, D.J. Sparse coding with an overcomplete basis set: A strategy employed by V1? *Vis. Res.* **1997**, *37*, 3311–3325. [CrossRef]
40. Shannon, C.E.; Weaver, W. *The Mathematical Theory of Communication*; University of Illinois Press: Urbana, IL, USA, 1949.
41. Cover, T.M.; Thomas, J.A. *Elements of Information Theory*; John Wiley & Sons: New York, NY, USA, 1991.
42. Pearl, J. *Probabilistic Reasoning in Intelligent Systems: Networks of Plausible Inference*; Morgan Kaufmann: San Fransisco, CA, USA, 1988.
43. Friston, K.J. Life as we know it. *J. R. Soc. Interface* **2013**, *10*, 20130475. [CrossRef] [PubMed]
44. Kullback, S.; Leibler, R.A. On information and sufficiency. *Ann. Math. Stat.* **1951**, *22*, 79–86. [CrossRef]
45. Arora, S.; Risteski, A. Provable benefits of representation learning. *arXiv* **2017**, arXiv:1706.04601.
46. Jaynes, E.T. Information theory and statistical mechanics. *Phys. Rev.* **1957**, *106*, 620–630. [CrossRef]
47. Jaynes, E.T. Information theory and statistical mechanics. II. *Phys. Rev.* **1957**, *108*, 171–190. [CrossRef]
48. Isomura, T.; Kotani, K.; Jimbo, Y. Cultured Cortical Neurons Can Perform Blind Source Separation According to the Free-Energy Principle. *PLoS Comput. Biol.* **2015**, *11*, e1004643. [CrossRef] [PubMed]
49. Xu, L. Least mean square error reconstruction principle for self-organizing neural-nets. *Neural Netw.* **1993**, *6*, 627–648. [CrossRef]
50. Amari, S.I.; Cichocki, A.; Yang, H.H. A new learning algorithm for blind signal separation. *Adv. Neural Inf. Proc. Syst.* **1996**, *8*, 757–763.
51. Oja, E. Neural networks, principal components, and subspaces. *Int. J. Neural Syst.* **1989**, *1*, 61–68. [CrossRef]
52. Bell, A.J.; Sejnowski, T.J. An information-maximization approach to blind separation and blind deconvolution. *Neural Comput.* **1995**, *7*, 1129–1159. [CrossRef] [PubMed]
53. Bell, A.J.; Sejnowski, T.J. The "independent components" of natural scenes are edge filters. *Vis. Res.* **1997**, *37*, 3327–3338. [CrossRef]
54. Hinton, G.E.; Salakhutdinov, R.R. Reducing the dimensionality of data with neural networks. *Science* **2006**, *313*, 504–507. [CrossRef] [PubMed]
55. LeCun, Y.; Bengio, Y.; Hinton, G. Deep learning. *Nature* **2015**, *521*, 436–444. [CrossRef] [PubMed]
56. Isomura, T.; Toyoizumi, T. Error-gated Hebbian rule: A local learning rule for principal and independent component analysis. *Sci. Rep.* **2018**, *8*, 1835. [CrossRef] [PubMed]
57. Hyvärinen, A.; Pajunen, P. Nonlinear independent component analysis: Existence and uniqueness results. *Neural Netw.* **1999**, *12*, 429–439. [CrossRef]
58. Yang, H.H.; Amari, S.I. Adaptive online learning algorithms for blind separation: Maximum entropy and minimum mutual information. *Neural Comput.* **1997**, *9*, 1457–1482. [CrossRef]
59. Latham, P.E.; Nirenberg, S. Synergy, redundancy, and independence in population codes, revisited. *J. Neurosci.* **2005**, *25*, 5195–5206. [CrossRef] [PubMed]
60. Amari, S.I.; Nakahara, H. Correlation and independence in the neural code. *Neural Comput.* **2006**, *18*, 1259–1267. [CrossRef] [PubMed]
61. Berkes, P.; Orbán, G.; Lengyel, M.; Fiser, J. Spontaneous cortical activity reveals hallmarks of an optimal internal model of the environment. *Science* **2011**, *331*, 83–87. [CrossRef] [PubMed]
62. Sengupta, B.; Stemmler, M.B.; Friston, K.J. Information and efficiency in the nervous system—A synthesis. *PLoS Comput. Biol.* **2013**, *9*, e1003157. [CrossRef] [PubMed]
63. Frémaux, N.; Gerstner, W. Neuromodulated Spike-Timing-Dependent Plasticity, and Theory of Three-Factor Learning Rules. *Front. Neural Circuits* **2016**, *9*. [CrossRef] [PubMed]
64. Isomura, T.; Toyoizumi, T. A Local Learning Rule for Independent Component Analysis. *Sci. Rep.* **2016**, *6*, 28073. [CrossRef] [PubMed]
65. Kuśmierz, Ł.; Isomura, T.; Toyoizumi T. Learning with three factors: Modulating Hebbian plasticity with errors. *Curr. Opin. Neurobiol.* **2017**, *46*, 170–177.
66. Isomura, T.; Sakai, K.; Kotani, K.; Jimbo, Y. Linking neuromodulated spike-timing dependent plasticity with the free-energy principle. *Neural Comput.* **2016**, *28*, 1859–1888. [CrossRef] [PubMed]
67. Markram, H.; Lübke, J.; Frotscher, M.; Sakmann, B. Regulation of synaptic efficacy by coincidence of postsynaptic APs and EPSPs. *Science* **1997**, *275*, 213–215. [CrossRef] [PubMed]

68. Bi, G.Q.; Poo, M.M. Synaptic modifications in cultured hippocampal neurons: Dependence on spike timing, synaptic strength, and postsynaptic cell type. *J. Neurosci.* **1998**, *18*, 10464–10472. [CrossRef] [PubMed]
69. Karnani, M.; Pääkkönen, K.; Annila, A. The physical character of information. *Proc. R. Soc. A Math. Phys. Eng. Sci.* **2009**, *465*, 2155–2175. [CrossRef]
70. Annila, A. On the character of consciousness. *Front. Syst. Neurosci.* **2016**, *10*, 27. [CrossRef] [PubMed]

entropy

MDPI

Article

Category Structure and Categorical Perception Jointly Explained by Similarity-Based Information Theory

Romain Brasselet [1,2,*] and **Angelo Arleo** [3]

[1] Cognitive Neuroscience Sector, SISSA, Via Bonomea 265, 34136 Trieste, Italy
[2] Center for Brain and Cognition, Computational Neuroscience Group, Department of Information and Communication Technologies, Universitat Pompeu Fabra, 08018 Barcelona, Spain
[3] Sorbonne Université, INSERM, CNRS, Institut de la Vision, 17 rue Moreau, F-75012 Paris, France; angelo.arleo@inserm.fr
* Correspondence: rbrasselet@sissa.it

Received: 27 April 2018; Accepted: 10 July 2018; Published: 14 July 2018

check for
updates

Abstract: Categorization is a fundamental information processing phenomenon in the brain. It is critical for animals to compress an abundance of stimulations into groups to react quickly and efficiently. In addition to labels, categories possess an internal structure: the goodness measures how well any element belongs to a category. Interestingly, this categorization leads to an altered perception referred to as categorical perception: for a given physical distance, items within a category are perceived closer than items in two different categories. A subtler effect is the perceptual magnet: discriminability is reduced close to the prototypes of a category and increased near its boundaries. Here, starting from predefined abstract categories, we naturally derive the internal structure of categories and the phenomenon of categorical perception, using an information theoretical framework that involves both probabilities and pairwise similarities between items. Essentially, we suggest that pairwise similarities between items are to be tuned to render some predefined categories as well as possible. However, constraints on these pairwise similarities only produce an approximate matching, which explains concurrently the notion of goodness and the warping of perception. Overall, we demonstrate that similarity-based information theory may offer a global and unified principled understanding of categorization and categorical perception simultaneously.

Keywords: goodness; categorical perception; perceptual magnet; information theory; perceived similarity

1. Introduction

Categorization is a cognitive process through which a large number of *items* (objects, events, stimuli; sometimes referred to as *instances* or *exemplars*) are grouped into a few classes. It is a bottleneck from an immensely complex world to relevant representations and actions [1] and thus it allows us to react quickly and communicate efficiently. Categorizing amounts to compressing the perceived world by putting the same label on many items, thereby preserving the relevant information and discarding the irrelevant one. Importantly, such a binary perception has been shown to be suboptimal [2], since categories to which the item may belong with weaker probabilities are discarded. From a relationist viewpoint, categorization consists in considering as *similar* two items in a category and as *different* two items in different categories. According to this view, Rosch [3] gives the following definition "To categorize a stimulus means to consider it (...) not only equivalent to other stimuli in the same category but also different from stimuli not in that category".

In addition, categories also have an internal structure: each item has its own measure of how well it represents its category, which is called *goodness* [4–6] (also referred to as *membership* or *typicality*). The item with the largest goodness in a category is called the prototype of this category [5]. This internal structure plays an important role in the speed of classification [4], exemplar production [7], or two-item discrimination [8]. The ontology of this graded internal structure is dependent on both the frequency of instantiation of an item as a member of the category [7,9,10] and the pairwise similarity structure [6,8]. The prototype has in general a large frequency and it is similar to other items of the category [5]. For instance, although it is a frequently mentioned bird, a chicken is not judged as being very similar to other birds and thus it has low goodness [11].

Interestingly, categorization is not only a bottom-up process as it bears effects on perception. One of these effects is called *categorical perception* [12]: items within a category are harder to discriminate than items in different categories, even if they are separated by the same physical distance (physical distance here means distance in the relevant physical space: frequency, amplitude, wavelength... or any metric space in which the items may be embedded). In other words, within-category discrimination is reduced while between-category discrimination is enhanced. Discrimination performance is only slightly better than category identification, though the within-category subtleties can be observed through the reaction times [13]. This effect has been observed on similarity between faces [14,15], colors [16,17], or speech sounds [12].

An additional effect is that prototypes of a category pull other items in the category toward themselves, which is called the *perceptual magnet* effect [18]. Items at the center of a category are perceived closer than at the border of a category. Iverson and Kuhl have shown the warping of the perceptual space using multi-dimensional scaling [19,20]. This effect is known to be asymmetrical [8,18,19]: in a two-stimuli discrimination task, if a prototype is presented before a non-prototype, the discrimination results are poorer than when the non-prototype is presented first. For a thorough review of these phenomena in natural and artificial categories, as well as an account of them through Bayesian inference, alternative to our explanation, we refer the reader to Feldman et al. [21].

The interactions between category boundaries, category structures, categorical perception, and perceptual magnet are still debated. It is so far unknown whether one of them is more fundamental and entails the others as consequences. We take here a holistic approach and attempt to show that they are all facets of categorization. All categories indeed have an internal structure [11], the notion of goodness thus appears inseparable from categorization. In addition, categorical perception has been observed commonly, although with some variability [22].

A large body of work already attempted at modeling categorical effects. Of particular interest to us here is the context theory of classification [23,24] that takes into account the similarities between items and proposes a measure comparing within-category similarities to between-category similarities. Our work will naturally lead us to consider a very similar measure, which we will interpret in information-theoretic terms. We will also consider the work by Bonnasse-Gahot and Nadal [25] that is the closest to ours in terms of explanations of categorical phenomena. They give an information-theoretic account of categorical perception and perceptual magnet as optimization of neural coding of categories.

Here, we aim at explaining altogether the structure of categories and the categorical perception phenomena by applying a recently introduced optimization principle for information processing. We start with well-defined categories whose items appear with uniform or bell-shaped frequencies. Following Tversky and Gati [8], we model human discrimination between two items with a notion of perceived similarity that can take any values in the range $[0,1]$: two items with a similarity of 1 are perceived as identical while two items with a similarity of 0 are perceived as different. Therefore, this work makes use of frequencies and pairwise similarities as fundamental features of our cognitive processes. We then use a new information-theoretic principle for optimizing the pairwise similarity values.

As information theory is a suitable tool and a very efficient framework to understand information processing in the brain [26–28] (and references therein), we account for categorical perception by applying information theory to categorization. We use a recently introduced version of information theory integrating pairwise similarities [29], whose formulation naturally merges reliability of information transmission and compression of the stimulus space. In this sense, the present work can be compared to that of Bonnasse-Gahot and Nadal [25] who used information theory to find optimal neural population codes to encode categories and account for some aspects of categorical perception.

In a general way, the process of learning categories can be described with two opposing strategies, as stated by Pothos [30] in the case of Artificial Grammar Learning, "the similarity/rules/association and the *information premise*", arguing that a shortcoming of information theory to understand cognition is its lack of tools for understanding representations. We hope to demonstrate the possibility to develop such tools. In this paper, we provide a theory that naturally encompasses the two approaches, thereby attempting to show that they may not be as much in opposition as they seem. We build on a recent modification of information theory that involves quantities for representations, namely, pairwise similarities between items considered.

Armed with this principle, we derive the internal structure of categories and categorical perception simultaneously. Indeed, the formulation naturally involves, for each element, an average of its similarities weighted by the probabilities of every other element within the category, which is readily interpreted as how similar one object is to the others on average, and it is shown to match the notion of goodness [5].

One could understand our method the following way: suppose a subject is trying to perfectly categorize an ordered set of stimuli. By perfectly, we mean that all specificities of the stimuli beyond the category are forgotten. This amounts exactly to maximizing the information while minimizing the equivocation. Now let us in addition suppose that there are limits to the ability of a subject to achieve this perfect categorization because of finite discrimination capacity. We choose to model the latter by similarity functions. The framework of similarity-based information theory (SBIT) is a very natural one to use here as it integrates similarities to information theory (IT).

2. Methods

2.1. Item Space and Categories

Items $S = \{s\}$ are considered on a one-dimensional axis and grouped into a set of categories $C = \{c\}$. These categories represent the pre-existing ideal categorization that the similarity measures have to emulate. These do not only depend on the observer: as in the case of colors, they can be very influenced by culture [31] (we come back to this issue in the discussion). As categorical perception effects are ubiquitous and appear in many modalities, we make no specific assumptions about the distribution of items inside a category. For example, it appears highly reasonable to consider light wavelengths to be uniformly distributed [32]. On the contrary, it is known that, in a given language, speech sounds are well defined but are modulated by noise or idiosyncrasies. This results in an ensemble of bell-shaped categories. Thus, for the sake of completeness, we will consider two extreme cases: one with contiguous uniformly distributed categories and another with bell-shaped categories. We also consider the case of bimodal categories. These three cases represent extreme cases in between which other one-dimensional cases will exist. In all cases, we will refer to their width (or, equivalently, the number of items) as W. We thus believe we exhibit exhaustively the phenomena of interest on all potential distributions of categories. The extension to higher-dimensional cases is straightforward and leads to qualitatively similar behaviors.

2.2. Similarity Functions

We define a similarity function $\phi_s(s')$, between the item s' and the reference item s, that takes values in the interval $[0,1]$ and describes how similar the item s' is to reference item s. We use a biologically reasonable constraint on the similarity function: at each point, it has to be a non-strictly decreasing function of the physical distance (e.g., Heaviside, bell-shaped, Gaussian, triangular). This is a very light constraint and seems, to the best knowledge of the authors, the only sensible behavior a similarity function can adopt. Unless we twist words and concepts heavily, two very different objects cannot be construed as more similar than two less different objects.

We choose a triangular pairwise similarity function given by $\phi_s(s') = 1 - \frac{|s-s'|}{\sigma(s)}$ when $|s - s'| < \sigma(s)$ and 0 otherwise. The width function $\sigma(s)$ takes a different value at each point in the item space and is the only free parameter to be tuned. The variation of the similarity width is akin (although not necessarily fully equivalent) to the attention-specific warping of the stimulus space in the context theory of classification or to the variation of the widths of the neural tuning curves in Bonnasse-Gahot and Nadal [25]. Note the potential asymmetry here—if $\sigma(s) \neq \sigma(s')$, then $\phi_s(s') \neq \phi_{s'}(s)$. The choice of a triangular function over another type of bell-shaped function is motivated by the empirical fact that the results are not qualitatively affected by the choice of similarity function, as long as it is not singular in any way (a condition akin to the one in [25] about the smoothness of the tuning curves), and by the ease of the mathematical treatment of the triangular function, which allow us to directly compare simulations and calculations.

Now that we are equipped with such similarity functions, we need to define an optimization principle. To do so, we make use of the similarity-based mutual information between the set of categories and the items with their pairwise similarities.

2.3. Similarity-Based Information Theory

Here, we recall the main concepts of similarity-based information theory (SBIT), a well-established theory with a versatile, albeit recent, history.

SBIT can be seen as an extension of IT, which is a framework to quantify statistical dependencies between variables, mainly through the definition of the entropy of a distribution, that quantifies its uncertainty. While IT takes into account only the probabilities of events or items, it discards entirely all other features of the dataset, in particular to what extent two items are similar or not. SBIT is precisely an attempt to extend IT by incorporating similarities in the very definition of entropy.

A similarity-based entropy was first introduced as a measure of biodiversity. In this field, the original concept was Rao's quadratic entropy that incorporates the distance between two species [33]. Ricotta and Szeidl [34] proposed a family of similarity-based Tsallis entropy, while Leinster and Cobbold [35] introduced a family of similarity-based Renyi entropy. Both Tsallis and Renyi entropies entail Rao's quadratic entropy as a special case. The similarity-based Renyi entropy reduces to a similarity-based Shannon entropy for particular values of its free parameter. The rationale for these concepts is that entropies only deal with probabilities but do not take into account potential similarities between items. Therefore, in the field of biodiversity, a population of canopy butterflies or a mixed population of canopy and understorey butterflies with the same probability distribution would have the same Shannon entropy, but not the same similarity-based entropy. Accounting for the similarities between species sheds a new light on the meaning of biodiversity.

Among all these entropies, the advantage of Shannon entropy is that there exists an unequivocal definition of conditional entropy [36,37], thereby allowing the mutual information between two variables to be defined. This mutual information is readily interpreted as a reduction in uncertainty about a variable when the other is known. This similarity-based mutual information was introduced and applied to neural coding [29], and we use it here. We wish to emphasize how the theory used here is grounded in other fields of study and applied without ad hoc extensions to the topic of cognition.

Our hypothesis is that information processing has to focus on thoroughly discriminating some pairs of items, thereby guaranteeing information transmission, but, simultaneously, it also has to overlook differences between other pairs, i.e., compress the stimulus set, which is assessed by the conditional entropy. This is what is referred to as categorization. This of course builds heavily on previous work and is not a new way of looking at categorization. Neither is the use of information theory to do so (see for example [25,38]). However, we address it in a new manner, using an extension of information theory that allows us to account for a set of phenomena that, to our knowledge, was never explained by a single model.

We first define (following Brasselet et al. [29]) the specific similarity-based entropy $h(s)$ (also known as *surprise*) :

$$h(s) \;=\; -\log g(s), \tag{1}$$

where $g(s) = \sum_{s'} p(s') \phi_s(s')$, which was defined similarly by Ricotta and Szeidl [34], Leinster and Cobbold [35]. Note here that this quantity $g(s)$ is a sum of probabilities weighted by similarities with s and is therefore always comprised between 0 and 1. It can therefore be thought of as "the probability of item s or another item s' similar to it". A high value of $h(s)$ means that it is surprising to observe the item s or another item s' similar to it. On the contrary, a small value tells us it is not surprising to observe it. This may happen when:

- the probability of s is itself large,
- another item s' with large probability is very similar to s,
- many low-probability items have large similarity with s.

In classical IT, only the first case exists.

Note that the corresponding entropy can be obtained by averaging the specific entropies over all items, $H(S) = \sum_s p(s) h(s)$.

Mathematically, the behavior of the similarity-based entropy is well-understood [29,35]. In the extreme case where the similarities are 0 everywhere, except for $s = s'$, we recover the probability of s and thus the original definition of Shannon surprise and Shannon entropy. Similarity-based information theory thus departs from Shannon theory that considers all items s to be different with no gradation. In such cases, the arguments of the logarithm cannot be taken as probabilities, since they do not sum to 1. However, the behavior of $h(s)$ is smooth as a function of the similarity matrix as it departs from the identity matrix, and eventually reaches $h(s) = 0$ when all the similarities are equal to 1, akin to the classical Shannon case where all items are indistinguishable (in the case of binning continuous variables for example). The similarity-based entropy thus takes values in the same range as Shannon entropy and has a natural interpretation. More properties are given in the references previously mentioned in this paragraph and we follow them in calling this quantity "entropy" as it meets all the criteria established by [39]. Researchers in ecology defined it to account for potential genetic similarities between species, while, in neuroscience, it was purposefully defined to account for similarities between percepts or representations.

Once we made the first step towards extending entropy with similarities, the definition of all the other quantities naturally follows. We can then define the specific conditional entropy $h(s|C)$ [40]:

$$h(s|C) \;=\; -\sum_c p(c|s) \log g(s|c), \tag{2}$$

where $g(s|c) = \sum_{s'} p(s'|c) \phi_s(s')$. This quantity can be thought of as "knowing the value of variable c, the probability of item s or another item s' similar to it". This specific conditional entropy can be thought as the uncertainty of items or items similar to them within a category c.

Note that, again, the corresponding conditional entropy can be obtained by averaging the specific entropies over all items $H(S|C) = \sum_s p(s) h(s|C)$. The conditional entropy is also known as *equivocation* because it measures how items are confused or, in other words, how the mapping from s to c is equivocal.

As usual, the specific similarity-based mutual information $i(s;C)$ is defined as the difference between the specific entropy and conditional entropy $i(s;C) = h(s) - h(s|C)$ and reads:

$$i(s;C) = \sum_c p(c|s) \log \frac{g(s|c)}{g(s)}. \tag{3}$$

This information increases with the argument of the logarithm, which is positive when an item is more probable within the category or more similar to other items within the category ($g(s|c)$) than it is probable overall or similar to other items items overall ($g(s)$). Therefore, for a given item with fixed probability, the information is large when the item is similar to items within the category but not with items outside. As usual, we recover the information by summing the specific information over the items: $I(S; C) = \sum_s p(s)i(s; C)$.

Importantly, all of these quantities reduce to Shannon specific entropies and specific mutual information in the case where the similarity function $\phi_s(s')$ is equal to 0 except for identity $s = s'$.

2.4. Optimization Principle

An important feature of the previously defined entropy $h(S)$, conditional entropy $h(s|C)$ and mutual information $i(s; C)$ is that they all depend on the similarity function $\phi_s(s')$. Large values of similarities, i.e., items are very much alike, will lead to low values of $h(s)$, $h(s|C)$ and $i(s; C)$. Conversely, low values of the similarities, i.e., all items appear as different, will increase the values of $h(s)$, $h(s|C)$ and $i(s; C)$. In our case, following Rosch's suggestion, we want to guarantee high similarities between objects from the same categories and low similarities between objects from different categories. High similarities inside categories amount to minimizing the conditional entropy while low similarities between categories amount to maximizing information between categories and items.

As is usually done in models of categorization, we implement the trade-off between maximizing information and minimizing conditional entropy by introducing an objective function involving a free trade-off parameter α:

$$q(s; C) = i(s; C) - \alpha \frac{h(s|C)}{h_0}. \tag{4}$$

(In the specific case of Brasselet et al. [29], α was chosen to be infinity). Note that this trade-off parameter is akin to the one we find in rate distortion theory (RDT) or in the information bottleneck (IB). The addition of our model is the integration of similarities. Just like in RDT, the objective is to minimize a cost function subject to an upper bound on the information, just like in IB, the objective is to maximize the information between two sets while minimizing that between one of these sets and an encoder, in SBIT, the objective is to maximize the information while minimizing the equivocation. To go further, we can compare this objective function with the one used by Sims et al. [38]:

$$\min_{p(y|x)} E[f(y - x) + P(C_x \neq C_y)], I(x, y) < C. \tag{5}$$

Maximizing the information between the item and its category is akin to not mistaking an item for another one. Minimizing the equivocation is akin to bounding the information.

We believe that a trade-off parameter is a necessary feature of any model of categorical perception. Indeed, there is a need for a compromise between compression (categorization) and information conveyance that may depend on the subject or the task at hand. This can only be captured by a quantity akin to a trade-off parameter.

A technical note about the quantity h_0 is in order here. In the present paper, we discuss the discrete case, but we aim at providing a framework that accomodates both discrete and continuous cases. Unlike information, conditional entropy is not independent from discretization. Therefore, we have to regularize it by a measure that depends commensurately on the discretization. This also allows the specific values of α to be independent from discretization. For a given problem, however, only the ratio between α and h_0 matters, so α can be redefined in units of h_0. In the sequel, we choose $h_0 = \log(W)$. Note that we are only concerned with positive values of α since we are looking at information maximization and conditional entropy minimization.

We apply this method to a categorization protocol and we use both a mathematical and computational approach. The problem is in general solvable analytically and we provide a solution in the case of uniform categories and triangular similarities in Appendix A. In the main body of the paper, we treat the problem computationally. As the maximization of the objective function $q(s; C)$ can be done independently at each point of the item space, we optimize the similarity measure by making an exhaustive search in the width space $\sigma(s)$ and by selecting the optimal value for the objective function.

3. Results

3.1. Non-Overlapping Uniform Categories

We first consider items $S = \{s\}$ characterized by a single parameter that is distributed uniformly over a single axis. Items S are grouped into N categories $C = \{c\}$ of width W (see Methods). For the sake of ease of explanation and interpretation, all the categories have the same size, though it is not a necessary condition. We optimize the objective function that maximizes information and minimizes the conditional entropy (see Methods). At each point, we compute the value of the similarity-based entropy, conditional entropy, information as well as objective function for each value of the similarity width $\sigma(s)$. We select the value of $\sigma(s)$ that maximizes the objective function. The particular results shown here are for $n = 10$ categories of width $W = 100$, for a total of 1000 items.

We provide the behavior of the different entropies at the center and at the boundary of categories as we explore the possible values of the similarity width. As the width of the similarity function increases, both the information and the conditional entropy decrease, although they do so at different paces. At the center of a category (Figure 1A,B), on the one hand, the conditional entropy starts at $\log(W)$, it undergoes a sharp drop as the width increases from 0 to $W/2$ and then decreases more slowly. On the other hand, for values of the width smaller than $W/2$, the information remains at its maximum of $\log(N)$ as all similarities between items from different categories remain zero. It only starts decreasing for values larger than $W/2$. Thus, the objective function always finds its maximum at a value larger than $W/2$. As the trade-off parameter α increases, the optimal width also does.

However, on the border of a category (Figure 1C,D), the behavior is radically different. The conditional entropy starts at $\log(W)$ and undergoes a drop as well as the width increases but less sharp than at the center as the similarity is increased with fewer items of the same category at a given similarity width. The information starts also at $\log(N)$ when the width is very small as there is no positive similarity between items of different category. However, as soon as the width increases, similarities between items of different categories increase and the information consequently drops and plateaus to $\log(N/2)$. This value comes from the fact that the similarity function essentially mixes two categories. Then, when the similarity width overpasses W, confusion arises with even more categories and the information decreases steadily even more. Thus, for low values of α, the optimal width is zero. As α increases, there is a sudden transition of the optimal width from 0 to W that then steadily keeps on increasing. More details and computations are given in Appendix A. We give only a summary of the final results here. We find that the optimal value of the width σ at the center of a category, i.e., at $(m + 1/2)W$, is:

$$\sigma_{opt} = \frac{W}{2}\left(1 + \frac{\alpha}{2\log W}\right). \tag{6}$$

This value is always larger than (or equal to) $W/2$. The intuitive reason for this is that, up to $W/2$, it only increases similarity with items within the category, and thus does not reduce information while decreasing equivocation. In addition, note that, to a good approximation, σ_{opt} is proportional to W, meaning that the similarity function scales with the category. Note that this value also grows linearly with α, for W fixed. This is because the more we focus on minimizing the equivocation, the larger the width has to be.

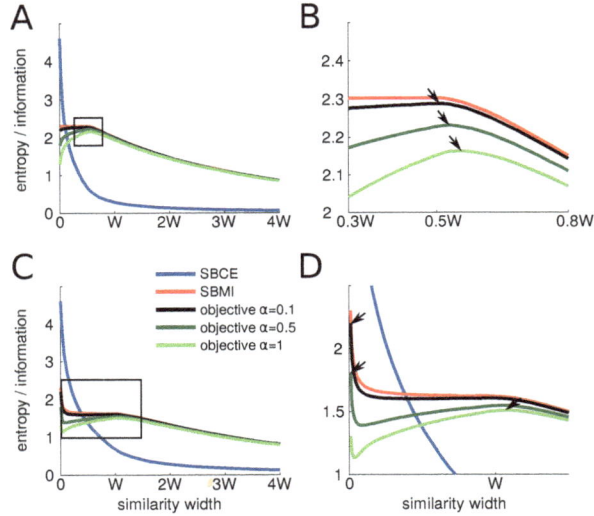

Figure 1. The mutual information SBMI, conditional entropy SBCE and objective function at the center (**A,B**) and border (**C,D**) of a category of width W depending on the similarity width with triangular similarity measures. Since information and entropies are always positive, larger values of α always lead to lower objective functions. The right panels (**B,D**) are magnifications of the rectangles depicted in the left panels (**A,C**). Arrows indicate the maxima of the objective functions. We observe that the maximum of the objective function at the center is always larger than $W/2$ and increases slowly with trade-off parameter α, while the maximum at the border is very small for small values of α and then undergoes an abrupt increase. These results are obtained for categories of width W.

We also find that the similarity width at the boundary of a category (i.e., the item closest to a boundary) is:

$$\sigma_{opt} = \quad 1, \qquad \text{if } \alpha < \alpha_{th}, \tag{7}$$

$$\sigma_{opt} = \quad (W-1)(1 + \tfrac{\alpha}{2\log W}), \quad \text{otherwise.} \tag{8}$$

When α is small, i.e., when little focus is on minimizing the equivocation, the optimal width is 1. Indeed, any departure from this would create confusion between the element s of category c and a neighboring category c'. When α is large, the optimal width is always larger than $W-1$. This is due to the possible reduction in equivocation yielded by extending the similarity function to encompass all the elements in category c and, collaterally, those of a neighbouring category c'.

Before turning to more complete computational simulations, we see already that the perceptual magnet effect will happen only for low values of α, when the optimal width at the center is larger than that at the boundary of a category.

We then assess the value of the similarity width across the stimulus space for a low value of the trade-off parameter, i.e., $\alpha = 0.5$ (Figure 2A). We recall that a low value of α means that the first objective is to maximize information and then, as a secondary objective, to minimize the conditional entropy. We observe that, in agreement with the previous results, the similarity measure width is much larger at the center of categories than at their boundaries. Examples of the similarity functions at selected places in the stimulus space are given in Figure 2B.

We also evaluate the functions $g(s|c)$ (see Equation (2)) for all members across the stimulus space for $\alpha = 0.5$ (Figure 2C). They exhibit a graded behavior: central members have high values while border members have low ones. This is due to the fact that central elements have large similarity widths and thus they are considered similar to other elements of the category. In the particular case of one-dimensional uniform categories, the $g(s|c)$ and the similarity width have the exact same behavior, but this will not be the case for other distributions, as we will see in following. The term $g(s|c) = \sum_{s'} p(s'|c)\phi_s(s')$ is readily interpreted as a measure of the goodness of the member s in category c.

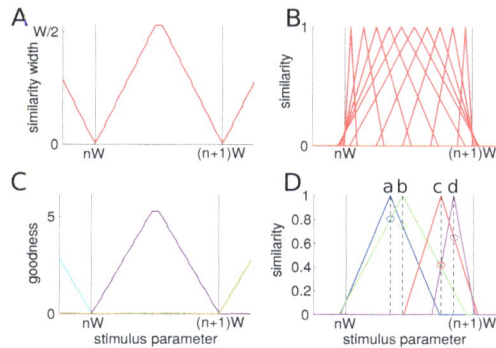

Figure 2. Results for categories of width W and trade-off parameter $\alpha = 0.5$. (**A**) similarity width for all members of a category and (**B**) similarity functions of selected members. The further away from the the center of the category, the smaller the similarity width and the narrower the similarity function; (**C**) goodness of all members of a category. Members at the center have larger goodnesses. In the case of uniform categories, they mimic the similarity widths; (**D**) similarities between two equidistant pairs of members either at the center or a border of a category. We see that a pair of items (a and b) near the center has larger similarities than a pair (c and d) closer to the border.

3.2. Preliminary Discussion

These results collectively match cognitive effects presented earlier:

- Categorical perception: two items at a given distance are perceived as more similar if they are within a category than if they are from two different categories. Indeed, if two items a and b from the same category are presented simultaneously, their similarities are 0.75 and 0.8, while two items c and d, which are at the same physical distance, have lower similarities 0.4 and 0.65 (Figure 2D).
- Perceptual magnet: the similarity measure is wider at the center of category, so items around it are perceived as more similar to each other than two other items at other locations within the category. Finally, we observe the asymmetry effect: starting from the prototype, an object on the border appears more similar than the other way around, or $\phi_{s_0}(s_1) > \phi_{s_1}(s_0)$.

However, the perceptual magnet effect depends on the value of the trade-off parameter α. For $\alpha < 0.79$ (see Appendix A), the perceptual magnet effect appears, but, for large values, the similarity measure peaks in-between categories, as shown in Figure 1. When the value of α is low, the objective is mainly to maximize information. Therefore, anything that reduces information is prohibited. More precisely, having a non-nil similarity between two items of two different categories reduces information. Thus, the maximal width of a similarity is the one that keeps the similarity low between items from two different categories. However, central items of a category are more remote to other categories than off-central items. Their similarity measure can therefore safely be wider.

As α increases, minimizing conditional entropy becomes more and more important, forcing the similarity measures to be wider. At $\alpha = 0.79$, there is a jump in the optimal width at the border of a category: it goes from $x_{opt} = 1$ to $x_{opt} > W$ (see Appendix A). At this point, the similarity width becomes larger at the border than at the center and the perceptual magnet effect disappears.

These opposite behaviors at low α and large α were found in all simulations no matter the shapes of category distribution. Because our main interest is in the ability of the model to produce a categorical perception effect, in the subsequent parts, we always choose a value of α below the critical value that sees a sudden jump in the similarity width at the boundaries of categories.

3.3. Gaussian Categories

We performed the same analysis on a set of items distributed on a one-dimensional space. These items are organized in Gaussian categories separated by a distance $W = 100$ and variance v:

$$p(s|c_i) \sim \mathcal{N}(W(i+\frac{1}{2}), v). \tag{9}$$

Again, we consider a triangular kernel whose width has to be optimized according to the similarity-based objective function. We carried the analysis with 10 values of α ranging from 5×10^{-10} to 5×10^{-1} equally spaced logarithmically.

The results are qualitatively similar to those obtained with flat categories. We observe that the similarity width (Figure 3, middle) is larger at the center of categories than at the border, although in this case, the width does not reach the minimum value as opposed to the flat category case. As for $g(s|c)$, we also observe a qualitatively similar behavior with items at the center having larger values than items on the border (Figure 3, bottom). The shape of the goodness curve $g(s|c)$ differs from the one of the similarity function since it also involves the Gaussian shape of categories. In particular, we observe that it drops faster when moving away from the prototype as the effects of the probabilities and distances multiply.

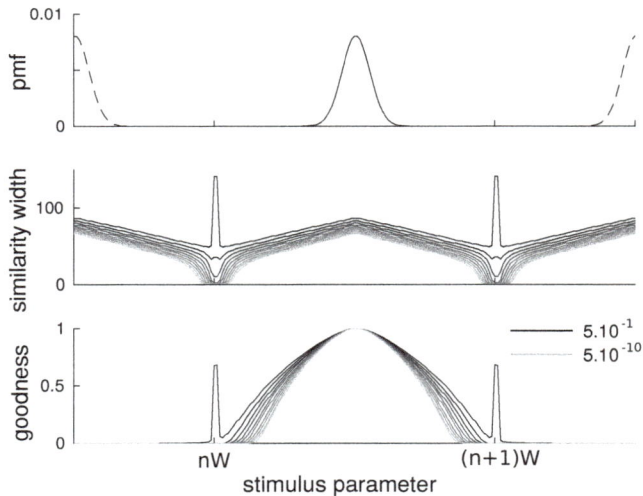

Figure 3. Results for Gaussian categories of variance $v = 5$ separated by a distance $W = 100$ and trade-off parameter α ranging from 5×10^{-10} to 5×10^{-1}. (**top**) probability mass function of the categories; (**middle**) similarity width for all members of a category. As in the uniform category case, the similarity is wider at the center than at the border, although the exact shape is more complicated; (**bottom**) goodness of all members of a category. The goodness results here from an interplay between frequencies and similarities, but it displays the expected behavior of larger goodnesses at the center.

3.4. Displacement Measure

In order to quantify the warping of the perceptual space, in the spirit of Feldman et al. [21], we use a measure of the displacement of each item within its category. We define the displacement of item s_0 as the average of the positions of the other items s_i within the category weighted by the similarity between s_0 and s_i. It reads:

$$D(s_0) = \sum_i \phi_{s_i}(s_0)s_i. \tag{10}$$

This is indeed a measure of where the average similar item is in the stimulus space. Therefore, this measure provides a good estimate of the position of the item in the perceptual space. For any value of the trade-off parameter α, we find qualitatively similar results, see Figure 4: the more the items are far from the center of a category, the more they are displaced towards its center. Items near the center are only slightly displaced.

The larger effect for Gaussian categories may account for the stronger perceptual magnet in discrete categories such as consonants compared to continuous, such as vowels.

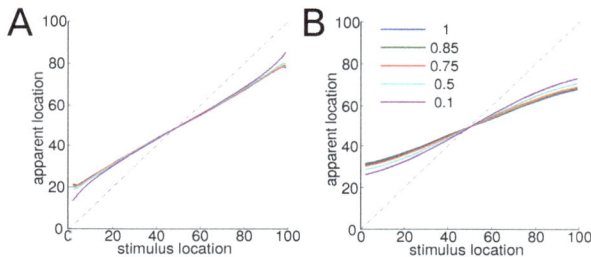

Figure 4. Measure of displacement of items within a category for different values of the trade-off parameter α. (**A**) uniform categories; (**B**) Gaussian categories. We observe that the more central element, the prototype, is not displaced while other items are all more displaced towards the center of the category, as they are far from the prototype. This result holds for any value of the trade-off parameter.

3.5. Bimodal Categories

To assess the effect of potential bimodality of distributions, the same analysis was made on bimodal categories. The items in each category are now distributed as the sum of two Gaussians:

$$p(s|c_i) \sim \mathcal{N}(W(i + \frac{1}{2}) - \delta, \nu) + \mathcal{N}(W(i + \frac{1}{2}) + \delta, \nu). \tag{11}$$

We kept the spacing of $W = 100$ between categories from the previous cases and chose a variance of $\nu = 5$ for the Gaussians to have clear bimodality. To study these cases, we chose a value of $\alpha = 5 \times 10^{-4}$ for which the perceptual magnet effect is very clear. We display results for δ from 0 to 30 (point at which the two peaks are further apart than they are to other categories) in increments of 5: namely, the category shapes, the similarity width and the goodnesses associated. See Figure 5.

We find that, overall, the similarity width keeps the same shape as in the monomodal Gaussian case, although its absolute value decreases. For a marked bimodal case, this is due to the the absence of a central mass in the distribution, compared to the bell-shaped case. In such a case, points close to one peak of the category do not decrease equivocation much by increasing their similarity width. The similarity width would have to be very large, but then the proximity of the other category prevents them from decreasing equivocation without decreasing information.

Note that, at some point, the similarity width becomes small and thus the goodness is mostly determined by the frequency. In a strong bimodal case, it naturally makes the goodness bimodal. We are not aware of an empirical case where this is observed. This leads us to venture that such unnatural cases where categories are bimodal with their modes further apart than they are from other categories may be unmanageable. However, it gives a prediction for artificial category learning in a laboratory setting.

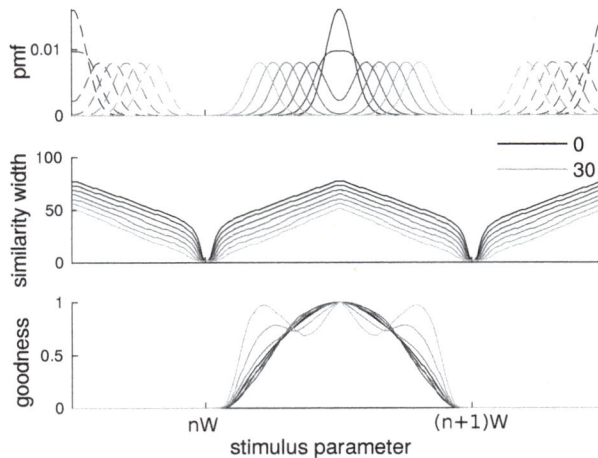

Figure 5. Results for bimodal categories made of two Gaussians of variance $v = 5$ separated by a distance δ ranging from 0 to 30. We used $\alpha = 0.0005$. (**top**) probability mass functions of the categories in the 7 cases; (**middle**) similarity width for all members of a category. As in the uniform and Gaussian category cases, the similarity is wider at the center than at the border; (**bottom**) goodness of all members of a category. The goodness has a more complicated behavior than in previous cases—from bell-shaped in the $\delta = 0$ case (i.e., Gaussian monomodal) to multimodal for large δ.

4. Discussion

First, the argument $g(s|c)$ that appears in the logarithm of the conditional entropy behaves naturally like the goodness of any given item s in the category c. The exemplar at the center of a category has the largest $g(s|c)$ and then qualifies as a prototype. Note that, in the uniform category case, the high goodness is only due to the internal similarity structure and not to frequencies. This matches the notion of prototype as having more attributes in common with other members of the category and less with members of other categories [5]. A posteriori, it appears natural to define the goodness of an exemplar as the average of its similarity with other exemplars: a good prototype should have features in common with most other items.

Second, among the theoretical advances brought by the present work, we note that the argument of the logarithm in the information formula (see Methods) is actually the function that appears in the context theory of classification [23,24] without the probabilities. We here give it an information-theoretic interpretation in terms of marginal and conditional similarity-based entropies. Indeed, in the context-theory of classification, a given item r is classified as belonging to a category c_1 based on the ratio between its similarity to other items in the category c_1 and its similarity with items in all categories $p(c_1|r) = \frac{\sum_i \phi_r(r_i^1)}{\sum_j \sum_i \phi_r(r_i^j)}$. This is similar to the term obtained in the logarithm of the similarity-based information theory, though the latter includes probabilities as well. It thus adds to the context-theory of classification by bringing together in a single formula the contributions of frequency and similarity [6], and by being grounded in the framework of information theory. The advantage of our formulation is that it includes both probabilities and similarity, so the resulting terms can be interpreted as goodnesses.

Third, among the other theoretical advances of this work is the ability to tune the similarity functions with an information-theoretic principle. It could be obvious to some readers that we can model perception with similarity functions. However, we then need an optimization principle that would have a clear interpretation and would not be ad hoc. This is exactly how we proceeded here, by taking without modification an existing theory and applying it directly to the issue of categorization.

Fourth, we show a warping of perceptual space. Similarity measures are wide at the center of categories and narrow their boundaries, akin to the poor within-category and high between-categories discriminability, known as the perceptual magnet effect. We can thus interpret the entropy as the expectation of the goodness surprise. The prototype is the least surprising member of a category. An important feature of our model is that, for a given trade-off α, if categories become larger, then the similarity functions scale likewise. Our model is therefore scale-invariant.

Fifth, as in the experiments mentioned in the introduction, there is an asymmetry between distinguishing two items depending on their order of presentation. When the prototype is presented first, the similarity measure used is wide and the discrimination of a following non-prototype becomes hard. However, if the non-prototype is presented first, the similarity measure is narrow and the discrimination of the prototype is easier. When a first item appears, it sets the comparison level. If it is close to the border of the category, it is much more stringent than when it is in the center.

It has been postulated that the magnitude of the categorical effects depends on the protocol used in the experiments [22]. Here, we suggest that the culprit of these variations may be the trade-off parameter α, which sets the relative importance of discrimination and categorization.

A large body of studies suggests that, in order to account for classification, discrimination, and categorization capacities of human brains, we need to consider them information-processing systems that rely on two essential features: probabilities and similarities. As stated by Quine [41]: "There is nothing more basic to thought and language than our sense of similarity; our sorting of things into kinds", as also stressed by Tversky and Gati [8]. Only by considering these two factors can we understand the way we perform at various tasks and our cognitive biases. This work attempts at bringing together these three elements: information, probabilities and similarities, thereby mapping Ref. [6]'s paper triptych: "Similarity, Frequency and Category Representation". It fits in the general attempt to phrase cognition into an information theoretical framework with the addition of a similarity structure.

There have already been attempts to explain categorical perception effects. Lacerda [42] proposed an explanation based on exemplar theory, which only accounts for between-category discrimination though. Following insights from Huttenlocher et al. [43], Feldman et al. [21] proposed a Bayesian model to explain the effect. However, according to the authors, the model only works in the case of unimodal categories and it cannot explain the effect on uniform categories. Our account thus finds its interest by fully explaining the perceptual magnet even in the case of uniform distributions as well as the internal structure of categories in a single framework using both probabilities and similarities.

Note the resemblance with the information-bottleneck method [44]. This approach aims at finding the codewords X that maximize the functional:

$$L = I(X; C) - \alpha I(X; S), \tag{12}$$

which is interpreted as maximizing the information of X about the category C, but trying to compress the stimulations S. The formulation bears a strong kinship with our method, but, as it is expressed only in Shannon information theoretic terms, it does not take into account the similarities between items. It would thus remain an unconstrained compression and would entail the optimal solution $X = C$ (for low values of α), which could not account for the category structure nor categorical perception.

It is now established that the information bottleneck and RDT are related [38,45]. The trade-off parameters in these two theories are linked to our α. In the field of categorization, this trade-off parameter is present in almost all models, or, more precisely, understanding categorization necessarily requires two opposing phenomena: a faithful encoding of the stimulus and a limiting/grouping phenomenon. For example, in Sims et al. [38], it naturally stems from the application of RDT. Although the link is not completely transparent: (a) minimizing Sims' cost function is akin to maximizing the similarity-base information, in the sense that it attempts at avoiding confusion or error between two different items; (b) the channel capacity constraint is akin to maximizing the similarity-based equivocation, as it allows elbow room for categories.

One of the strongest points of the work presented here is that our solution is abstract, as we work with similarities between items, without mentioning the type of neural substrates that could actually implement it. However, the nature of the neurophysiological substrates of the similarity remains an important question to address. Bonnasse-Gahot and Nadal [25] approached the problem with Shannon information theory and, although they focused on partially overlapping categories, our results conceptually matches theirs. In particular, they reached the conclusion that a neural population code optimized for categorization should display a higher resolution of the sensory input near the boundary of categories. This could be done by having more neurons with tuning curves centered near the boundary and/or to have neurons with narrower tuning curves near the boundary. Their actual implementation of a neural code matches our more abstract solution since, more neurons with narrow tuning curves will naturally yield larger neural distances between otherwise similar stimuli. It is interesting to observe that they take the neural code as a fundamental object. An exact quantification of the difference (or lack thereof) between the predictions of the models would require a mapping between population neural activity and item similarities, using, for example, common spike-train or firing rate distances [46].

As for the decision, similarity values can be seen as a level of evidence per unit of time in a drift model of decision-making [47]. This could explain why it takes longer to react in some cases [9,13,48]. This would bring this work closer to the drifting models of decision-making.

Limitations

One of the main limitations of our work is that we do not explain the origin and birth of categories. We just assumed that categorization is a needed feature of sensory information processing. We take categories for granted and then we try to maximize an objective information theoretical function. Other studies have already attempted to explain the sizes of categories. Komarova et al. [49] proposed an evolutionary mechanism to explain the birth of categories from discrimination. They used a parameter k_{sim} that is very close to our definition of similarity width and that is eventually generalized as a similarity matrix comparable to our family of similarity functions. Regier et al. [32] also attempted to explain categories as an optimal partitioning of a complex irregular perceptual space. It is so far uncertain whether our framework could also bring light on the birth of categories, but it is compatible with the former approaches. We hope in the near future to extend our framework to include the birth of categories. However, it will necessarily involve a measure of the allocation of brain resources to the discrimination of items and thus another trade-off. We believe that the simplicity (one objective function with one trade-off paramater) of the current framework can make it stand alone, notwithstanding these potential extensions.

Another limitation of our work is that it does not include hierarchy of categories, i.e., some items may be part of a subordinate category itself inside a superordinate category (a typical example being "a rabbit is a mammal is an animal"). Corter and Gluck [50] already used information theory to compute what hierarchical level of categorization is optimal to describe an item. Their goal being different from ours, their framework does not include similarities and does not attempt to explain categorical perception. This problem could be dealt with by adding a similarity structure on the set of categories, a feature that can be naturally integrated in the similarity-based information-theoretic framework. However, we are not aware of any experimental studies showing categorical perception or perceptual magnet effects on sub- and super-ordinate categories.

5. Conclusions

In this work, we account for both the internal structure of categories and its effects on perception by using a single information-theoretic framework. This framework integrates the notion of pairwise similarity between items into information measures. It provides an optimization principle that, at the same time, maximizes information transmission and compression of the stimulus space. Applying this framework to pre-existing categories to be learned, we derive the notion of goodness of items belonging to a category, categorical perception and the perceptual magnet effect.

The main point of this work is to set forth a single hypothesis about human discrimination in terms of perceived similarity as well as frequencies. There is no assumption on the shape of categories, which can be either bell-shaped or uniform. Nonetheless, we are able to derive the internal structure of categories through a naturally occurring quantity matching goodness, and the main effects of categorical perception with their potential asymmetry, thereby expliciting an ontological link between category structure and categorical perception. In other words, we show that both the goodness and warped perception are consequences of the interaction between pairwise similarities and frequencies, thereby expliciting a previously unknown link between the structure of categories and the perceptual effects of categorization. We show that the discrimination level can be accounted through a parameter α that implements a trade-off between discriminating and confusing pairs of stimulations. Empirical categorical perception is accounted for at low values of this parameter, i.e., when discrimination remains the main objective.

Author Contributions: All authors contributed in a significant way to the manuscript, have read and approved the final manuscript.

Acknowledgments: This work was granted by the EC project SENSOPAC, IST-027819-IP and by the French Medical Research Foundation. The authors thank Alessandro Treves for reading the manuscript and offering suggestions, as well as an anonymous reviewer for his insightful comments.

Conflicts of Interest: The authors declare no conflict of interests.

Appendix A

Here, we will give an intuition for the behavior of the similarity width in the case of uniform categories with a triangular kernel (that we use instead of a Gaussian kernel for ease of computation). Let's look in particular at the optimal similarity width at the center and at the boundary of a category.

Appendix A.1. Center of Categories

The computations go as follows in the case of N uniform categories of width W and a triangular kernel of width x. Note that $p(s|c) = Np(s)$:

$$g(s) = p(s)x. \tag{A1}$$

If $x < W/2$ (in blue in Figure A1):

$$g(s|c) = p(s|c)x, \tag{A2}$$

so we can compute the objective function:

$$q(s;c) \propto -\log p(s)x + (1 + \frac{\alpha}{\log W})\log p(s|c)x, \tag{A3}$$

which, upon inspection, is an ever-increasing function of x. Therefore, the optimal value of x must be larger than or equal to $\frac{W}{2}$.

If $x > W/2$ (in red in Figure A1):

$$g(s|c) = p(s|c)(W - \frac{W^2}{4x}),$$ (A4)

so we can compute the objective function if $x > W/2$:

$$q(s;c) \propto -\log p(s)x + (1 + \frac{\alpha}{\log W})\log p(s|c)(W - \frac{W^2}{4x}).$$ (A5)

Deriving this function and equating it to 0 yields:

$$x_{opt} = \frac{W}{4}(2 + \frac{\alpha}{\log W}),$$ (A6)

and we find perfect agreement with simulations.

As long as the kernel is positive only within the category, the information remains unaltered, only the conditional entropy is reduced compared to the Kronecker kernel. By increasing the width of the kernel, the conditional entropy is further reduced, but the information is reduced even more. There is thus a trade-off appearing.

If the trade-off parameter α is nil, then the objective is merely to maximize information. In that case, any value of the width smaller than the width of a category is suitable. As soon as α increases and thus minimizing conditional entropy starts to matter, then larger values of the width are optimal.

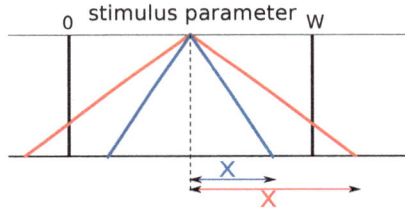

Figure A1. Similarity functions.

Appendix A.2. Boundary of Categories

The argument of the logarithm in the entropy is:

$$g(s) = p(s)x,$$ (A7)

and in the conditional entropy if $x < W$ (in blue in Figure A2):

$$g(s|c) = p(s|c)(\frac{1}{2} + \frac{x}{2}).$$ (A8)

Note that the $\frac{1}{2}$ comes from the fact that the item s considered is fully in category c.
We can compute the objective function:

$$q(s;c) \propto -\log[p(s)x] + (1 + \frac{\alpha}{\log W})\log[p(s|c)(\frac{1}{2} + \frac{x}{2})].$$ (A9)

The minimum of this function is at $x = \frac{\log W}{\alpha}$, so that, in the $0 < x < W$ part, there are two local maxima at $x = 1$ and $x = W$.

If $x > A$ (in red in Figure A2):

$$g(s|c) = p(s|c)W(1 - \frac{W}{2x} + \frac{1}{2x}) \tag{A10}$$

so we can compute the objective function:

$$q(s;c) \propto -\log[p(s)x] + (1 + \frac{\alpha}{\log W})\log[p(s|c)W(1 - \frac{W}{2x} + \frac{1}{2x})]. \tag{A11}$$

Deriving this function and equating it to 0 yields:

$$x_{opt} = \frac{(2 + \frac{\alpha}{\log W})(W - 1)}{2}. \tag{A12}$$

Note that the value of the objective function at x_{opt} is always larger than that at $x = W$. Thus, it only remains to know when the local maximum at $x = 1$ is larger than that at $x = x_{opt}$.

At $x = 1$:

$$q1 = q(s|c) \propto -\log p(s) + (1 + \frac{\alpha}{\log W})\log p(s|c). \tag{A13}$$

At $x = x_{opt}$:

$$q2 = q(s|c) \propto -\log[p(s)x_{opt}] \tag{A14}$$

$$+ (1 + \frac{\alpha}{\log W})\log[p(s|c)W(1 + \frac{1 - W}{2x_{opt}})]. \tag{A15}$$

The difference is:

$$q2 - q1 = -\log x_{opt} + (1 + \frac{\alpha}{\log W})\log[W(1 - \frac{W}{2x_{opt}} + \frac{1}{2x_{opt}})]. \tag{A16}$$

This cancels when:

$$\alpha = \log W(\frac{\log x_{opt}}{\log[W(1 + \frac{1-W}{2x_{opt}})]} - 1). \tag{A17}$$

We solved the last equation numerically and found for $W = 100$ a value of $\alpha_{th} = 0.796$. In the end, the optimal value of x depending on α is:

$$x_{opt} = \qquad 1, \qquad \text{if } \alpha < \alpha_{th}, \tag{A18}$$

$$x_{opt} = \quad \frac{(2 + \frac{\alpha}{\log W})(W-1)}{2}, \quad \text{otherwise.} \tag{A19}$$

With this, we obtain the values of the optimal x for all values of α and we observe an exact matching with the numerical results from the core of the article. We see that the optimal value of x scales with $W - 1$ (and not with W, which creates small departures from scale-independence at very low Ws).

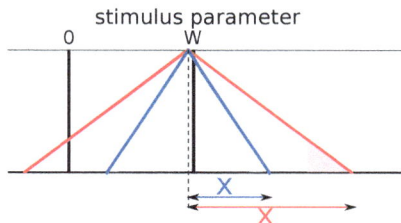

Figure A2. Similarity functions.

References

1. Harnad, S. To cognize is to categorize: Cognition is categorization. In *Handbook of Categorization in Cognitive Science*; Elsevier: New York, NY, USA, 2005; pp. 20–45.

2. Fleming, S.M.; Maloney, L.T.; Daw, N.D. The irrationality of categorical perception. *J. Neurosci.* **2013**, *33*, 19060–19070. [CrossRef] [PubMed]

3. Rosch, E.H. Principles of Categorization. In *Concepts: Core Readings*; MIT Press: Cambridge, MA, USA, 1999; pp. 189–206.

4. Rips, L.J.; Shoben, E.J.; Smith, E.E. Semantic distance and the verification of semantic relations. *J. Verbal Learn. Verbal Behav.* **1973**, *12*, 1–20. [CrossRef]

5. Rosch, E.H.; Mervis, C.B. Family resemblances: Studies in the internal structure of categories. *Cognit. Psychol.* **1975**, *7*, 573–605. [CrossRef]

6. Nosofsky, R.M. Similarity, frequency, and category representations. *J. Exp. Psychol. Learn. Mem. Cognit.* **1988**, *14*, 54. [CrossRef]

7. Freedman, J.L.; Loftus, E.F. Retrieval of words from long-term memory. *J. Verbal Learn. Verbal Behav.* **1971**, *10*, 107–115. [CrossRef]

8. Tversky, A.; Gati, I. Studies of similarity. *Cognit. Categ.* **1978**, *1*, 79–98.

9. Wilkins, A.J. Conjoint frequency, category size, and categorization time. *J. Verbal Learn. Verbal Behav.* **1971**, *10*, 382–385. [CrossRef]

10. Kuhl, P.K.; Williams, K.A.; Lacerda, F.; Stevens, K.N.; Lindblom, B. Linguistic experience alters phonetic perception in infants by 6 months of age. *Science* **1992**, *255*, 606–608. [CrossRef] [PubMed]

11. Barsalou, L.W. Ideals, central tendency, and frequency of instantiation as determinants of graded structure in categories. *J. Exp. Psychol. Learn. Mem. Cognit.* **1985**, *11*, 629. [CrossRef]

12. Liberman, A.M.; Harris, K.S.; Hoffman, H.S.; Griffith, B.C. The discrimination of speech sounds within and across phoneme boundaries. *J. Exp. Psychol.* **1957**, *54*, 358–368. [CrossRef] [PubMed]

13. Pisoni, D.B.; Tash, J. Reaction times to comparisons within and across phonetic categories. *Percept. Psychophys.* **1974**, *15*, 285–290. [CrossRef] [PubMed]

14. Etcoff, N.L.; Magee, J.J. Categorical perception of facial expressions. *Cognition* **1992**, *44*, 227–240. [CrossRef]

15. Campanella, S.; Quinet, P.; Bruyer, R.; Crommelinck, M.; Guerit, J.M. Categorical perception of happiness and fear facial expressions: An ERP study. *J. Cogn. Neurosci.* **2002**, *14*, 210–227. [CrossRef] [PubMed]

16. Franklin, A.; Drivonikou, G.V.; Clifford, A.; Kay, P.; Regier, T.; Davies, I.R. Lateralization of categorical perception of color changes with color term acquisition. *Proc. Natl. Acad. Sci. USA* **2008**, *105*, 18221–18225. [CrossRef] [PubMed]

17. Holmes, A.; Franklin, A.; Clifford, A.; Davies, I. Neurophysiological evidence for categorical perception of color. *Brain Cognit.* **2009**, *69*, 426–434. [CrossRef] [PubMed]

18. Kuhl, P.K. Human adults and human infants show a "perceptual magnet effect" for the prototypes of speech categories, monkeys do not. *Percept. Psychophys.* **1991**, *50*, 93–107. [CrossRef] [PubMed]

19. Iverson, P.; Kuhl, P.K. Mapping the perceptual magnet effect for speech using signal detection theory and multidimensional scaling. *J. Acoust. Soc. Am.* **1995**, *97*, 553–562. [CrossRef] [PubMed]

20. Iverson, P.; Kuhl, P.K. Influences of phonetic identification and category goodness on American listeners' perception of /r/ and /l/. *J. Acoust. Soc. Am.* **1996**, *99*, 1130–1140. [CrossRef] [PubMed]

21. Feldman, N.H.; Griffiths, T.L.; Morgan, J.L. The influence of categories on perception: Explaining the perceptual magnet effect as optimal statistical inference. *Psychol. Rev.* **2009**, *116*, 752–782. [CrossRef] [PubMed]

22. Gerrits, E.; Schouten, M.E.H. Categorical perception depends on the discrimination task. *Percept. Psychophys.* **2004**, *66*, 363–376. [CrossRef] [PubMed]

23. Medin, D.L.; Schaffer, M.M. Context theory of classification learning. *Psychol. Rev.* **1978**, *85*, 207. [CrossRef]

24. Nosofsky, R.M. Choice, similarity, and the context theory of classification. *J. Exp. Psychol. Learn. Mem. Cognit.* **1984**, *10*, 104. [CrossRef]

25. Bonnasse-Gahot, L.; Nadal, J.P. Neural coding of categories: Information efficiency and optimal population codes. *J. Comput. Neurosci.* **2008**, *25*, 169–187. [CrossRef] [PubMed]

26. Bialek, W.; Rieke, F.; de Ruyter van Steveninck, R.R.; Warland, D. Reading a neural code. *Science* **1991**, *252*, 1854–1857. [CrossRef] [PubMed]

27. Borst, A.; Theunissen, F.E. Information theory and neural coding. *Nat. Neurosci.* **1999**, *2*, 947–957. [CrossRef] [PubMed]

28. Quiroga, R.Q.; Panzeri, S. Extracting information from neuronal populations: Information theory and decoding approaches. *Nat. Rev. Neurosci.* **2009**, *10*, 173–185. [CrossRef] [PubMed]

29. Brasselet, R.; Johansson, R.S.; Arleo, A. Quantifying neurotransmission reliability through metrics-based information analysis. *Neural Comput.* **2011**, *23*, 852–881. [CrossRef] [PubMed]

30. Pothos, E.M. An entropy model for artificial grammar learning. *Front. Psychol.* **2010**, *1*, 16. [CrossRef] [PubMed]

31. Davidoff, J.; Davies, I.; Roberson, D. Colour categories in a stone-age tribe. *Nature* **1999**, *398*, 203. [CrossRef] [PubMed]

32. Regier, T.; Kay, P.; Khetarpal, N. Color naming reflects optimal partitions of color space. *Proc. Natl. Acad. Sci. USA* **2007**, *104*, 1436–1441. [CrossRef] [PubMed]

33. Radhakrishna, R.C. Diversity and dissimilarity coefficients: A unified approach. *Theor. Popul. Biol.* **1982**, *21*, 24–43.

34. Ricotta, C.; Szeidl, L. Towards a unifying approach to diversity measures: Bridging the gap between the Shannon entropy and Rao's quadratic index. *Theor. Popul. Biol.* **2006**, *70*, 237–243. [CrossRef] [PubMed]

35. Leinster, T.; Cobbold, C.A. Measuring diversity: The importance of species similarity. *Ecology* **2012**, *93*, 477–489. [CrossRef] [PubMed]

36. Teixeira, A.; Matos, A.; Antunes, L. Conditional Renyi Entropies. *IEEE Trans. Inf. Theory* **2012**, *58*, 4273–4277. [CrossRef]

37. Berens, S. Conditional Renyi Entropy. Ph.D. Thesis, Mathematisch Instituut, Universiteit Leiden, Leiden, The Netherlands, 2013.

38. Sims, C.R.; Ma, Z.; Allred, S.R.; Lerch, R.A.; Flombaum, J.I. Exploring the cost function in color perception and memory: An information-theoretic model of categorical effects in color matching. In Proceedings of the 38th Annual Conference of the Cognitive Science Society, Philadelphia, PA, USA, 10–13 August 2016; pp. 2273–2278.

39. Shannon, C.E. A Mathematical Theory of Communication. *Bell Syst. Tech. J.* **1948**, *27*, 379–423. [CrossRef]

40. DeWeese, M.R.; Meister, M. How to measure the information gained from one symbol. *Network* **1999**, *10*, 325–340. [CrossRef] [PubMed]

41. Quine, W. *Ontological Relativity and Other Essays*; Number 1; Columbia University Press: New York, NY, USA, 1969.

42. Lacerda, F. The perceptual-magnet effect: An emergent consequence of exemplar-based phonetic memory. In Proceedings of the XIIIth International Congress of Phonetic Sciences, Stockholm, Sweden, 14–19 August 1995; Volume 2, pp. 140–147.

43. Huttenlocher, J.; Hedges, L.V.; Vevea, J.L. Why do categories affect stimulus judgment? *J. Exp. Psychol. Gen.* **2000**, *129*, 220. [CrossRef] [PubMed]

44. Tishby, N.; Pereira, F.C.; Bialek, W. The information bottleneck method. *arXiv* **2000**, arXiv:0004057.

45. Harremoës, P.; Tishby, N. The information bottleneck revisited or how to choose a good distortion measure. In Proceedings of the 2007 IEEE International Symposium on Information Theory (ISIT 2007), Nice, France, 24–29 June 2007; pp. 566–570.

46. Houghton, C.; Sen, K. A new multineuron spike train metric. *Neural Comput.* **2008**, *20*, 1495–1511. [CrossRef] [PubMed]

47. Brunton, B.W.; Botvinick, M.M.; Brody, C.D. Rats and humans can optimally accumulate evidence for decision-making. *Science* **2013**, *340*, 95–98. [CrossRef] [PubMed]

48. Bonnasse-Gahot, L.; Nadal, J.P. Perception of categories: From coding efficiency to reaction times. *Brain Res.* **2012**, *1434*, 47–61. [CrossRef] [PubMed]

49. Komarova, N.L.; Jameson, K.A.; Narens, L. Evolutionary models of color categorization based on discrimination. *J. Math. Psychol.* **2007**, *51*, 359–382. [CrossRef]

50. Corter, J.E.; Gluck, M.A. Explaining basic categories: Feature predictability and information. *Psychol. Bull.* **1992**, *111*, 291. [CrossRef]

entropy

MDPI

Article

The Identity of Information: How Deterministic Dependencies Constrain Information Synergy and Redundancy

Daniel Chicharro [1,2,*], Giuseppe Pica [2] and Stefano Panzeri [2]

1   Department of Neurobiology, Harvard Medical School, Boston, MA 02115, USA
2   Neural Computation Laboratory, Center for Neuroscience and Cognitive Systems@UniTn, Istituto Italiano di Tecnologia, Rovereto (TN) 38068, Italy; giuseppe.pica@iit.it (G.P.); stefano.panzeri@iit.it (S.P.)
*   Correspondence: daniel_chicharro@hms.harvard.edu or daniel.chicharro@iit.it; Tel.: +39-0464-808696

Received: 13 November 2017; Accepted: 28 February 2018; Published: 5 March 2018

Abstract: Understanding how different information sources together transmit information is crucial in many domains. For example, understanding the neural code requires characterizing how different neurons contribute unique, redundant, or synergistic pieces of information about sensory or behavioral variables. Williams and Beer (2010) proposed a partial information decomposition (PID) that separates the mutual information that a set of sources contains about a set of targets into nonnegative terms interpretable as these pieces. Quantifying redundancy requires assigning an identity to different information pieces, to assess when information is common across sources. Harder et al. (2013) proposed an identity axiom that imposes necessary conditions to quantify qualitatively common information. However, Bertschinger et al. (2012) showed that, in a counterexample with deterministic target-source dependencies, the identity axiom is incompatible with ensuring PID nonnegativity. Here, we study systematically the consequences of information identity criteria that assign identity based on associations between target and source variables resulting from deterministic dependencies. We show how these criteria are related to the identity axiom and to previously proposed redundancy measures, and we characterize how they lead to negative PID terms. This constitutes a further step to more explicitly address the role of information identity in the quantification of redundancy. The implications for studying neural coding are discussed.

Keywords: information theory; mutual information decomposition; synergy; redundancy

MSC: 94A15; 94A17

1. Introduction

The characterization of dependencies between the parts of a multivariate system helps to understand its function and its underlying mechanisms. Within the information-theoretic framework, this problem can be investigated by breaking down into parts the joint entropy of a set of variables [1–3] or the mutual information between sets of variables [4–6]. These approaches have many applications to study dependencies in complex systems such as gene networks (e.g., [7–9]), neural coding and communication (e.g., [10–12]), or interactive agents (e.g., [13–15]).

An important aspect of how information is distributed across a set of variables concerns whether different variables provide redundant, unique or synergistic information when combined with other variables. Intuitively, variables share redundant information if each variable carries individually the same information carried by other variables. Information carried by a certain variable is unique if it is not carried by any other variables or their combination, and a group of variables carries synergistic information if some information arises only when they are combined. The presence of these different types of information has implications for example to determine how the information can be decoded [16], how robust it is to disruptions of the system [17], or how the variables' set can be compressed without information loss [18].

Characterizing the distribution of redundant, unique, and synergistic information is especially relevant in systems neuroscience, to understand how information is distributed in neural population responses. This requires identifying the features of neural responses that represent sensory stimuli and behavioral actions [19,20] and how this information is transmitted and transformed across brain areas [21,22]. The breakdown of information into these different types of components can determine the contribution of different classes of neurons and of different spatiotemporal components of population activity [23,24]. Moreover, the identification of synergistic or redundant components of information transfer may help to map dynamic functional connectivity and the integration of information across neurons or networks [25–28].

Although the notions of redundant, unique, and synergistic information seem at first intuitive, their rigorous quantification within the information-theoretic framework has proven to be elusive. Synergy and redundancy have traditionally been quantified with the measure called interaction information [29] or co-information [30], but this measure does not quantify them separately, and the presence of one or the other is associated with positive or negative values, respectively. Synergy has also been quantified using maximum entropy models as the information that can only be retrieved from the joint distribution of the variables [1,31,32].

However, a recent seminal work of [33] introduced a framework, called Partial Information Decomposition (PID), to more precisely and simultaneously quantify the redundant, unique, and synergistic information that a set of variables (or primary sources) S has about a target X. This decomposition has two cornerstones. The first is the definition of a general measure of redundancy following a set of axioms that impose desirable properties, in agreement with the corresponding abstract notion of redundancy [34]. The second is the construction of a redundancy lattice, structured according to these axioms, which reflects a partial ordering of redundancies for different sets of variables [33].

The PID framework has been further developed by others (e.g., [35–40]). However, the properties that the PID should have continue to be debated [38,41]. In particular, properly quantifying redundancy is inherently difficult because it requires assigning an identity to different pieces of information. This is needed to assess when different sources carry the same information about the target. The work in [35] argued that the original redundancy measure of [33] quantifies only quantitatively equal amounts of information and not information that is qualitatively the same. They introduced a new axiom, namely the identity axiom, which states that, for the concrete case of a target that is a copy of two sources, redundancy should correspond to the mutual information between the sources, and thus vanish for independent sources. Several redundancy measures that fulfill the identity axiom have been subsequently proposed [35–37]. However, although this axiom imposes a necessary condition to capture qualitatively common information, the question of how to generally determine the identity of different pieces of information to assess redundancy has not yet been solved, and information identity criteria are implicit in the axioms and measures used. Furthermore, the identity axiom is incompatible with ensuring the nonnegativity of the PID terms when there are more than two sources (multivariate case). This was proven by [42] with a counterexample that involves deterministic target-source dependencies, just like the target-source copy example used to motivate the axiom.

In this work, we examine in more detail how assumptions on the assignment of information identity determine the properties of the PIDs. We study in a general way the form of the PID terms for systems with deterministic target-source dependencies. These dependencies are particularly relevant to address the question of information identity because they allow exploring the consequences of alternative assumptions about how target-source identity associations constrain the existence of information synergistic contributions. These target-source identity associations naturally occur for example when the same variable appears both as a source and in the target: if some piece of information is assumed to be only associated with a variable that appears both as a source and as part of the target, this identity association would imply that there is no need to combine that source with any other to retrieve that piece of information. In other words, the corresponding synergy should be zero. Importantly, the deterministic relationships between the target and sources allow us to analyze how information identity criteria constrain the properties of the PIDs without the need to rely on any specific definition of PID measures.

To formalize the effect of deterministic target-source dependencies on the PID terms, we enunciate and compare the implications of two axioms that propose two alternative ways in which deterministic dependencies can constrain synergistic contributions because of assumptions on target-source identity associations. These axioms impose constraints to synergy for any (possibly multivariate) system with deterministic target-source dependencies, while the identity axiom only concerns a particular class of bivariate systems. We prove that the fulfillment of these axioms implies the fulfillment of the identity axiom and that several measures that fulfill the identity axiom also comply with one of the synergy axioms in general [36], or at least for a wider class of systems [35,43] than the one addressed by the identity axiom. The proof of the existence of negative terms when adopting the identity axiom was based on a concrete counterexample [41,42,44]. Oppositely, the stricter conditions of our synergy axioms allow us to explain in general how negative PID terms result from the specific information identity criteria underlying these axioms. More concretely, we derive, specifically for each of the two axioms, general expressions for deterministic components of the PID terms, which occur in the presence of deterministic target-source dependencies.

The comparison of the two axioms allows us to better understand the role of information identity in the quantification of redundancy. When the target contains a copy of some primary sources, an important difference between the redundancy measures derived from the two axioms regards their invariance, or lack thereof, to a transformation that reduces the target by removing all variables within it that are deterministically determined by this copy. This transformation does not alter the entropy of the target, and thus, a redundancy measure not invariant under it depends on semantic aspects regarding the identity of the variables within the target. We discuss why, in contrast to the mutual information itself, the PID terms may not be invariant to this transformation and depend on semantic aspects, as a consequence of the assignment of identity to the pieces of information, which is intrinsic to the notion of redundancy. In particular, we indicate how the overall composition of the target can affect the identity of the pieces of information and also can determine the existence of redundancy for independent sources (mechanistic redundancy [35]). Furthermore, based on this analysis, we identify the minimal set of assumptions that when added to the original PID axioms [33,34] can lead to negative PID terms. We indicate that this set comprises the assumption of the target invariance mentioned above. Overall, we conclude that if the redundancy lattice of [33] is to remain as the backbone of a nonnegative decomposition of the mutual information, a new criterion of information identity should be established that is compatible with the identity axiom, considers the semantic aspects of redundancy and results in less restrictive constraints on synergy in the presence of deterministic target-source dependencies than the two synergy axioms herein studied. Alternatively, the redundancy lattice should have to be modified to preserve nonnegativity.

We start this work by reviewing the PIDs (Section 2). We then introduce two alternative axioms that impose constraints on the value of synergistic terms in the presence of deterministic target-source dependencies, following an information identity criterion based on target-source identity associations (Section 3). Using these axioms, we derive general expressions that separate each PID term into a stochastic and a deterministic component for the bivariate (Section 4.1) and trivariate (Section 5.1) case. We show how these axioms constitute two alternative extensions of the identity axiom (Section 4.2) and examine if several previously-proposed redundancy measures conform to our axioms (Section 4.3). We reconsider the examples used by [42], characterizing their bivariate and trivariate decompositions and illustrating how in general negative PID terms can occur as a consequence of the information identity criteria underlying the synergy axioms (Sections 4.4 and 5.2). The comparison between the two axioms allows us to discuss the implications of using an information identity criterion that, in the presence of deterministic target-source dependencies, identifies pieces of information in the target by assuming that their identity is related to specific sources. More generally, we discuss how our results constitute a further step to more explicitly address the role of information identity in the quantification of information (Sections 4.5, 4.6 and 5.3).

2. A Review of the PID Framework

The seminal work of [33] introduced a new approach to decompose the mutual information into a set of nonnegative contributions. Let us consider first the bivariate case. Assume that we have a target X formed by one variable or by a set of variables and two variables (primary sources) 1 and 2 from which we want to characterize the information about X. The work in [33] argued that the mutual information of each variable about the target can be expressed as:

$$I(X;1) = I(X;1.2) + I(X;1\backslash 2), \tag{1}$$

and similarly for $I(X;2)$. The term $I(X;1.2)$ refers to a redundancy component between variables 1 and 2, which can be obtained either by observing 1 or 2 separately. The terms $I(X;1\backslash 2)$ and $I(X;2\backslash 1)$ quantify a component that is unique to 1 and to 2, respectively, that is, the information that can be obtained from one of the variables alone, but that cannot be obtained from the other alone. Furthermore, the joint information of 12 can be expressed as:

$$I(X;12) = I(X;1.2) + I(X;1\backslash 2) + I(X;2\backslash 1) + I(X;12\backslash 1,2), \tag{2}$$

where the term $I(X;12\backslash 1,2)$ refers to the synergistic information of the two variables, that is information that can only be obtained when combining the two variables. Therefore, given the standard information-theoretic chain rule equalities [45]:

$$\begin{aligned} I(X;12) &= I(X;1) + I(X;2|1) \\ &= I(X;2) + I(X;1|2), \end{aligned} \tag{3}$$

the conditional mutual information $I(X;2|1)$, that is the average information that 2 provides about X once the value of 1 is known, is decomposed as:

$$I(X;2|1) = I(X;2\backslash 1) + I(X;12\backslash 1,2), \tag{4}$$

and analogously for $I(X;1|2)$. Conditioning removes the redundant component, but adds the synergistic component so that conditional information is the sum of the unique and synergistic terms.

In this decomposition, a redundancy and a synergy component can exist simultaneously. The work in [33] showed that the measure of co-information [30] that previously had been used to quantify synergy and redundancy, defined as:

$$C(X;1;2) = I(i;j) - I(i;j|k) = I(i;j) + I(i;k) - I(i;j,k) \tag{5}$$

for any assignment of $\{X,1,2\}$ to $\{i,j,k\}$, corresponds to the difference between the redundancy and the synergy terms of Equation (2):

$$C(X;1;2) = I(X;1.2) - I(X;12\backslash 1,2). \tag{6}$$

More generally, [33] defined decompositions of the mutual information about a target X for any multivariate set of variables S. This general formulation relies on the definition of a general measure of redundancy and the construction of a redundancy lattice. In more detail, to decompose the information $I(X;S)$, [33] defined a *source* A as a subset of the variables in S and a collection α as a set of sources. They then introduced a measure of redundancy to quantify for each collection the redundancy between the sources composing the collection, and constructed a redundancy lattice, which reflects the relation between the redundancies of all different collections. Here, we will generically refer to the redundancy of a collection α by $I(X;\alpha)$. Furthermore, following [46], we use a more concise notation than in [33]: for example, instead of writing $\{1\}\{23\}$ for the collection composed by the source containing variable 1 and the source containing variables 2 and 3, we write 1.23, that is, we save the curly brackets that indicate for each source the set of variables and we use instead a dot to separate the sources. We will also refer to the single variables in S as *primary* sources when we want to specifically distinguish them from general sources that can contain several variables.

The work in [34] argued that a measure of redundancy should comply with the following axioms:

- Symmetry: $I(X;\alpha)$ is invariant to the order of the sources in the collection.
- Self-redundancy: The redundancy of a collection formed by a single source is equal to the mutual information of that source.
- Monotonicity: Adding sources to a collection can only decrease the redundancy of the resulting collection, and redundancy is kept constant when adding a superset of any of the existing sources.

The monotonicity property allows introducing a partial ordering between the collections, which is reflected in the redundancy lattice. Self-redundancy links the lattice to the joint mutual information $I(X;S)$ because at its top there is the collection formed by a single source including all the variables in S. Furthermore, the number of collections to be included in the lattice is limited by the fact that adding a superset of any source does not change redundancy. For example, the redundancy between the source 12 and the source 2 is all the information $I(X;2)$. The set of collections that can be included in the lattice is defined as:

$$\mathcal{A}(S) = \{\alpha \in \mathcal{P}(S) - \{\varnothing\} : \forall A_i, A_j \in \alpha, A_i \not\subseteq A_j\}, \tag{7}$$

where $\mathcal{P}(S) - \{\varnothing\}$ is the set of all nonempty subsets of the set of nonempty sources that can be formed from S. This domain reflects the symmetry axiom in that it does not distinguish the order of the sources. For this set of collections, [33] defined a partial ordering relation to construct the lattice:

$$\forall \alpha, \beta \in \mathcal{A}(S), (\alpha \preceq \beta \Leftrightarrow \forall B \in \beta, \exists A \in \alpha, A \subseteq B), \tag{8}$$

that is, for two collections α and β, $\alpha \preceq \beta$ if for each source in β there is a source in α that is a subset of that source. This partial ordering relation is reflexive, transitive, and antisymmetric. In fact, the consistency of the redundancy measures with the partial ordering of the collections, that is that $I(X;\alpha) \leq I(X;\beta)$ if $\alpha \preceq \beta$ represents a stronger condition than the monotonicity axiom. This is because the monotonicity axiom only considers the cases in which α is obtained from β adding more sources (e.g., $\alpha = 1.2.3$ and $\beta = 1.2$), while the partial ordering comprises also the removal of variables from sources (e.g., $\alpha = 1.2$ and $\beta = 1.23$, or $\alpha = 1$ and $\beta = 12.13$).

The mutual information multivariate decomposition was constructed in [33] by implicitly defining partial information measures $\Delta(X;\alpha)$ associated with each node α of the redundancy lattice, such that redundancy measures are obtained from the sum of partial information measures:

$$I(X;\alpha) = \sum_{\beta \in \downarrow\alpha} \Delta(X;\beta), \tag{9}$$

where $\downarrow \alpha$ refers to the set of collections lower than or equal to α in the partial ordering, and hence reachable descending from α in the lattice. The partial information measures are obtained inverting Equation (9) by applying the Möbius inversion to the terms in the lattice [33]. Redundancy lattices for S being bivariate and trivariate are shown in Figure 1. As studied in [46], a mapping exists between the terms of the trivariate and bivariate PIDs, as indicated by the colors and labels.

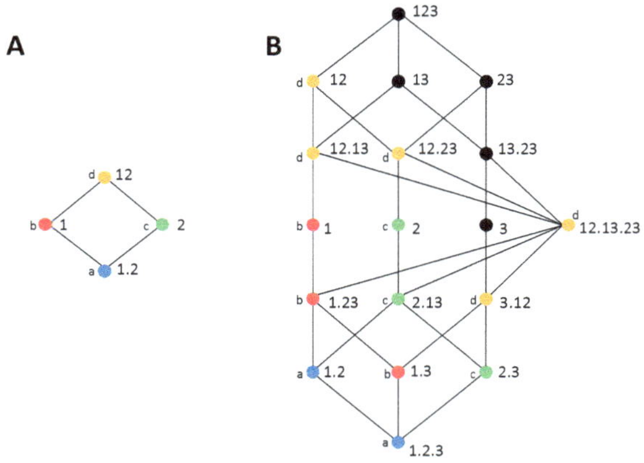

Figure 1. Redundancy lattices of [33]. The lattices reflect the partial ordering defined by Equation (8). (**A**) Bivariate lattice corresponding to the decomposition of $I(X; 12)$. (**B**) Trivariate lattice corresponding to the decomposition of $I(X; 123)$. The color and label of the nodes indicate the mapping of partial information decomposition (PID) terms from the trivariate to the bivariate lattice; in particular, nodes with the same color in the trivariate lattice are accumulated in the corresponding node in the bivariate lattice.

An extra axiom, called the identity axiom, was later introduced by [35] specifically for the bivariate redundancy measure:

- Identity axiom: For two sources A_1 and A_2, $I(A_1 \cup A_2; A_1.A_2)$ is equal to $I(A_1; A_2)$.

The work in [35] pointed out that with the original measure of redundancy of [33] a nonzero redundancy is obtained for two independent variables and a target being a copy of them, while a measure quantifying the amount of qualitatively common information and not the quantitatively equal amount of information should be zero in this case. The work in [38] has specifically differentiated between the identity axiom, which states the form of redundancy for any degree of dependence between the primary sources when the target is a copy of them, and a more specific property, namely the independent identity property:

- Independent identity property: For two sources A_1 and A_2, $I(A_1; A_2) = 0 \Rightarrow I(A_1 \cup A_2; A_1.A_2) = 0$.

This means that the independent identity property is fulfilled when the identity axiom is fulfilled, but the fulfilling of the independent identity property does not necessarily imply fulfilling the identity axiom. Several alternative measures have been proposed that fulfill the identity axiom [35–37]. The properties of the PID terms have been characterized, either based on the axioms and the structure of the redundancy lattice [46,47], or also considering the properties of specific measures [37,41,42,44,48,49]. However, only for specific cases such as multivariate Gaussian systems with univariate targets, it has been shown that several of the proposed measures are actually equivalent [50,51].

3. Stochasticity Axioms for Synergistic Information

In this section, we analyze the consequences of information identity criteria that in the presence of deterministic target-source dependencies identify pieces of information in the target by assuming that their identity is related to specific primary sources. As a first step, we formulate two axioms that impose constraints on synergistic information due to the presence of identity associations between variables within the target and the primary sources. Both axioms assume that, when a subset $X(S_k)$ of the target X can be completely determined by one primary source S_k from the set of primary sources $S = \{S_1, ..., S_n\}$, the identity of the bits of $X(S_k)$ is associated with S_k. This target-source identity association between $X(S_k)$ and S_k then imposes constraints on the synergistic information that S_k can provide about X combined with the other variables in S, because S_k alone already can provide all the information about $X(S_k)$. That is, the amount of synergy is constrained by the degree of stochasticity of the target variables with respect to the sources. The strength of the synergy constraints varies between the two axioms, as we will describe below, and we distinguish them as the weak and strong axiom. We start formulating the constraints that each axiom imposes on synergy conceptually, and subsequently, we will propose concrete constraints for the PID synergistic terms following from these axioms.

Weak axiom on stochasticity requirements for the presence of synergy: *Any primary source S_k that completely determines a subset $X(S_k)$ of variables of the target X does not provide information about $X(S_k)$ synergistically, since S_k alone provides all the information about $X(S_k)$.*

Strong axiom on stochasticity requirements for the presence of synergy: *Any primary source S_k that completely determines a subset $X(S_k)$ of variables of the target X does not provide information about $X(S_k)$ synergistically, since S_k alone provides all the information about $X(S_k)$. Furthermore, S_k can only provide synergistic information about the rest of the target, $X - X(S_k)$, to the extent that there is some remaining uncertainty in both $X - X(S_k)$ and S_k after determining $X(S_k)$.*

Both axioms impose a common conceptual constraint on the presence of synergy, and the strong axiom imposes an extra constraint. The difference between the two axioms, as we will see in Section 4, can be understood in terms of the order in which S_k is used to obtain information about the target. According to the weak axiom, there are no constraints on the synergistic contributions of S_k to the information about $X - X(S_k)$. Oppositely, the logic of the strong axiom is that, because S_k already provides alone the information associated to the entropy $H(X(S_k))$, only if $H(X - X(S_k)|X(S_k)) > 0$ and $H(S_k|X(S_k) > 0)$, then S_k can still provide some extra information about X, and only in this case can this information possibly be synergistic.

The axioms constrain synergy on the basis that an identity is assigned to the bits of information related to the uncertainty $H(X(S_k))$ as corresponding to source S_k. In general, in the presence of dependencies between the variables constituting the target, bits cannot be associated univocally to specific variables within the target. Therefore, the identification of the bits of $H(X(S_k))$ with source S_k does not follow univocally from the joint distribution of the variables. The assignment of an identity to the different pieces of information determines the assessment of whether different sources provide the same information and thus determines the quantification of redundant, unique, and synergistic information. This means that this quantification will also in general depend on the criterion used to assign identity, and will not be reducible to an analysis of the dependencies that are present in the joint distribution.

For simplicity, we will from now on refer to the axioms as the weak or strong stochasticity axioms, or simply the weak or strong axiom. In order to render these axioms operative, we have to formalize their conceptual formulation into sets of constraints imposed on the synergistic PID terms. We now propose the concrete formalization of the axioms. For the weak axiom, we will propose constraints on synergistic PID terms resulting from the existence of functional dependencies of target variables on primary sources (Section 3.1). For the strong axiom, we will also propose constraints resulting from these general functional dependencies, and moreover, we will propose extra constraints specific for the case in which some of the sources themselves are contained in the target (Section 3.2). Finally, we will briefly discuss the motivation to study these axioms in our subsequent analyses, namely as a way to examine how information identity criteria determine the PID terms (Section 3.3).

3.1. Constraints on Synergistic PID Terms That Formalize the Weak Axiom

We propose the following constraints to formalize the weak axiom:

Constraints imposed by functional dependencies of target variables on primary sources: *For a target X and a set of n variables (primary sources) $S = \{S_1, ..., S_n\}$, consider the subsets $X(S_i)$ of X, $i = 1...n$, such that $X(S_i)$ can be determined completely by the single primary source S_i. Define $X' = \bigcup_i X(S_i)$ as the subset of X determined by single primary sources, then:*

$$\Delta(X; \alpha) = \Delta(X - X'; \alpha) \ \ \forall \alpha \notin \bigcup_{i \in S} \downarrow i, \tag{10}$$

where $\downarrow i$ indicates the collections reachable by descending the lattice from node i, corresponding to primary source S_i.

The above means that the synergy about X is equal to the synergy in the lattice associated with the decomposition of the mutual information $I(X - X'; S)$ about a target $X - X'$ that does not include the variables in X' determined by single primary sources alone. This implies that the primary sources cannot have synergistic information about a part X' of the target that is deterministically related to any of them. However, if we define S' as the subset of S comprising any primary source S_k that determines some of the target variables (i.e., S_k having a nonempty $X(S_k)$), the weak axiom does not constrain that the variables in S' may provide information about other parts of the target in a synergistic way. Conversely, the strong axiom imposes that the variables in S' can only provide synergistic information to the extent that they are not themselves deterministically related to the variables in X'.

3.2. Constraints on Synergistic PID Terms that Formalize the Strong Axiom

We propose the following constraints as a formalization of the strong axiom. First, we propose general constraints for any system with functional dependencies of target variables on primary sources:

Constraints imposed by functional dependencies of target variables on primary sources: *For a target X and a set of n variables (primary sources) $S = \{S_1, ..., S_n\}$, consider the subsets $X(S_i)$ of X, $i = 1...n$, such that $X(S_i)$ can be determined completely by the single primary source S_i. Define $X' = \bigcup_i X(S_i)$ as the subset of X determined by single primary sources, then:*

$$\Delta(X; \alpha) = \Delta(X - X'; \alpha | X') \ \ \forall \alpha \notin \bigcup_{i \in S} \downarrow i. \tag{11}$$

That is, the synergy about X is equal to the synergy in the lattice associated with the decomposition of the mutual information $I(X - X'; S | X')$ that S has about $X - X'$ conditioned on X'. Note that the PID of $I(X - X'; S | X')$ is the same as the one of $I(X; S | X')$, and thus, $\Delta(X - X'; \alpha | X') = \Delta(X; \alpha | X')$.

Comparing Equations (10) and (11), we can outline an important difference of the PIDs derived from each axiom. Define X'' as the variables in $X - X'$ that can be determined as a function of X'. Because in Equation (11) the synergistic PID terms are related to the decomposition of $I(X; S | X')$, given the conditioning on X', these terms are invariant to a transformation of the target that removes from it all variables X'', i.e., $\Delta(X; \alpha) = \Delta(X - X''; \alpha)$. Note that $I(X; S | X')$ and also $I(X; S)$ are themselves invariant to this transformation. Oppositely, according to Equation (10), the synergistic PID terms are related to the decomposition of $I(X - X'; S)$, which is not invariant to the removal of X'' from $X - X'$. As we will discuss in detail in Section 4.6, the invariance, or lack thereof, to this transformation plays an important role in the characterization of the notion of redundancy that underpins the PIDs, and in particular determines the sensitivity of the PID terms to the overall composition of the target, comprising semantics aspects beyond the statistical properties of the joint distribution of the target variables.

In general, Equation (11) only expresses the synergistic terms of the PID of $I(X; S)$ in terms of the synergistic terms of the PID of $I(X - X'; S | X')$. However, these latter PID terms are themselves only specified after the definition of particular measures to implement the PID. However, in the more specific case where the primary sources in S' are in fact contained in the target (i.e., $X' = S'$), the logic of the conceptual formulation of the strong axiom leads to more specific constraints on the synergistic terms. In particular, because $X(S_k) = S_k$, then $H(S_k|X(S_k)) = 0$, and the primary sources in S' cannot provide other information about the target than the information about themselves. Since such information is already available without combining the sources in S' with any other source, this implies that the primary sources in S' do not provide any information about X synergistically. Therefore, we propose the following extra constraints for the synergistic PID terms specifically for the case in which $X' = S'$.

Constraints imposed by copies of the primary sources within the target: *For a target X and a set of n variables (primary sources) $S = \{S_1, ..., S_n\}$, consider the subset X' formed by all variables in X, which are a copy of one of the primary sources. Similarly, consider the subset S' formed by all primary sources with a copy within the target, i.e., $X' = S'$. Then:*

$$\Delta(X; \alpha) = 0 \ \forall \alpha \notin \bigcup_{i \in S} \downarrow i \ : \ \exists A \in \alpha, S' \cap A \neq \varnothing. \tag{12}$$

That is, there is no synergy for those nodes whose collection α has a source A containing a variable S_k from S'.

Since we have separately proposed the constraints of Equations (11) and (12) from the conceptual formulation of the strong axiom, they constitute, for the case of $X' = S'$, complementary requirements that should be fulfilled by a PID to be compatible with the strong axiom. However, we will show that for those previously proposed measures that at least for some class of systems comply with Equation (12), [35,36,43], Equation (11) is consistently fulfilled (Appendix D). More generally, the constraints of Equation (12) can be derived from Equation (11) in the case of $X' = S'$ if an extra desirable property is imposed to construct the PIDs (Appendix A). This extra property requires that the PID of $I(X; S|S')$ is equivalent to the PID of $I(X; S - S'|S')$.

Furthermore, although the distinction between a *weak* and a *strong* axiom is motivated by the fact that the strong axiom conceptually imposes an extra requirement for the presence of synergy, this hierarchical relation is not conferred by construction to the concrete constraints imposed to the synergistic PID terms. The PIDs depend on the specific definition of the measures used to construct them, and these measures are expected to comply with one or the other axiom, so that PIDs complying with the axioms cannot be compared on the same measures. However, in agreement with the conceptual formulation, synergistic PID terms are expected to be smaller under the strong axiom because, in $\Delta(X - X'; \alpha|X')$ (Equation (11)), for primary sources S_k with a nonempty $X(S_k)$, the synergy that other primary sources may have with S_k will already be partially accounted by the combination of these other primary sources with $X(S_k)$, which is part of X'. See Appendix A for further details on the effect of conditioning on X' for the specific case of $X' = S'$.

3.3. Using the Stochasticity Axioms to Examine the Role of Information Identity Criteria in the Mutual Information Decomposition

In this work we will study how, based on the constraints imposed to the synergistic PID terms following the two stochasticity axioms, bivariate and trivariate PIDs are affected by deterministic relations between the target and the primary sources. Before focusing on that analysis, we complete the general formulation of the axioms with some considerations about their role in this work and their generality.

Regarding their role, we remark that we do not introduce the stochasticity axioms *per se*, to propose that they should be added to the set of axioms that PID measures should satisfy. Instead, the axioms are introduced to study the implications of identifying different pieces of information based on the target-source identity associations that result from deterministic target-source dependencies. The final objective is to better understand the role of information identity criteria in the quantification of redundancy. As we will see, these axioms are instrumental to characterize how the underlying information identity criterion can lead to negative PID terms. This characterization is also relevant in relation to previous studies because we prove that several previously proposed measures conform to the strong axiom generally [36] or at least for a wider class of systems [35,43] than the one concerned by the identity axiom.

Since our intention is not to formulate these axioms in their most general form, we have only considered their conceptual formulation, and propose concrete constraints to synergistic PID terms following from them, for the case in which there exist functional relations of target variables on single primary sources, that is, $X(S_k)$. The same logic could be applied to formulate the axioms more generally and propose further constraints regarding functional relations of target variables on a subset of S. The PID terms affected by these other functional relations differ from the ones involved in Equations (10)–(12). For example, the existence of a functional relation $X(1,2)$ depending on sources 1 and 2, would constrain the synergy of 12 with other variables, but not the synergy between 1 and 2. We will not pursue this more general formulation of synergy constraints. Conversely, to further simplify the derivations, we will focus on cases where the target X contains some of the primary sources themselves, that is, when the target overlaps with the sources as $X' = S'$. The more general formulation that considers target variables determined as a function of primary sources leads to the same main qualitative conclusions. All the general derivations in the rest of this work follow from the relations characteristic of the redundancy lattice (Equation (9)) and from the constraints to synergistic PID terms proposed following the axioms (Equations (10)–(12)). We do not need to select any specific measure of redundant, unique or synergistic information. For simplicity, from now on we will not distinguish between the conceptual formulation of the axioms and the constraints to synergistic PID terms proposed following from them, and we will refer to them as the weak and strong stochasticity axioms.

4. Bivariate Decompositions with Deterministic Target-Source Dependencies

We start with the bivariate case. Consider that the target X may have some overlap $X \cap 12$ with the sources 1 and 2. Following the weak stochasticity axiom (Equation (10)), synergy is expressed as:

$$I(X; 12\backslash 1, 2) = I(X - 12; 12\backslash 1, 2). \tag{13}$$

On the other hand, the strong stochasticity axiom (Equation (12)) implies that:

$$I(X; 12\backslash 1, 2) = \begin{cases} I(X - 12; 12\backslash 1, 2) & \text{if } X \cap 12 = \varnothing \\ 0 & \text{if } X \cap 12 \neq \varnothing \end{cases}. \tag{14}$$

From these expressions of the synergistic terms, we will now derive how deterministic relations affect the other PID terms.

4.1. General Formulation

For both forms of the stochasticity axiom, we will derive expressions of unique and redundant information in the presence of a target-source overlap. These derivations follow the same procedure: First, given that unique and synergistic information are related to conditional mutual information by Equation (4), the synergy stochasticity axioms determine the form of the unique information terms. Second, once the unique information terms are derived, their relation to the mutual information together with the redundancy term (Equation (1)) allows identifying redundancy. For both unique and redundant information terms this procedure separates the PID term into stochastic and deterministic components. These stochastic and deterministic components quantify contributions associated with the information that the sources provide about the non-overlapping part of the target, $X - 12$, and the overlapping part, $X \cap 12$, respectively. However, how these components are combined depends on the order in which stochastic and deterministic target-source dependencies are partitioned. In particular, using the chain rule [45] of the mutual information, we can separate the information about the target in two different ways:

$$I(X; 12) = I(X - 12; 12) + I(X \cap 12; 12 | X - 12) \tag{15a}$$
$$= I(X \cap 12; 12) + I(X - 12; 12 | X \cap 12). \tag{15b}$$

The first case considers first the stochastic dependencies and after the conditional deterministic dependencies. In the second case, this order is reversed. We will see that for each axiom only one of these partitioning orders leads to expressions that additively separate stochastic and deterministic components for each PID term. Oppositely, the other partitioning order leads to cross-over components across PID terms, in particular to some PID terms being expressed in terms of the stochastic component of another PID term.

4.1.1. PIDs with the Weak Axiom

We start with the PID of $I(X; 12)$ derived from the weak axiom (Equation (13)). Consider the mutual information partitioning order of Equation (15a), which can be re-expressed as:

$$I(X; 12) = I(X - 12; 12) + H(X \cap 12 | X - 12), \tag{16}$$

that is, the second summand corresponds to the conditional entropy of the overlapping target variables given the non-overlapping ones. We now proceed analogously for the PID terms. Since conditional mutual informations are the sum of a unique and a synergistic information component (Equation (4)), we have that:

$$I(X; 1 \backslash 2) = I(X; 1 | 2) - I(X; 12 \backslash 1, 2)$$
$$= I(X - 12; 1 | 2) + I(X \cap 12; 1 | 2, X - 12) - I(X - 12; 12 \backslash 1, 2). \tag{17}$$

The first equality indicates that unique information is conditional information minus synergy. The second equality uses the chain rule to separate the conditional mutual information stochastic and deterministic components, and applies the stochasticity axiom to remove the overlapping part of the target in the synergy term. Using again the relation between conditional mutual information and unique and synergistic terms (Equation (4)), but now, for the target $X - 12$ we get:

$$I(X; 1 \backslash 2) = I(X - 12; 1 \backslash 2) + H(X \cap 1 | 2, X - 12), \tag{18}$$

where we also used that $I(X \cap 12; 1 | 2, X - 12)$ equals the entropy $H(X \cap 1 | 2, X - 12)$. Accordingly, the unique information of 1 can be separated into a stochastic component, the unique information about target $X - 12$, and a deterministic component, the entropy $H(X \cap 1 | 2, X - 12)$. This last term is zero if the target does not contain source 1. If it does, it quantifies the entropy that only 1 as a source can explain about itself as part of the target, which is thus an extra contribution to the unique information.

Once we have identified the stochastic and deterministic components of the unique information we can use Equation (1) to characterize the redundancy. Combining Equations (1) and (18), we obtain:

$$I(X; 1.2) = I(X - 12; 1.2) + \begin{cases} 0 \text{ if } X \cap 12 = \emptyset \\ I(1; 2 | X - 12) \text{ if } X \cap 12 \neq \emptyset \end{cases}. \tag{19}$$

Therefore, it suffices that one of the two primary sources overlaps with the target so that their conditional mutual information given the non-overlapping target variables contributes to redundancy. Note that when $X \cap 12 = \emptyset$ then $X = X - 12$; hence, the axiom has no effect on the redundancy.

Following the same procedure, it is possible to derive expressions for the unique and redundant information terms, but applying the other mutual information partitioning order of Equation (15b). The resulting terms can be compared in Table 1 and are derived in more detail in Appendix B, where we also show the consistency between the expressions obtained with each partitioning order. We present in the upper part of the table the decompositions into stochastic and deterministic contributions for each PID term and for the two partitioning orders. To simplify the expressions, their form is shown only for the case of $X \cap i \neq \emptyset$. With the alternative partitioning order, both the expressions of unique information and redundancy contain a cross-over component, namely the synergy about $X - 12$, instead of being expressed in terms of the unique information and redundancy of $X - 12$, respectively. Furthermore, the separation of the deterministic and stochastic components is not additive. This indicates that, while the chain rule holds for the mutual information, it is not guaranteed that the same type of separation holds separately for each PID term. Only for a certain partitioning order, when stochastic dependencies are considered first, unique and redundant information terms derived from the weak axiom can both be separated additively into a stochastic and a deterministic component without cross-over terms. We individuate in the lower part of the table the deterministic PID components obtained from the partitioning order for which each PID term is separated additively into a stochastic and deterministic component.

Table 1. Decompositions of synergistic, unique, and redundant information terms into stochastic and deterministic contributions obtained assuming the weak stochasticity axiom. For each term we show the decompositions resulting from two alternative mutual information partitioning orders (Equation (15)), which are consistent with each other (see Appendix B). For the partitioning order leading to an additive separation of each partial information decomposition (PID) term into a stochastic and deterministic component we also individuate the deterministic contributions $\Delta_d(X; \beta)$. Synergy has only a stochastic component, according to the axiom (Equation (13)). Expressions of unique information come from Equations (18) and (A3), and the ones of redundancy from Equations (19) and (A5). The expressions have been simplified with respect to the equations, indicating their form for the case $X \cap i \neq \emptyset$. The terms $\Delta_d(X; \beta)$ have analogous expressions for $X \cap j \neq \emptyset$ when a symmetry exists between i and j and are zero otherwise.

Term	Decomposition
$I(X; ij \backslash i, j)$	$I(X - ij; ij \backslash i, j)$
$I(X; i \backslash j)$	$I(X - ij; i \backslash j) + H(i\|j, X - ij)$ $H(i\|j) - I(X - ij; ij \backslash i, j)$
$I(X; i.j)$	$I(X - ij; i.j) + I(i; j\|X - ij)$ $I(i; j) + I(X - ij; ij \backslash i, j)$

Term	Measure
$\Delta_d(X; ij)$	0
$\Delta_d(X; i)$	$H(i\|j, X - ij)$
$\Delta_d(X; i.j)$	$I(i; j\|X - ij)$

4.1.2. PIDs with the Strong Axiom

The procedure to derive the unique and redundant PID terms is the same if the strong stochasticity axiom is assumed, but determining synergy with Equation (14) instead of Equation (13). To simplify the expressions we indicate in advance that if $X \cap 12 = \emptyset$ each PID term with target X is by definition equal to the one with target $X - 12$ and we only provide expressions derived with some target-source overlap. In contrast to the weak axiom, with the strong axiom an additive separation of stochastic and deterministic components is obtained with the partitioning order of Equation (15b). See Appendix B for details about the other partitioning order. For the unique information the strong axiom implies that:

$$I(X; 1 \backslash 2) = \begin{cases} I(X - 12; 1 \backslash 2) + I(X - 12; 12 \backslash 1, 2) & \text{if } X \cap 1 = \emptyset \\ H(1\|2) & \text{if } X \cap 1 \neq \emptyset \end{cases}, \tag{20}$$

and for the redundancy:

$$I(X; 1.2) = I(1; 2). \tag{21}$$

As before, we summarize the PIDs in Table 2. Comparing Tables 1 and 2, we see that the expressions obtained with the weak and strong axiom differ because of a cross-over contribution, corresponding to the synergy about $X - 12$, which is transferred from redundancy to unique information. This is due to the synergy constraints imposed by each axiom: the strong axiom imposes that there is no synergy, and hence this part of the information has to be transferred to the unique information because the sum of synergy and unique information is constrained to equal the conditional mutual information. As a consequence, redundancy is reduced by an equivalent amount to comply with the constraints that relate unique informations and redundancy to mutual informations (Equation (1)). Furthermore, like for the weak axiom, the chain rule property does not generally hold for each PID term separately. This has been previously proven for specific measures. The work in [42] provided a counterexample for the original redundancy measure of [33] (I_{min}) and for the one of [35] (I_{red}). The work in [44] provided counterexamples for the redundancy and synergy measures of the decomposition based on maximum conditional entropy [36]. Our results prove this for any measure conforming to the stochasticity axioms. In particular, they show that the PID terms are consistent with the mutual information decompositions obtained applying the chain rule, but that, depending on the partitioning order and on the version of the axiom assumed, information contributions are redistributed between different PID terms, and between their stochastic and deterministic components.

Table 2. Decompositions of synergistic, unique, and redundant information terms into stochastic and deterministic contributions obtained assuming the strong stochasticity axiom. The table is analogous to Table 1. Synergy is null according to the axiom (Equation (14)). Expressions of unique information come from Equations (A8) and (20), and the ones of redundancy from Equations (A9) and (21). Again, expressions are shown for the case $X \cap i \neq \emptyset$, with the corresponding symmetries holding for $X \cap j \neq \emptyset$ and with terms $\Delta_d(X; \beta)$ equal to zero otherwise.

Term	Decomposition
$I(X; ij \backslash i, j)$	0
$I(X; i \backslash j)$	$I(X - ij; i \backslash j) + I(X - ij; ij \backslash i, j) + H(i|j, X - ij)$
	$H(i|j)$
$I(X; i.j)$	$I(i; j|X - ij) + I(X - ij; i.j) - I(X - ij; ij \backslash i, j)$
	$I(i; j)$

Term	Measure
$\Delta_d(X; ij)$	0
$\Delta_d(X; i)$	$H(i|j)$
$\Delta_d(X; i.j)$	$I(i; j)$

4.2. The Relation between the Stochasticity Axioms and the Identity Axiom

In the previous section, we derived how the two stochasticity axioms imply different expressions for the redundancy term. We now examine how these expressions are related to the redundancy term stated by the identity axiom [35]. It is straightforward to show that the identity axiom is subsumed by both stochasticity axioms:

Proposition 1. *The fulfillment of the synergy weak or strong stochasticity axioms implies the fulfillment of the identity axiom.*

Proof. If $X = 12$ then $X \cap 12 = 12$ and $X - 12 = \emptyset$. If the weak stochasticity axiom holds, redundancy (Equation (19)) reduces to $I(12; 1.2) = I(1; 2)$. If the strong stochasticity axiom holds, Equation (21) is already $I(12; 1.2) = I(1; 2)$. □

Therefore, the stochasticity axioms represent two alternative extensions of the identity axiom: First, they do not only consider a target that is a copy of the primary sources, but a target with any degree of overlap or functional dependence with the primary sources. Second, they are not restricted to the bivariate case but are formulated for any number of primary sources. This means that their fulfillment imposes stricter conditions to the redundancy measures. Redundancy terms derived from each axiom coincide for the particular case that is addressed by the identity axiom, but more generally they differ. We will further discuss these differences below based on concrete examples.

4.3. How Different PID Measures Comply with the Stochasticity Axioms

We now investigate whether several proposed measures conform to the predictions of the stochasticity axioms. We examine the original redundancy measures of [33] (I_{min}), the one based on the pointwise common change in surprisal of [38] (I_{ccs}), the one based on maximum conditional entropy of [36] (SI), the one based on projected information of [35] (I_{red}), and the one based on dependency constraints of [43] (I_{dep}).

It is well-known that the redundancy measure I_{min} does not comply with the identity axiom [35]. Even if $I(1;2) = 0$, a redundancy $I_{min}(12;1.2) > 0$ can be obtained. Nor does I_{ccs} comply with the identity axiom. Since the fulfillment of the stochasticity axioms implies the fulfillment of the identity axiom, none of these measures complies with the stochasticity axioms.

On the other hand, SI, I_{dep}, and I_{red} fulfill the identity axiom. We will show that SI always conforms to the strong stochasticity axiom. For I_{dep}, we will show that it complies with the strong axiom at least when some primary source is part of the target, i.e., $X \cap 12 \neq \emptyset$. We will also show that I_{red} complies with the strong axiom at least for the case of $X \cap 12 = 12$. This latter case is particularly relevant to examine nonnegativity (Section 5.3).

We proceed as follows: For SI, we now prove that it complies with Equation (12) specifically for the case in which some primary sources are also part of the target, which is the case considered throughout this work. The longer complete proof of compliance with Equation (11) for systems comprising any type of functional relation between parts of the target and single primary sources is left for Appendix C. For I_{dep}, we also provide here the proof of compliance with Equation (12) for the case of primary sources being part of the target. Again because of length reasons, the proof for I_{red} is left for Appendix C. In all cases, we also prove in Appendix D that, for those cases herein studied in which these measures comply with Equation (12), Equation (11) is consistently fulfilled. We start with SI:

Proposition 2. *The PID associated with the redundancy measure SI [36] conforms to the synergy strong stochasticity axiom when some primary source is part of the target.*

Proof. Consider a target X and two sources 1 and 2. The redundancy measure SI is defined as:

$$SI(X;1,2) = \max_{Q \in \Delta(p)} C_Q(X;1;2), \tag{22}$$

where the co-information $C_Q(X;1;2)$ is maximized within the family of distributions $\Delta(p)$ that preserves the marginals $p(X,1)$ and $p(X,2)$. We will now show that $SI(X;1,2)$ conforms to Equation (21) when $X \cap 12 \neq \emptyset$. It is a general property following from the definition of the co-information (Equation (5)) that, if either 1 or 2 are in X, that is, $\exists i \in \{1,2\} : X \cap i = i$, then $C_Q(X;1;2) = I_Q(1;2)$. Further, because $\Delta(p)$ preserves $p(X,1)$ and $p(X,2)$, it suffices that $X \cap 12 \neq \emptyset$ so that $p(X,1,2)$ is preserved. This means that $p(1,2)$ is preserved and $I_Q(1;2) = I(1;2)$ for all $Q \in \Delta(p)$. This leads to $SI(X;1,2) = I(1;2)$. Given that for any valid bivariate PID one of the four PID terms already determines the other PID terms, because they have to comply with Equations (1) and (4), this shows that the PID equals the one derived from the strong axiom. $\square$

We now continue with the proof for I_{dep}:

Proposition 3. *The PID associated with the redundancy measure I_{dep} [43] conforms to the synergy strong stochasticity axiom when some primary source is part of the target.*

Proof. The work in [43] defined unique information based on the construction of a dependency constraints lattice in which constraints to maximum entropy distributions are hierarchically added. The unique information $I(X; 1\backslash 2)$ is defined as the least increase in the information $I_Q(X; 12)$ when adding the constraint of preserving the distribution $p(X, 1)$ to the list of constraints imposed to the maximum entropy distribution Q. This results in the following expression for $I(X; 1\backslash 2)$, according to Appendix B of [43]:

$$I(X; 1\backslash 2) = \min\{I(X; 1), I_{X1,X2}(X; 1|2), I_{X1,X2,12}(X; 1|2)\}, \tag{23}$$

where $I_{X1,X2}(X; 1|2)$ indicates the conditional mutual information for the maximum entropy distribution preserving $p(X, 1)$ and $p(X, 2)$, and analogously for $I_{X1,X2,12}(X; 1|2)$. Now, consider that some source is part of the target, in particular, without loss of generality, that $X \cap 1 = 1$. In this case $I(X, 1) = H(1)$, and preserving $p(X, 2)$ and $p(X, 1)$ implies preserving the joint distribution $p(X, 1, 2)$, given that 1 is part of X. This means that $I_{X1,X2}(X; 1|2) = I_{X1,X2,12}(X; 1|2) = I(X; 1|2)$. Furthermore, $I(X; 1|2) = H(1|2)$. Since $H(1|2) \le H(1)$ the unique information is $I(X; 1\backslash 2) = H(1|2)$, which already determines the PID, and in particular leads to the redundancy being $I(X; 1.2) = I(1; 2)$. $\square$

4.4. Illustrative Systems

So far, we have derived the predictions for the PIDs according to each version of the stochasticity axiom, pointed out the relation with the identity axiom, and checked how different previously proposed measures conform to these predictions. We now analyze concrete examples to further examine the implications of our axioms on the PIDs. In particular, we reconsider two examples that have been previously studied in [42,44], namely the decompositions of the mutual information about a target jointly formed by the inputs and the output of a logical XOR operation or of an AND operation (see Figures 2A and 3A, respectively). We first describe below the decompositions obtained, then in Section 4.5, we will discuss these decompositions in the light of the underlying assumptions on how to assign an identity to different pieces of information of the target. The deterministic components for these examples are derived without assuming any specific measure of redundancy, unique, or synergistic information. The stochastic components have already been previously studied and some of the terms depend on the measures selected to compute the PID terms. We will indicate previous work examining these terms when required.

4.4.1. XOR

We first examine the XOR system. Consider an output variable 3 determined through the operation $3 = 1$ XOR 2, resulting in the joint probability displayed in Figure 2A. We also indicate the values of the information-theoretic measures needed to calculate the PID bivariate decompositions studied here and that will also serve for the trivariate decompositions addressed in Section 5.2. We want to examine the decomposition of $I(123; 1, 2)$, where the target is composed by the three variables. For each version of the stochasticity axiom we will focus on the mutual information partitioning order that allows separating additively a stochastic and a deterministic component of each PID term.

Since $X - 12 = 3$, for the weak axiom the PID (Figure 2B) can be obtained by implementing the decomposition of $I(3; 12)$ and separately calculating the deterministic PID components $\Delta_d(123; \beta)$ as collected in Table 1. As indicated in [37], the decomposition of $I(3; 12)$ for the XOR operation can be derived without adopting any particular redundancy measure, just using Equations (1) and (4) and the axioms of [34] described in Section 2. Furthermore, the same value is obtained with I_{ccs}, which does not fulfill the nonnegativity axiom. There is no stochastic component of redundancy or unique information because $I(3; i) = 0$ for $i = 1, 2$, and synergy contributes one bit of information. Regarding the deterministic components, redundancy has 1 bit because $I(1;2|3) = 1$. The deterministic unique information components are zero because $H(i|jk) = 0$ for $i = 1, 2$, and according to the axiom, there is no deterministic component of synergy.

In the case of the strong axiom (Figure 2C), since both primary sources overlap with the target, only deterministic components are larger than zero in the decomposition when selecting the partitioning order that additively separates stochastic and deterministic contributions, as indicated in Table 2. By assumption, there is no synergy. Since $I(1;2) = 0$, the redundancy is also zero and all the information is contained in the unique information terms. As pointed out for the generic expressions, the two decompositions differ in the transfer of the stochastic component of synergy to unique information, which in turns forces an equivalent transfer from redundancy to unique information.

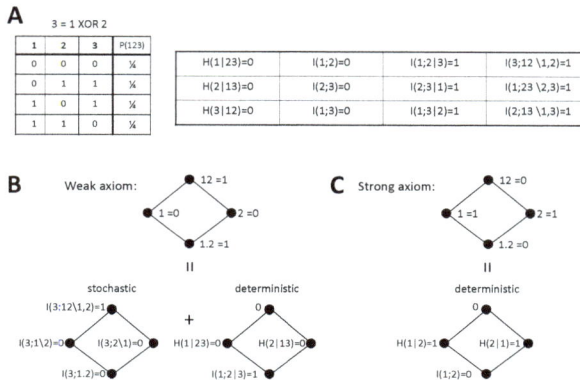

Figure 2. Bivariate decomposition of $I(123; 12)$ for the XOR system. (**A**) Joint distribution of the inputs 1 and 2 and the output 3 for the XOR operation. We also collect the value of the information-theoretic quantities used to calculate this bivariate decomposition and the trivariate decomposition $I(123; 123)$ in Section 5.2. (**B**) Bivariate decomposition derived from the weak stochasticity axiom. Stochastic and deterministic components are separated in agreement with Table 1. (**C**) Bivariate decomposition derived from the strong axiom. Only deterministic components are present, following Table 2.

4.4.2. AND

As a second example, we now consider the AND system. Following the weak axiom, again the decomposition can be obtained by implementing the PID of $I(3; 12)$ and separately calculating the deterministic PID components from Table 1, using the joint distribution of inputs and output displayed in Figure 3A. The PID of $I(3; 12)$ for the AND operation has also been already characterized and coincides for I_{min} [33], I_{red} [35], and SI [36]. However, in contrast to the XOR case, this decomposition depends on the redundancy measure used and for example differs for I_{ccs}. Each PID term contributes half a bit to $I(123; 12)$. Unique contributions come exclusively from the deterministic components. Each unique information amounts to half a bit because the output and one input determine the other input only when not both have a value of 0. Redundancy is also 0.5 bit, but it comes in part from a stochastic component and in part from a deterministic one. The stochastic component appears intrinsically because of the AND mechanism, even if the inputs are independent. This type of redundancy has been called mechanistic redundancy [35]. The deterministic component appears because, although the inputs are independent, conditioned on the output $I(1;2|3) > 0$. The synergy $I(3; 12\backslash 1, 2) = 0.5$ was also previously determined [33,35,36]. This PID differs from the one obtained with the weak axiom for the XOR example. Conversely, with the strong axiom the decomposition is the same as for the XOR example, because it is completely determined by $I(1;2) = 0$. This latter decomposition is again in agreement with the arguments of [42,44] based on the identity axiom.

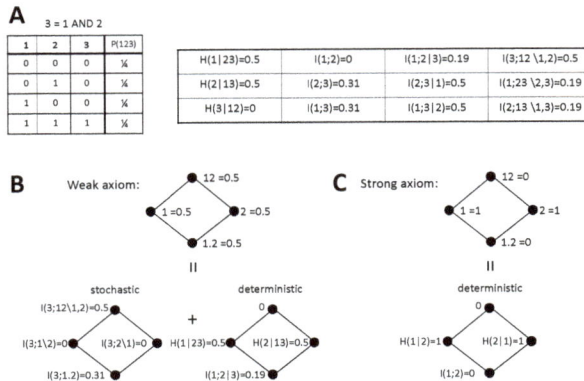

Figure 3. Bivariate decomposition of $I(123;12)$ for the AND system. The structure of the figure is analogous to Figure 2. (**A**) Joint distribution of the inputs 1 and 2 and the output 3 for the AND operation. (**B**) Bivariate decomposition derived from the weak stochasticity axiom. (**C**) Bivariate decomposition derived from the strong axiom.

4.5. Implications of Target-Source Identity Associations for the Quantification of Redundant, Unique, and Synergistic Information

Each version of the stochasticity axiom implies a different quantification of redundancy. We now examine in more detail how these different quantifications are related to the notion of redundancy as common information about the target that can be obtained by observing either source alone. The key point is how identity is assigned to different pieces of information in order to assess which information about the target carried by the sources is qualitatively common to the sources. In particular, the logic of the strong axiom is that if a source is part of the target it cannot provide other information about the target than the information about itself. As a consequence, if the other source does not contain information about the former source, this information is unique. This logic rests on the assumption that when there is a copy of a primary source in the target we can identify and separate the bits of information about that copy from the information about the rest of the target. The idea of assigning an identity to bits of information in the target by associating them with specific variables also motivated the introduction of the identity axiom. Although this axiom was formulated for sources with any degree of dependence, its motivation [35] was mainly based on the case of independent sources, that is, the particular case considered by the independent identity property. For that case, we can identify the bits of information associated with variable 1 and the ones with variable 2, and thus redundancy, that should quantify the qualitatively equal information that is shared among the sources and not only common amounts of information, has to be null.

However, assigning an identity to pieces of information in the target is in general less straightforward. For example, in the XOR system, with target 123 and sources 1 and 2, we have two target-source identity associations, namely between each source and its copy in the target. However, the two bits of 123 cannot be identified as belonging to a certain variable, because of the conditional dependencies between the variables. The only information identity criterion that seems appropriate in this case to identify the two bits is the following: the bit that any first variable provides alone, and the bit that a second variable provides combined with the first. This lack of correspondence between pieces of information and individual variables is incompatible with the identification of the pieces of information based on the target-source identity associations that are formalized by the stochasticity axioms. To show this, we now consider different combinations of mutual information partitioning orders for $I(123;1)$ and $I(123;2)$ and show how, if the assignment of identity to the bits in the target 123 is based on target-source identity associations, the interpretation of redundant and unique information is ambiguous. First, consider that we decompose the information of each primary source as follows:

$$I(123;1) = I(1;1) + I(2;1|1) + I(3;1|12) = I(1;1) = H(1)$$
$$I(123;2) = I(2;2) + I(1;2|2) + I(3;2|12) = I(2;2) = H(2). \tag{24}$$

If we assume that we can identify the bit of information carried by each primary source about the target using the target-source identity associations, these decompositions would suggest that there is no redundant information. This is because each source only carries one bit of information about its associated copy within the target and $I(1;2) = 0$ for the XOR system. However, keeping the same decomposition of $I(123;1)$, we can consider alternative decompositions of $I(123;2)$:

$$I(123;2) = I(3;2) + I(1;2|3) + I(2;2|13) = I(1;2|3) = H(1) \tag{25a}$$
$$= I(1;2) + I(3;2|1) + I(2;2|13) = I(3;2|1) = H(3). \tag{25b}$$

The redundancy and unique information terms should not depend on how we apply the chain rule to $I(123;2)$. However, in contrast to Equation (24), the first decomposition of Equation (25a) suggests, based on the target-source identity associations, that there is redundancy between sources 1 and 2. In particular, $I(123;2) = I(1;2|3)$ in Equation (25a) can be interpreted as information that source 2 provides about the copy of source 1 within the target, thus redundant with the information $I(123;1) = I(1;1)$ in Equation (24) that source 1 has about its copy. The second decomposition in Equation (25b) further challenges the interpretation of redundancy and unique information based on the assignment of an identity to bits of information in the target given their association with the overlapping target variables. Given $I(123;2) = I(3;2|1)$, source 2 provides information about 3. However, the bit of 3 is shared with the copies of 1 and 2 within the target, given the conditional dependencies of the XOR system. Moreover, the information in $I(3;2|1)$ is information that source 2 provides about 3 after conditioning on the copy of source 1 within the target, so that the target-source identity association of 1 suggests that both sources are combined to retrieve this information. Note that for both $I(1;2|3)$ and $I(3;2|1)$ in Equation (25), we expressed the information in terms of the entropy of the target variable, 1 and 3, respectively, because it is the identity of the pieces of information within the target what determines their assignment to a certain PID term.

In summary, when using the target-source identity associations to identify pieces of information, different partitioning orders of the mutual information ambiguously suggest that the same information can be obtained uniquely, redundantly, or even in a synergistic way. These problems arise because, in contrast to the case of $I(12;1,2)$ with independent sources, in the XOR system the two bits of 123 cannot be identified as belonging to a certain variable, and thus the target-source identity associations between the variables cannot identify the bits unambiguously.

The differences in the quantification of redundancy with each stochasticity axiom are related to the alternative interpretations of identity discussed for Equations (24) and (25). A notion of redundancy compatible with the weak axiom considers the common information about the target that can be obtained by observing either source alone or conditioned on variables in the target, which means that redundancy depends on the overall composition of the target. Indeed, the deterministic component of redundancy comprises the conditional dependence of the sources given the rest of the target, $I(1;2|X-12)$, when there is a target-source overlap, and thus fits to Equation (25a), where the term $I(1;2|3)$ appears. Conversely, with the strong axiom, when there is a target-source overlap, redundancy equals $I(1;2)$ independently of $X-12$, in agreement with Equation (24). We will now further discuss the implications of this independence or dependence of redundancy on the overall composition of the target.

4.6. The Notion of Redundancy and the Identity of Target Variables

We showed above that enforcing the identification of the bits of 123 based on target-source identity associations between the variables leads to ambiguous interpretations of whether this information is retrieved redundantly, uniquely, or synergistically, depending on the partitioning order used to decompose the target. That is, the ambiguity arises because we consider 1, 2, and 3 as three separate variables within the target, which furthermore can be observed sequentially in any order, and not only simultaneously. The possibility to separately observe these variables is not relevant to quantify their entropy $H(123)$ or the mutual information $I(123;1,2)$, but, as we will argue below, it is potentially relevant to determine the PID terms.

In particular, for both the XOR and AND systems, 3 is completely determined by 12, so that $H(123) = H(12)$ and $I(123;1,2) = I(12;1,2)$. That is, the entropy and mutual information do not depend on whether we consider 3 as a separate variable or it is removed from the target. We can then ask how the assignment of the two bits to the PID terms depends on reducing the target 123 to 12. We repeat the comparison of different partitioning orders of $I(123;1)$ and $I(123;2)$ of Section 4.5 but now after this reduction. For $I(12;1)$ the only possible partitioning orders are:

$$I(12;1) = I(1;1) + I(2;1|1) = I(1;1) = H(1)$$
$$= I(2;1) + I(1;1|2) = I(1;1) = H(1), \tag{26}$$

and analogously for $I(12,2)$. Since $I(1;2) = 0$, in all cases each source retrieves information about its associated copy in the target, and thus all information is contained in the unique information terms. Therefore, with the reduction of 123 to 12, the decompositions obtained are consistent with the ones derived from the strong axiom, which effectively also reduces 123 to 12 since the decomposition is independent of $X - 12$ when 12 is part of the target. Indeed, [44] derived for the AND system the same decomposition as with the strong axiom using the measure SI and the reduction of 123 to 12.

The consistency between the strong axiom and the reduction of 123 to 12 is also reflected in the equality between the PIDs derived from the strong axiom for the XOR and AND systems. This is because the distributions of the targets 123 of these systems are isomorphic, i.e., one can be mapped to the other by relabeling the states, and are indistinguishable after the reduction to 12. However, the decompositions of the XOR and AND systems differ when derived with the weak axiom, and are not consistent with the reduction of 123 to 12. This is because with the weak axiom the PID terms have components in which 12 is explicitly separated from 3, in particular the redundancy contains the terms $I(1;2|3)$ and $I(3;1.2)$ (Table 1).

Therefore, an important difference between the redundancy measures derived from the two stochasticity axioms regards their invariance to transformations of the target consisting on the removal of the variables within it that are completely determined by copies of the primary sources contained in the target. We will in general call this type of invariance as TSC (target to sources copy) reduction invariance. Because the removal of these variables does not alter the entropy of the target, the mutual information is TSC reduction invariant. The lack of TSC reduction invariance implies that the redundancy depends on semantic aspects of the joint probability distribution of the target, related to the identity of the variables. The reason why the redundancy measure following from the weak axiom depends on these semantic aspects, while the measure following from the strong axiom does not, can be understood from how the identification of pieces of information based on the target-source variables associations is later used to constrain synergy in each case. With the weak axiom, the bits of $H(X(S_k))$ identified with the primary source S_k due to the presence of $X(S_k)$ within the target, are constrained to be non-synergistic in nature when the primary sources provide information about $X(S_k)$. Oppositely, the weak axiom imposes no restriction on synergy about $X - X(S_k)$. However, if there is some dependence between the variables $X(S_k)$ and $X - X(S_k)$ (i.e., $I(X(S_k); X - X(S_k)) > 0$), part of the bits of $X(S_k)$ are shared by the variables $X - X(S_k)$. This means that the same bits that are constrained to be non-synergistic in nature when the primary sources provide information about $X(S_k)$ are still allowed to be synergistic when the primary sources provide information about $X - X(S_k)$. Therefore, it is the identity of the variables about which the primary sources provide information what determines whether the same bits are subjected to the synergy constraints or not. The dependence of synergistic terms on the semantic aspects of the probability distribution determines that also redundancy terms inherit this dependence, because of their relation as terms of the PID. Oppositely, in the case of the strong axiom, once the identity of the bits of $H(X(S_k))$ is associated with the primary source S_k due to the presence of $X(S_k)$ within the target, the fact that due to $I(X(S_k); X - X(S_k)) > 0$ these bits can also be associated with $X - X(S_k)$ is not considered, and they are constrained to be non-synergistic in nature without any consideration of the identity of the target variables about which the primary sources provide information.

The fact that redundancy is invariant or not to the TSC reduction has implications to determine other properties of the PIDs. In particular, it plays a crucial role in the counterexamples provided by [42,44] to prove that nonnegativity and left monotonicity are not compatible with the independent identity property. We will address in detail the counterexample of nonnegativity after studying trivariate PIDs with the stochasticity axioms, in Section 5. With regard to left monotonicity, it is useful to remind that [44] assumed the invariance of SI when reducing 123 to 12 to prove that left monotonicity is violated in the decomposition of $I(123; 1, 2)$ of the AND system, because $SI(3; 1.2) > SI(123; 1.2)$. As can be seen in Figure 3, although we have that $I(3; 1.2) > I(123; 1.2)$ with the strong axiom, the opposite holds with the weak axiom, for which the invariance under reduction of 123 to 12 does not hold.

More generally, the TSC reduction is just one type of isomorphism of the target to which entropy and mutual information are always invariant. The comparison of the decompositions obtained with the two stochasticity axioms raises the question of whether we should expect the PIDs to be invariant to isomorphisms of the target, as the entropy and mutual information are. This question is intrinsically related to the role assigned to information identity in the notion of redundancy. Two aspects of this notion would justify a lack of invariance. First, the assessment of redundancy implies assigning an identity to pieces of information, and this identity can change depending on the variables included in the target. For example, for the target 123 in the XOR and AND systems, if 1, 2, and 3 are taken as variables that can be observed separately and sequentially, the bits of 123 cannot be identified as belonging to a certain variable, because of the conditional dependence $I(1; 2|3)$. However, after the reduction of 123 to 12, the two bits can be associated each to a single variable of the target because $I(1; 2) = 0$. Second, mechanistic redundancy can only be assessed when explicitly considering the mechanism of the input-output deterministic relation generating 3 from 12. This mechanism is not preserved under isomorphic transformations, and the information about it is lost when reducing 123 to 12 for the XOR and AND systems. These two arguments highlight the role of information identity to quantify redundancy, and indicate that requiring or not that the redundancy measures should be invariant to target isomorphisms implies further specifications of which is the underlying notion of redundancy that is quantified.

5. Trivariate Decompositions with Deterministic Target-Source Dependencies

We now extend the analysis to the trivariate case. This is particularly relevant because, in contrast to the bivariate case, it has been proven that, in the multivariate case, the PIDs that jointly comply with the monotonicity and the identity axioms do not guarantee the nonnegativity of the PID terms [42]. In particular, [42] used the XOR example we reconsidered above as a counterexample to show that PID terms can be negative. The work in [41] reexamined this counterexample indicating that the independent identity property, which is a weaker condition than the identity axiom, already implies the existence of negative terms. Therefore, we would like to be able to extend the general formulation of Section 4.1 to the trivariate case, and thus apply it to further examine the XOR and AND examples by identifying each component of the trivariate decomposition of $I(123;123)$ and not only of the decomposition of $I(123;12)$.

5.1. General Formulation

While in the bivariate lattice there is a single PID term that involves synergistic information, in the trivariate lattice of Figure 1B, all nodes that are not reached descending from 1, 2, or 3 imply by definition synergistic information, and the nodes of the form $i.jk$ too. This is because these nodes correspond to collections containing sources composed by several primary sources, and hence quantify information only obtained by combining primary sources. The weak and strong axioms impose constraints on these terms given Equations (10) and (11), respectively.

5.1.1. PIDs with the Weak Axiom

We begin with the weak stochasticity axiom, for a target X and three primary sources 1, 2, and 3. Expressing the general constraints of the weak axiom (Equation (10)) particularly for the trivariate case, and separating stochastic and deterministic components of the PID terms as in Table 1, that is, as $\Delta(X;\alpha) = \Delta(X - X';\alpha) + \Delta_d(X;\alpha)$, Equation (10) can be expressed as:

$$\Delta_d(X;\alpha) = 0 \ \forall \alpha \notin \bigcup_{i=1,2,3} \downarrow i. \tag{27}$$

To characterize the remaining deterministic contributions to PID terms, analogously to the bivariate case, we apply the mutual information chain rule to separate stochastic and deterministic dependencies. Again we focus on the partitioning order that considers first the stochastic dependencies, since only this order leads to an additive separation of stochastic and deterministic components for each PID term. With this partitioning order, we obtain:

$$I(X;123) = I(X - 123;123) + I(X \cap 123;123|X - 123)$$
$$= I(X - 123;123) + H(X \cap 123|X - 123). \tag{28}$$

Following derivations analogous to the ones of Section 4.1 (see Appendix E), if a certain primary source i does not overlap with the target, the nodes that can only be reached descending from its corresponding node will not have a deterministic component. Accordingly, deterministic contributions are further restricted by:

$$\Delta_d(X;\alpha) = 0 \ \forall \alpha \notin \bigcup_{i \in X \cap \{1,2,3\}} \downarrow i. \tag{29}$$

This can be understood examining the term $H(X \cap 123|X - 123)$ in Equation (28). For example, suppose that the target includes 1 and 2 but not 3. Then the entropy in Equation (28) is $H(12|X - 123)$, corresponding to $I(12;123|X - 123)$. The PID terms that can be reached descending from 3 and not from 1 or 2 are $\Delta(X;3)$ and $\Delta(X;3.12)$ (see Figure 1B). The first quantifies information that can only be obtained from 3, and not from 12. The second is information that can be obtained from 3 or from 12, but not from 1 or 2 alone. However, the information $I(12;123|X - 123)$ can be obtained from either 1 or 2 alone, so there is no information exclusive of 12. This means that $\Delta(X;3)$ and $\Delta(X;3.12)$ do not contribute to the decomposition of $H(12|X - 123)$.

Using the condition of Equation (29), we can use the same procedure as in Section 4.1 to derive the expressions of all the deterministic PID trivariate components. These terms are collected in Table 3 and we leave the detailed derivations and discussion for Appendix E. Their expressions are indicated for the case in which variable i is part of the target and are symmetric with respect to j or k when this symmetry is characteristic of a certain PID term, or vanish otherwise, consistently with Equation (29).

The first two terms $\Delta_d(X;i)$ and $\Delta_d(X;i.jk)$ are nonnegative, the former because it is an entropy and the latter because according to the axiom adding a new source can only reduce synergy. However, for the terms $\Delta_d(X;i.j)$ and $\Delta_d(X;i.j.k)$ it is not guaranteed that they are nonnegative.

For $\Delta_d(X;i.j)$, we will see examples of negative values below. For $\Delta_d(X;i.j.k)$, the conditional co-information can be negative if there is synergy between the primary sources when conditioning on the non-overlapping target variables, and this can happen when there is no synergy about the target, leading to a negative value. Therefore, following the weak stochasticity axiom, the PID cannot ensure the nonnegativity of all terms when deterministic target-source dependencies are in place. We will further discuss this limitation after examining the full trivariate decomposition for the XOR and AND examples.

Table 3. Deterministic components of the PID terms for the trivariate decomposition derived from the weak stochasticity axiom. All terms not included in the table have no deterministic component due to the axiom. These expressions correspond to the case in which the primary source i overlaps with the target. If i does not overlap, $\Delta_d(X;i)$ and $\Delta_d(X;i.jk)$ are zero, while the other terms depend on their characteristic symmetry for the other variables j and k, and vanish if none of the variables with the corresponding symmetry overlaps with the target. See the main text and Appendix E for details.

Term	Measure
$\Delta_d(X;i)$	$H(i\|jk, X-ijk)$
$\Delta_d(X;i.jk)$	$I(X-jk;jk\backslash j,k) - I(X-ijk;jk\backslash j,k)$
$\Delta_d(X;i.j)$	$I(i;j\|k, X-ijk) - [\Delta_d(X;i.jk) + \Delta_d(X;j.ik)]$
$\Delta_d(X;i.j.k)$	$C(i;j;k\|X-ijk) + \Delta_d(X;i.jk) + \Delta_d(X;j.ik) + \Delta_d(X;k.ij)$

5.1.2. PIDs with the Strong Axiom

With the strong axiom, not only deterministic but stochastic components of synergy are restricted. There cannot be any synergistic contribution that involves a source overlapping with the target. Equation (12) can be applied with $S = 123$. Furthermore, since synergistic terms have to vanish not only for the terms $\Delta(X;\alpha)$ of the trivariate lattice but also of any bivariate lattice associated with it, given the mapping of PID terms between these lattices (Figure 1), this implies that in the trivariate lattice also the PID terms of the form $i.jk$ are constrained. There is only one case in which synergistic contributions can be nonzero if there is any target-source overlap for the trivariate case, and this is when only one variable overlaps. Consider that only variable 1 is part of the target. Since there cannot be any synergy involving 1, all synergistic PID terms contained in $I(X;1|2)$, $I(X;1|3)$, or $I(X;1|23)$ have to vanish, and also $\Delta(X;2.13)$ and $\Delta(X;3.12)$. It can be checked that this includes all synergistic terms except $\Delta(X;23)$ and $\Delta(X;1.23)$. The former quantifies synergy about other target variables and the latter synergy redundant with the information of 1 itself. With more than one primary source overlapping with the target all synergistic terms have to vanish for the trivariate case.

Like for the weak axiom, we now leave the derivations for Appendix E. The PID deterministic terms are collected in Table 4, again for simplicity showing their expressions for the case in which i overlaps with the target. The form of the expressions respects the symmetries of each term. For example, if j instead of i overlaps with the target then $\Delta_d(X;i.j) = I(i;j|k) - \Delta_d(X;j.ik)$. Note however that, because $\Delta_d(X;j.ik) = 0$ when i overlaps, if both i and j overlap then $\Delta_d(X;i.j) = I(i;j|k)$. See Appendix E for further details.

In comparison to the deterministic components derived from the weak axiom there are two differences: First, the lack of conditioning on $X - ijk$ is due to the reversed partitioning order selected. Like for the bivariate case, the deterministic PID components are independent of the non-overlapping target variables when adopting the strong stochasticity axiom. Second, assuming the strong axiom the terms $\Delta_d(X;i.jk)$ can only be nonzero if j and k are not contained in the target and when more than one source overlaps all terms of the form $\Delta_d(X;i.jk)$ vanish. In that case it is clear that $\Delta_d(X;i.j.k)$ can be negative, since the co-information can be negative. Therefore, also the PID derived from the strong axiom does not ensure nonnegativity. We will now show examples of negative terms for both PIDs.

Table 4. Deterministic components of the PID terms for the trivariate decomposition derived from the strong stochasticity axiom. All terms not included in the table have no deterministic component due to the axiom. Again, the expressions shown here correspond to the case in which the source i overlaps with the target. For $\Delta_d(X; i.jk)$ we further consider that neither j nor k overlap with the target, and otherwise this term vanishes. If i does not overlap, $\Delta_d(X; i)$ is zero, while the other terms depend on their characteristic symmetry for the other variables j and k and vanish otherwise. See the main text and Appendix E for details.

Term	Measure
$\Delta_d(X; i)$	$H(i\|jk)$
$\Delta_d(X; i.jk)$	$I(i; jk\backslash j, k)$
$\Delta_d(X; i.j)$	$I(i; j\|k) - \Delta_d(X; i.jk)$
$\Delta_d(X; i.j.k)$	$C(i; j; k) + \Delta_d(X; i.jk)$

5.2. Illustrative Systems

We now continue the analysis of the XOR and AND examples by decomposing $I(123; 123)$. Since now $X - 123 = \varnothing$ the decompositions are completely deterministic and are obtained calculating the PID components described in Tables 3 and 4. Accordingly, given that deterministic and joint PID terms are equal, we will use $\Delta(X; \beta)$ instead of $\Delta_d(X; \beta)$ to refer to them. As discussed in Section 4.4, the decompositions of the XOR system can be derived without assuming any particular redundancy measure. For the AND system, according to Tables 3 and 4, only the terms $\Delta_d(X; i.jk)$ require selecting a particular measure. As before we assign to these terms the value that is equally obtained with I_{min}, I_{red}, and SI.

5.2.1. XOR

We start with the XOR example and the decomposition derived from the weak stochasticity axiom (Figure 4A). We show the trivariate decomposition of $I(123; 123)$ and also again the decomposition of $I(123; 12)$, now indicating the mapping of the nodes with the trivariate decomposition. For the trivariate lattice we only show the nodes lower than the ones of the primary sources because for all others the corresponding terms are zero (Equation (27)). The PID terms are calculated considering Table 3 and the information-theoretic quantities displayed in Figure 2A.

The trivariate terms $\Delta(X; i)$ are all zero, because any two variables determine the third. This is also reflected in the terms $\Delta(X; i.jk)$ having 1 bit. The terms $\Delta(X; i.j)$ are all equal to -1 bit. These terms should quantify the redundant information between two variables, which is unique with respect to the third, but their interpretation is impaired by the negative values. Furthermore, $\Delta(X; i.j.k) = 2$, so that not only negative values exist but also the monotonicity axiom is violated, since $I(X; i.j.k) > I(X; i.j)$. However, it can be verified that the values obtained are consistent from the point of view of the constraints linking PID terms and mutual informations (Equation (9)). Similarly, the calculated PID components are consistent between the bivariate and trivariate decompositions. In particular, the sum of the nodes with the same color or label in the trivariate lattice equals the corresponding node in the bivariate lattice. This equality holds for the joint bivariate lattice, and not for the deterministic lattice alone, even if in the trivariate case the lattice is uniquely deterministic. This reflects a transfer of stochastic synergy in the bivariate case to deterministic redundancy in the trivariate case (see yellow nodes labeled with d).

We now consider the decomposition derived from the strong axiom (Figure 4B). In this case also $\Delta(X; i)$ are all zero because any two variables determine the third, but now also $\Delta(X; i.jk)$ are zero. This is because the axiom assumes that there is no synergy involving any of the primary sources overlapping with the target. $\Delta(X; 3.12) = 0$ is consistent with the lack synergy for the decomposition of $I(123; 12)$, as indicated by the mapping of the yellow nodes labeled with d. Furthermore, the mapping of all other

PID terms is consistent. In particular, the 1 bit corresponding to the unique informations of the bivariate decomposition is contained in the terms $\Delta(X; i.j) = I(i; j|k)$ of the trivariate one. In comparison to the decomposition from the weak axiom, these terms are not negative, but instead, a negative value is obtained for $\Delta(X; i.j.k)$. Therefore nonnegativity is neither fulfilled for this decomposition.

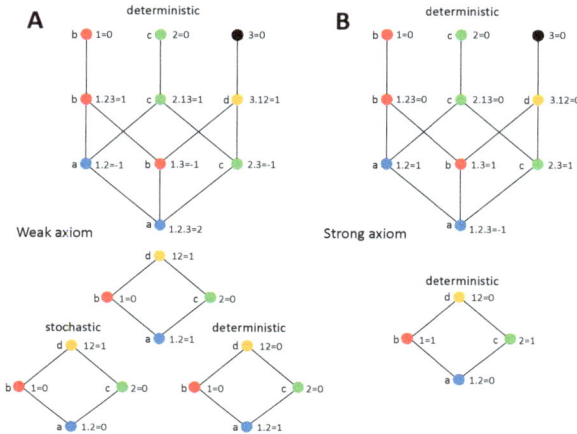

Figure 4. Trivariate decompositions of $I(123; 123)$ for the XOR system. (**A**) Decomposition derived from the weak stochasticity axiom. The trivariate redundancy lattice is displayed only for nodes lower than the single source nodes because all upper PID terms are zero. The bivariate decomposition of $I(123; 12)$ is shown again now indicating the mapping of the PID terms with colors and labels as in Figure 1. In particular, nodes with the same color in the trivariate lattice are accumulated in the corresponding node in the bivariate lattice. (**B**) Same as (**A**) but for the decomposition derived from the strong axiom.

We mentioned in Section 4.5 that, due to conditional dependencies, the only information identity criterion that seems appropriate for target 123 of the XOR system is to identify the two bits as follows: the bit that any first variable provides alone, and the bit that a second variable provides combined with the first. Oppositely, on one hand, the strong axiom assumes that each source alone can uniquely provide a bit, corresponding to its own identity, as reflected in the decomposition of $I(123; 12)$ (see Figure 4B). On the other hand, with the weak axiom, the second bit is classified as synergy, consistently with the idea that retrieving it requires the combination of two variables (Figure 4A). However, because the weak axiom still assumes that any information about an overlapping variable has to be redundant or unique, it imposes that the synergy is contained in the terms $\Delta(X; i.jk)$ in the trivariate decomposition and not in terms corresponding to nodes above the ones of single variables. Therefore, the weak axiom is still not compatible with that identification of the two bits as the one that can be obtained from a single variable and the one that can only be obtained from the combination of two variables.

5.2.2. AND

We present the AND decomposition as a further example (Figure 5). All PID terms are derived using the information-theoretic quantities of Figure 3A in combination with Tables 3 and 4. Like for the XOR case, the mapping of trivariate to bivariate decompositions is consistent. Again, both trivariate decompositions contain some negative term. With the strong axiom, while the bivariate decompositions for the XOR and AND example are equal because of the invariance reducing target 123 to 12, the trivariate PID terms differ substantially, reflecting the different symmetries of each operation.

This is because in the trivariate decomposition 3 explicitly appears as a primary source and cannot be removed even if determined by 12.

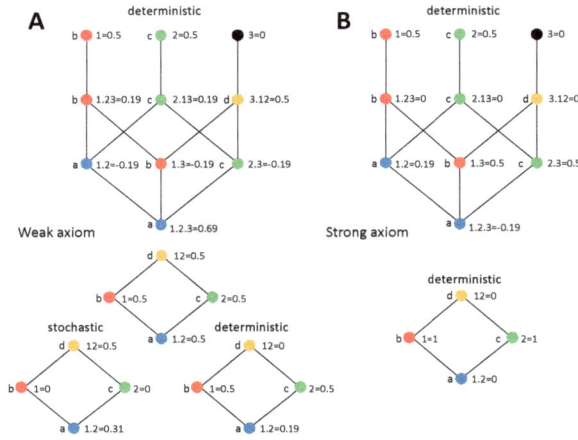

Figure 5. Trivariate decompositions of $I(123;123)$ for the AND system. The structure of the figure is the same as in Figure 4. (**A**) Decompositions derived from the weak stochasticity axiom. (**B**) Decompositions derived from the strong axiom. Nodes with the same color in the trivariate lattice are accumulated in the corresponding node in the bivariate lattice.

5.3. PID Terms' Nonnegativity and Information Identity

The decomposition of $I(123;1,2,3)$ for the XOR system was used by [42,44] as a counterexample to show that with more than two sources there is no decomposition that can simultaneously comply with the monotonicity axiom and the identity axiom and also lead to global nonnegativity of the PID terms. The work in [41] recently pointed out that negative terms appear just when assuming the independent identity property, and not necessarily the identity axiom. However, the existence proofs in [41,42,44] only indicate that a negative term exists, without finding exactly which is the negative term and without determining all the PID terms (see Appendix F for more details).

Our results complement their proofs because the combination of the stochasticity axioms with the relations of the form of Equation (9) allows us to derive the complete PIDs, as shown in Tables 3 and 4. The negative terms can be explained as a consequence of the deterministic components of the PID terms, which result from deterministic target-source dependencies. This is particularly relevant because, as proven in Section 4.3 and in the Appendix C, several proposed measures (i.e., SI, I_{red}, and I_{dep}) comply with the strong axiom (at least) when the primary sources are part of the target. Furthermore, the derivations from the stochasticity axioms relate the existence of negative terms to specific assumptions made to assign an identity to different pieces of information (Sections 3 and 4.5). In more detail, the stochasticity axioms enforce that certain pieces of information are attributed to redundancy or unique information terms because of the target-source identity associations. As a consequence, deterministic components of the decomposition are bounded to the non-synergistic part of the redundancy lattice, which leads to negative terms in order to conform to the lattice structure and to the relations between PID terms and mutual informations (Equation (9)). Furthermore, as argued by [41], if the PID terms are to depend continuously on the probability distributions, the same problem of obtaining negative PID terms is expected to occur not only when deterministic target-source dependencies exist, but also in the limit of strong dependencies tending to be deterministic.

More generally, it is important to identify the minimal assumptions that, when added to the original core ingredients of [33,34], can lead to negative PID terms. These original ingredients are the three axioms of symmetry, self-redundancy, and monotonicity, and the relations of the measures in the redundancy lattice (Equation (9)). The work in [42,44] found that negative terms follow from adding the identity axiom, and [41] showed that they already follow from the weaker independent identity property. Furthermore, the comparison of the two stochasticity axioms and the discussion of information identity criteria (Section 4.6) allow a deeper appreciation of an extra assumption used in the proofs of [41,42,44] (see Appendix F), namely the TSC reduction invariance discussed in Section 4.6. This invariance was assumed, but not motivated in terms of what is expected from the notion of redundancy. Instead, it was assumed as inherited from the mutual information.

These two additional assumptions are less restrictive than adding the strong axiom, because the fulfillment of the strong axiom is a sufficient condition to fulfill the independent identity property (Section 4.2) and also to fulfill the TSC reduction invariance (Section 4.6). In contrast, the weak axiom does not imply this invariance. However, the decomposition of the XOR system derived from this axiom (Figure 4A) shows that it leads to negative terms, but also that it contradicts the monotonicity axiom. Therefore, if we want to preserve the original ingredients of the PID framework, the minimal additional assumptions that lead to negative terms are the independent identity property and the TSC reduction invariance.

We now assess these two assumptions in the light of the discussion of the role of information identity in the quantification of redundancy. Regarding the TSC reduction invariance, in Section 4.6, we indicated that this reduction can affect the identity of the pieces of information and remove information about the mechanisms that would result into mechanistic redundancy. This provides some arguments suggesting that the TSC reduction invariance should not be imposed to the redundancy measure. Regarding the independent identity property, when the target is a copy of two independent sources, the bits of information can be identified with each variable within the target and the target-source identity associations can be applied to assess that there is no redundancy. In fact, for the more general case of dependent sources, the identification of the bits is also consistent with the identity axiom. The bits are shared by the copies of the two sources in the target, and the target-source identity associations can be used to assess that redundancy equals the mutual information between the sources. Therefore, the considerations about information identity suggest that both the independent identity property and the identity axiom should be required. Altogether, this suggests that, from the two assumptions, only the independent identity property, and more generally the identity axiom, should be preserved.

We now review how several proposed redundancy measures comply or not with these two assumptions. SI, I_{red}, and I_{dep} follow both assumptions (see [35,36,43], Section 4.3, and Appendix C). This leads to negative PID terms in the multivariate case. Oppositely, for the measures I_{min} [33] and its simplification I_I [42], it is straightforward to check that they fulfill the TSC reduction invariance, and nonnegativity has been proven [33], but they do not comply with the identity axiom [35]. This means that they only quantify common amounts of information, but not information that is qualitatively common to the sources. Furthermore, I_{ccs} complies with the TSC reduction invariance, which it inherits from the co-information, but this measure was not defined to be nonnegative. Therefore, none of all these proposed redundancy measures complies with the independent identity property and does not comply with the TSC reduction invariance.

If the redundancy lattice and the axioms of [33,34] are to remain as the backbone of a nonnegative PID, we would require a new information identity criterion compatible with the identity axiom but leading to different assumptions about how deterministic target-source dependencies constrain the synergistic terms. The common assumption of the weak and strong axioms that information about an overlapping variable can only be redundant or unique may be too restrictive. As mentioned above, the analysis of how negative terms originate from the stochasticity axioms shows that they are produced by the accumulation of deterministic PID components in non-synergistic terms due to the constraints imposed based on target-source identity associations. Since the TSC reduction invariance only holds for the strong axiom, we expect that an information criterion that further relaxes these constraints can be compatible with the PID terms being non-TSC-invariants.

6. Discussion

6.1. Implications for the Theoretical Definition of Redundant, Synergistic and Unique Information

The proposal of [33] of decomposing mutual information into nonnegative redundant, unique, and synergistic components has been a fruitful and influential conceptual framework. However, a concrete implementation consistent with a set of axioms formalizing the notions for such types of information has proven to be elusive. The main difficulty stems from determining if redundant sources contain the same qualitative information, which requires assigning an identity to pieces of information in the target. The work in [35] pointed out that the redundancy defined by [33] only captures quantitatively the common amounts of information shared by the sources. They introduced the identity axiom to ensure that two independent variables cannot have redundant information about a copy of themselves. The lack of redundancy for this particular case has been enunciated as the independent identity property by [38]. However, Ref. [42] provided a counterexample showing that nonnegativity of the PID terms is not ensured when the identity axiom is assumed. This counterexample also involved a target constituted as a copy of the primary sources, in particular as the inputs and output variables of the XOR logical operation.

Although the identity axiom provides a necessary condition to capture qualitatively common information in the redundancy measure, previous studies have not directly examined how to assign an identity to different pieces of information in order to assess which information is redundant between sources. Since systems with deterministic target-source dependencies have been investigated for the introduction of the identity axiom [35] and to prove its incompatibility with the nonnegativity of the PID terms [36], here we systematically studied how specific information identity criteria constrain the PIDs of such systems. In particular, we examined the PIDs resulting from two information identity criteria that impose constraints on synergistic terms based on identity associations between the target and source variables. These associations result generically from deterministic target-source dependencies and, more concretely for the case we mainly studied, from the overlap between the target and the primary sources. We enunciated (Section 3) two stochasticity axioms that impose constraints of different strength on the synergistic terms. The weak axiom states that there cannot be synergistic information about the overlapping target variables. The strong axiom further constrains synergy assuming that the overlapping sources cannot provide other information than about themselves, and thus cannot contribute synergistic information about the non-overlapping part of the target.

We derived (Section 4.1) general formulas for the PID terms in the bivariate case, following each version of the stochasticity axiom. We showed that the PID terms can be separated into a stochastic and a deterministic component, which account for the information about the non-overlapping and overlapping target variables, respectively. We indicated that the stochasticity axioms subsume the identity axiom and provide two alternative extensions to characterize redundancy for any multivariate system with any degree of target-source overlap (Section 4.2). We showed that several previously proposed measures conform to the strong axiom either in general [36] or at least for classes of systems with target-source overlaps [35,43] wider than the ones considered by the identity axiom (Section 4.3). We then examined (Section 4.4) two concrete examples based on the XOR and AND logical operations, with variables 1 and 2 as inputs and variable 3 as output, calculating the PID of the mutual information $I(123;12)$.

Using these examples, we showed how the identification of pieces of information based on target-source identity associations leads to an ambiguous determination of redundancy and unique information. This ambiguity is reflected in the possible selection of the two alternative stochasticity axioms, and is associated with different partitioning orders of the mutual information (Section 4.5). When using the weak axiom, each source can be combined with some target variables to provide information about other target variables, even in the presence of a target-source overlap. Conversely, the strong axiom assumes that any overlapping variable only provides information about itself, and thus redundant information is equal to the mutual information between the primary sources when there is some target-source overlap, independently of the non-overlapping target variables.

Therefore, a crucial difference between redundancy derived from the two axioms is its invariance to isomorphisms of the target. In the XOR and AND examples, since the output variable is completely determined by the inputs, the mutual information is invariant under the isomorphic reduction of the target 123 to 12, which we called the TSC (target to sources copy) reduction. This invariance of the mutual information holds without imposing any constraint on how the target variables are observed, either simultaneously or sequentially. However, it has been already shown that the PID terms can be sensitive to properties of the joint probability distribution of the target and sources to which the mutual information is not [52]. Here, we discussed two ways in which the overall composition of the target may affect redundancy (Section 4.6). First, the addition of new target variables can change the identity of the pieces of information associated with the previous variables by introducing new conditional dependencies. Second, the TSC reduction invariance is not compatible with the quantification of mechanistic redundancy [35], which has been recognized as the origin of redundancy between independent sources (e.g., [35,36,38,43]). This is because, in examples such as the XOR and AND systems with target 123, the TSC reduction erases the information about the mechanism generating the output variable 3, which is necessary to assess mechanistic redundancy. This means that the notion of redundancy itself depends on whether all the target variables can be observed separately or not. If it is accepted that redundancy should depend on the overall composition of the target, the corresponding redundancy measure cannot comply with the TSC reduction invariance and will depend on semantic aspects of the joint probability distribution of the target related to the identity of the variables.

In Section 5.1, we extended the general derivations to the trivariate case. This allowed us to understand what originates negative PID terms. While with the identity axiom the counterexample of [42] provides a proof of the existence of negative terms, under the stricter conditions of the stochasticity axioms, we could derive the complete PIDs, showing that several PID terms have a deterministic component that is not non-negatively defined (Section 5.2). This analysis is particularly relevant for the previously proposed measures that comply with the strong axiom for the XOR and AND systems with target 123 (Section 4.4). We have thus exposed the relation between the assumptions on information identity and the lack of nonnegativity. In particular, imposing that certain pieces of information can only be attributed to redundancy or unique information terms, based on the premise that their identity is associated with the sources, enforces that deterministic components of the mutual information are bounded to the non-synergistic part of the redundancy lattice. This leads to negative terms in order to conform to the lattice structure and to the lattice inherent relations between PID terms and mutual informations.

Although the notion of redundancy as information shared about the same pieces of information is intuitive in plain language, its precise implementation within the information-theoretic framework is not straightforward. The measure of mutual information has applications in many fields, such as communication theory and statistics [45]. Accordingly, a certain decomposition in terms of redundant, unique, and synergistic contributions may be compatible only with one of its interpretations. Indeed, if information is understood in the context of a communication channel [53], nonnegativity is required from its operational interpretation as the number of messages that can be transmitted without errors. Furthermore, semantic content cannot be attributed, and thus, information identity should rely only on the statistical properties of the distribution of the target variables. For example, in the case of the target composed by two independent variables, identity is assigned based on independence. Alternatively, if mutual information is used as a descriptor of statistical dependencies [54], nonnegativity is not required since locally negative information, or misinformation [55], simply reflects a certain change in the probability distribution of one variable due to conditioning on another variable. With this interpretation of information based on local dependencies, a criterion of information identity can introduce semantic content in association with the specific value of the variables, and common information of two sources can be associated with dependencies that induce coherent modifications of the probability distribution of the target variables [38]. These local measures of information may be interpreted operationally in terms of changes in beliefs, or in relation to a notion of information more associated with ideal observer analysis than with communication theory [55,56]. In this work, we have not considered local versions of mutual information, and we adopted the premise that nonnegativity is a desirable property for the PID terms.

With this aim, we identified the minimal extra assumptions that when added to the original axioms [33,34] lead to negative PID terms. Combining our analysis of the stochasticity axioms and the counterexample of [41,42,44], we pointed out that negativity appears due to the combined assumption of TSC reduction invariance and of the independent identity property (Section 5.3). Following our discussion of the role of information identity in the quantification of redundancy, we suggested that only the latter should be preserved in the search of a desirable redundancy measure.

6.2. Implications for Studying Neural Codes

Determining the proper criterion of information identity to evaluate when information carried by different sources is qualitatively common is essential to interpret the results of the PID in practical applications, such as in the analysis of the distribution of redundant, unique, and synergistic information in neural population responses. For example, when examining how information about a multidimensional sensory stimulus is represented across neurons, the decomposition should identify information about different features of the stimulus, and not only common amounts of information. The PID terms should reflect the functional properties of the neural population so that we can properly characterize the neural code. On the other hand, nonnegativity of the PID terms facilitates their interpretation not only as a description of statistical dependencies, but as a breakdown of the information content of neural responses, for example to assess the intersection information between sensory and behavioral choice representations [20,47,57].

The underlying criterion of information identity for the PID is also important when examining information flows among brain areas because, only if redundant and unique information terms correctly separate qualitatively the information, we can interpret the spatial and temporal dynamics of how unique new information is transmitted across areas. It is common to apply dynamic measures of predictability such as Granger causality [58] to characterize information flows between brain areas [21]. The effect of synergistic and redundant information components in the characterization of information flows with Granger causality has been studied [59,60], and ref. [61] applied their PID framework to decompose the information-theoretic measure of Granger causality, namely Transfer entropy [62,63], into terms separately accounting for state-independent and state-dependent components of information transfer. Furthermore, they also indicated which terms of the PIDs can be associated with information uniquely transmitted at a certain time or information transfer about a specific variable, such as a certain sensory stimulus [64]. These applications of the PID framework identify meaningful PID terms based on the redundancy lattice, and thus can be applied for any actual definition of the measures, but our considerations highlight the necessity to properly determine information identity in order to fully exploit their explanatory power.

Furthermore, our discussion of how the interpretation of information identity depends on the dependencies between the variables composing the target indicates that the analysis of how redundant, unique, and synergistic information components are distributed across neural population responses can be particularly useful in combination with interventional approaches [20,65]. In particular, the manipulation of neural activity with optogenetics techniques [66,67] can disentangle causal effects from other sources of dependencies such as common factors. Although this work illustrates the principled limitations of current PID measures, their combination with these powerful experimental techniques can help to better probe the functional meaning of the PID terms.

6.3. Concluding Remarks

We investigated the implications for the quantification of redundant, unique, and synergistic information of information identity criteria that, in the presence of deterministic target-source dependencies, assign an identity to pieces of information based on identity associations between the target and sources variables. Our analysis suggests that, if the redundancy lattice of [33] is to remain as the backbone of a nonnegative decomposition of the mutual information, a new criterion of information identity should be established that, while conforming to the identity axiom, it is less restrictive in the presence of deterministic target-source dependencies than the ones herein studied.

Acknowledgments: This work was supported by the Fondation Bertarelli. We are grateful to the anonymous reviewers that provided helpful insights to substantially improve this work.

Author Contributions: All authors contributed to the design of the research. The research was carried out by Daniel Chicharro. The manuscript was written by Daniel Chicharro with the contribution of Stefano Panzeri and Giuseppe Pica. All authors have read and approved the final manuscript.

Conflicts of Interest: The authors declare no conflict of interest.

Appendix A. The Relations between the Constraints to Synergy Resulting from the Strong Axiom for the General Case of Functional Dependencies and for the Case of Sources Being Part of the Target

In Section 3, we argued that when any primary source S_k that deterministically explains a subset $X(S_k)$ of the target is in fact part of the target (i.e., $X' = S'$), the conceptual formulation of the strong axiom implies extra constraints that cancel certain synergistic terms (Equation (12)). These constraints specific of the case $X' = S'$ were not derived as a subcase of the general constraints associated with the strong axiom for any type of functional dependence (Equation (11)). Oppositely, we separately proposed these specific constraints following the logic of the strong axiom that indicates that any source that is part of the target cannot be involved in synergistic contributions because it can only contribute to provide the information about itself and it can do so without being combined with any other. We here examine the consistency of Equation (12) with Equation (11), showing that, if imposing an extra desirable property to the PID terms, Equation (12) can be derived from Equation (11). Without this further condition, the fulfillment of Equation (11) and of Equation (12) can be seen as separate requirements that should be fulfilled by a PID compatible with the strong axiom.

In general, Equation (11) only establishes a relation between the synergistic terms of the PID of $I(X; S)$ and the synergistic terms of the PID of $I(X - X'; S|X')$, or equivalently of $I(X; S|X')$, as indicated in Section 3. When $X' = S'$, the synergistic terms of the decomposition of $I(X; S)$ are expressed in terms of the synergistic terms of the decomposition of $I(X - S'; S|S')$, or $I(X; S|S')$. Both decompositions have to comply with the relations between mutual informations and partial informations of Equation (9). However, without any further assumption on how the PID terms should be defined, we cannot further specify the form of the PID terms $\Delta(X; S|S')$. We will now show under which further conditions imposed to the PID terms $\Delta(X; S|S')$ Equation (12) is obtained as a subcase of Equation (11).

Before addressing the general case, we start with an example to understand the constraints that Equation (9) imposes to the decomposition of $I(X; S|S')$. In particular, consider three primary sources 1, 2, and 3 and a target $X = \{X - X', 3\}$, that is $X' = S' = 3$. In this case, Equation (11) links the decomposition of $I(X; 123)$ to the decomposition of $I(X; 123|3)$. In Figure A1A,B, we show the redundancy lattices of $I(X; 123)$ and $I(X; 123|3)$, respectively. The PID terms of both lattices should be consistent with the equations that relate them to mutual information quantities (Equation (9)). In particular, using the chain rule to break down $I(X; 123|3)$, we have that:

$$I(X; 123|3) = I(X; 12|3) + I(X; 3|12, 3) = I(X; 12|3) + 0 \tag{A1a}$$
$$I(X; 13|3) = I(X; 1|3) + I(X; 3|1, 3) = I(X; 1|3) + 0 \tag{A1b}$$
$$I(X; 23|3) = I(X; 2|3) + I(X; 3|2, 3) = I(X; 2|3) + 0 \tag{A1c}$$
$$I(X; 3|3) = 0. \tag{A1d}$$

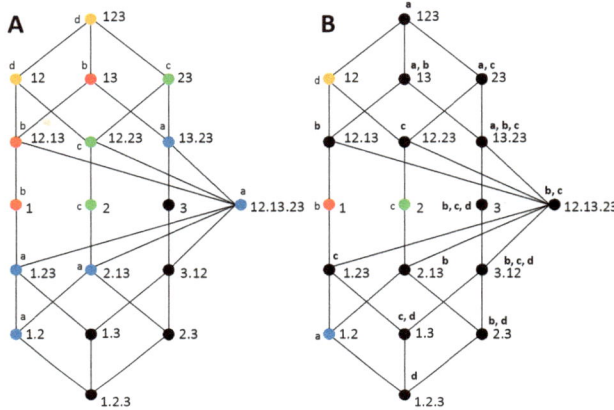

Figure A1. Redundancy lattices of the decomposition of $I(X; 123)$ (**A**) and of $I(X; 123|3)$ (**B**). (**A**) Black nodes correspond to PID terms decomposing $I(X; 3)$. Colored labeled nodes correspond to the decomposition of $I(X; 12|3)$. Nodes labeled with *a* contribute to the redundancy $I(X; 1.2|3)$, labeled with *b* to the unique information $I(X; 1\backslash 2|3)$, with *c* to the unique information $I(X; 2\backslash 1|3)$, and with *d* to synergy $I(X; 12\backslash 1, 2|3)$. (**B**) Colored labeled nodes correspond to the decomposition of $I(X; 12|3)$. Colors and labels indicate the mapping to the same decomposition in the lattice of (**A**). Black nodes correspond to PID terms involved in the constraints of Equation (A1a–d). The bold letters refer to the subequations constraining each node.

Note that the mutual information quantities in which 3 appears both as a source and as a target variable can only be related to the decomposition of $I(X; 123|3)$, while the others are related to both decompositions. In particular, given Equation (9), we should be able to recover, as sums of PID terms from any of the lattices, the measures $I(X; 12|3)$, $I(X; 1|3)$, and $I(X; 2|3)$. The other measures are only related to the PID of $I(X; 123|3)$. Because of the conditioning on the target variable 3, the source 3 cannot contribute any information, since $H(3|3) = 0$. This is reflected in the fact that $I(X; 3|12, 3)$, $I(X; 3|1, 3)$, $I(X; 3|2, 3)$, and $I(X; 3|3)$ are all zero. Given Equation (9), this imposes four constraints on the PID terms $\Delta(X; \alpha|3)$, as sums of terms that should cancel.

Without any further assumption on the properties of these PID terms these constraints cannot determine the PID terms, since the cancellation could be achieved by a combination of positive and negative PID terms. However, if we assume the nonnegativity of the terms $\Delta(X; \alpha|3)$, this implies that all PID terms summed by Equation (9) to obtain $I(X; 3|12, 3)$, $I(X; 3|1, 3)$, $I(X; 3|2, 3)$, or $I(X; 3|3)$ should vanish. In Figure A1B we indicate the PID terms that vanish with black colored nodes and with bold letters referring to the set of sub-equations of Equation (A1) that imply their cancellation. For example, from Equation (9), the fact that $I(X; 3|1, 3) = 0$ implies that PID terms of Figure A1B associated with the nodes in $(\downarrow 13) - (\downarrow 1)$ have to vanish. These are the nodes that can be reached descending from 13 but that cannot be reached descending from 1. Similarly, $I(X;3|12,3) = 0$ indicates that the terms associated with nodes in $(\downarrow 123) - (\downarrow 12)$ have to vanish. Each of the four constraints cancels a set of PID terms identified as $(\downarrow \{W,3\}) - (\downarrow W)$, where $W \subseteq S - S'$.

Altogether, given the constraints of Equation (A1), only the PID terms related to nodes 1.2, 1, 2, and 12 of the lattice decomposing $I(X;123|3)$ can be nonzero. This is consistent with the fact that $I(X;123|3) = I(X;12|3)$. The resulting decomposition of $I(X;123|3)$ is in fact equivalent to the decomposition of $I(X;12|3)$, which has only four terms. This is a desired property for the decomposition of $I(X;123|3)$ because source 3 has no information after conditioning on target variable 3, and thus it has no information that can be classified as redundant, unique, or synergistic. Therefore, instead of assuming the nonnegativity of the terms $\Delta(X;\alpha|3)$, we can alternatively directly impose that the PID of $I(X;123|3)$ should equal the PID of $I(X;12|3)$, as a desirable property of the PID terms. This already implies the fulfillment of the constraints of Equation (A1). This extra requirement is specific of the case $X' = S'$, since for the case $I(X;123|X(3))$, with $X(3)$ a part of the target functionally determined by 3, we have $H(3|X(3)) > 0$. The logic of imposing the equivalence between the decomposition of $I(X;123|3)$ and $I(X;12|3)$ can be further understood examining the mapping of the four PID terms that can be nonzero (1.2, 1, 2, and 12) of the decomposition of $I(X;123|3)$ (Figure A1B) to the PID terms of the decomposition of $I(X;123)$ (Figure A1A). These four terms are mapped to different sums of terms that represent all the information other than $I(X;3)$, which is not available after the conditioning on 3. Finally, the only synergistic term that can be nonzero in Figure A1B is 12, which is consistent with Equation (12), since the collection 12 is the only above the nodes corresponding to single primary sources such that none of its sources contains 3. There is no synergy between any source $W \subseteq S - 3$ and the primary source 3 because $W \subseteq S - 3$ already can provide, in combination with the conditioning copy of 3 from the target, any information obtained by jointly having $\{W,3\}$. Furthermore, also the terms 2.13 and 1.23 vanish in Figure A1B. This is consistent with the expression of $\Delta_d(X;i.jk)$ in Table 4, although Equation (12) does not concern these terms.

We can now resume the general case. In general, the equality of Equation (11) between the synergistic terms of $I(X;S)$ and $I(X;S|X')$ cannot be further specified. For the case of $X' = S'$, $I(X;S|X') = I(X;S|S')$. In this case, since $I(X;S|S') = I(X;S - S'|S')$, we can impose as a desirable property of the PID of $I(X;S|S')$ that it is equivalent to the PID of $I(X;S - S'|S')$. This is because $H(S'|S') = 0$, and thus the sources in S' have no information after conditioning on their target copies. We can now see that this extra condition guarantees the consistency of the constraints of Equation (11) and Equation (12) proposed following the conceptual formulation of the strong axiom. Because the sources in S' do not appear as sources in $I(X;S - S'|S')$, the PID of $I(X;S|S')$ can only be equal to the one of $I(X;S - S'|S')$ if all PID terms $\Delta(X;S|S')$ corresponding to a collection with a source A containing $S_i \in S'$ vanish. This corresponds to the condition $\exists A \in \alpha, S' \cap A = \varnothing$ of Equation (12). Accordingly, since $\Delta(X - S';S|S') = \Delta(X;S|S')$, Equation (12) is subsumed by Equation (11). The equivalence of the PID of $I(X;S - S'|S')$ and the PID of $I(X;S|S')$ is also consistent with the constraints of the form of Equation (A1) related to Equation (9). This is because, as shown for the example above, given the partial ordering relations of Equation (8), canceling the PID terms of all collections containing a source A such that $S' \cap A = \varnothing$ is equivalent to canceling the PID of all collections within $(\downarrow \{W,S'\}) - (\downarrow W)$, for any $W \subseteq S - S'$. Altogether, although in Section 3, we proposed the constraints of Equation (11) and of Equation (12) as separately following from the conceptual formulation of the strong axiom, if imposing as a further desirable condition to the PID of $I(X;S|S')$ its equivalence to the PID of $I(X;S - S'|S')$, Equation (12) can be derived from Equation (11).

Appendix B. Alternative Partitioning Orders for the Bivariate Decomposition with Target-Source Overlap

We here derive in more detail the alternative expressions for the unique and redundant information terms collected in Table 1, which are obtained applying the other mutual information partitioning order of Equation (15b). Using the relation decomposing conditional mutual information into unique information and synergy, we get:

$$
\begin{aligned}
I(X;1\backslash 2) &= I(X;1|2) - I(X;12\backslash 1,2)\\
&= I(X \cap 12;1|2) + I(X - 12;1|2, X \cap 12) - I(X - 12;12\backslash 1,2).
\end{aligned} \tag{A2}
$$

This leads to express the unique information of 1 as:

$$
I(X;1\backslash 2) = \begin{cases} I(X - 12;1\backslash 2) & \text{if } X \cap 1 = \emptyset \\ H(1|2) - I(X - 12;12\backslash 1,2) & \text{if } X \cap 1 \neq \emptyset \end{cases}. \tag{A3}
$$

In this case, the unique information is separated into nonadditive terms and involves the synergy about $X - 12$. This cross-over may seem at odds with the expression obtained with the other partitioning order (Equation (18)), but on the contrary it reflects the internal consistency of the relations between the information-theoretic quantities: Equations (18) and (A3) coincide if 1 is not part of the target. For $X \cap 1 \neq \emptyset$, their equality:

$$
H(1|2) - I(X - 12;12\backslash 1,2) = I(X - 12;1\backslash 2) + H(X \cap 1|2, X - 12) \tag{A4}
$$

is consistent with the definition $I(X - 12;1|2) = H(1|2) - H(1|2, X - 12)$, taking into account that conditional information is the sum of the unique and synergistic components.

Proceeding as with the other partitioning order, once we have the expression of the unique information we can use the relation with the mutual information to determine redundancy:

$$
I(X;1.2) = \begin{cases} I(X - 12;1.2) & \text{if } X \cap 12 = \emptyset \\ I(1;2) + I(X - 12;12\backslash 1,2) & \text{if } X \cap 12 \neq \emptyset \end{cases}. \tag{A5}
$$

Also here, internal consistency with Equation (19) holds. In particular, the equality:

$$
I(1;2) + I(X - 12;12\backslash 1,2) = I(X - 12;1.2) + I(1;2|X - 12) \tag{A6}
$$

reflects that:

$$
\begin{aligned}
C(X - 12;1;2) &= I(1;2) - I(1;2|X - 12)\\
&= I(X - 12;1.2) - I(X - 12;12\backslash 1,2)
\end{aligned} \tag{A7}
$$

because the co-information is invariant to permutations (Equation (5)) and also corresponds to the difference of the redundancy and synergistic PID components.

Also following the strong axiom the alternative partitioning order, in this case the one considering first stochastic dependencies with the non-overlapping target variables, can be derived. With overlap, Equation (14) implies that $I(X;1\backslash 2) = I(X;1|2)$. For the unique information, we get:

$$
I(X;1\backslash 2) = I(X - 12;1\backslash 2) + I(X - 12;12\backslash 1,2) + H(X \cap 1|2, X - 12), \tag{A8}
$$

and for the redundancy:

$$
I(X;1.2) = I(X - 12;1.2) + I(1;2|X - 12) - I(X - 12;12\backslash 1,2). \tag{A9}
$$

Like with the weak axiom, internal consistency holds for the expressions obtained with the two partitioning orders.

Appendix C. The Fulfillment of the Strong Axiom by the Measures SI, and I_{red}

We first show that the maximum conditional entropy-based decomposition of [36] conforms to the strong axiom:

Proposition A1. *The PID associated with the redundancy measure SI [36] conforms to the synergy strong stochasticity axiom for any system with a target X containing a subset $\{X(1), X(2)\}$ of variables that are completely determined by one of the primary sources 1 or 2.*

Proof. Consider a target X and two sources 1 and 2. The measure of synergy associated to SI is defined as:

$$CI(X; 1, 2) = I(X; 1, 2) - \min_{Q \in \Delta(p)} I_Q(X; 1, 2), \tag{A10}$$

where information is minimized within the family of distributions $\Delta(p)$ that preserves the marginals $p(X, 1)$ and $p(X, 2)$. Define $X(i)$, as the subset of variables in X that are completely determined by the source $i \in \{1, 2\}$. Define $X^* = X - \{X(1), X(2)\}$. We can now re-express $CI(X; 1, 2)$ as:

$$
\begin{aligned}
CI(X; 1, 2) &= I(X^* X(1) X(2); 1, 2) - \min_{Q \in \Delta(p)} I_Q(X^* X(1) X(2); 1, 2), \\
&= [I(X(1) X(2); 1, 2) + I(X^*; 1, 2 | X(1), X(2))] - \min_{Q \in \Delta(p)} [I_Q(X(1) X(2); 1, 2) + I_Q(X^*; 1, 2 | X(1), X(2))].
\end{aligned} \tag{A11}
$$

We now show that $I_Q(X(1) X(2); 1, 2) = I(X(1) X(2); 1, 2) \; \forall Q \in \Delta(p)$:

$$
\begin{aligned}
I_Q(X(1) X(2); 1, 2) &= I_Q(X(1); 1, 2) + I_Q(X(2); 1, 2 | X(1)) \\
&= I_Q(X(1); 1) + I_Q(X(1); 2|1) + I_Q(X(2); 2 | X(1)) + I_Q(X(2); 1 | 2, X(1)) \\
&= I_Q(X(1); 1) + I_Q(X(2); 2 | X(1)) = I(X(1); 1) + I(X(2); 2 | X(1)),
\end{aligned} \tag{A12}
$$

where the last equality holds because $p(X(1), 1)$ and $p(X(2), 2, X(1))$ are preserved in $\Delta(p)$. This means that Equation (A11) can be further simplified to:

$$
\begin{aligned}
CI(X; 1, 2) &= I(X^*; 1, 2 | X(1), X(2)) - \min_{Q \in \Delta(p)} I_Q(X^*; 1, 2 | X(1), X(2)) \\
&= CI(X^*; 1, 2 | X(1), X(2)).
\end{aligned} \tag{A13}
$$

This corresponds to the synergy of the bivariate decomposition of $I(X^*; 1, 2 | X(1), X(2))$, in agreement with the strong axiom formulated in Equation (11). □

We now show that the measure I_{red} also complies with the strong axiom, at least for a wider class of systems than the one concerned by the identity axiom:

Proposition A2. *The PID associated with the redundancy measure I_{red} [35] conforms to the synergy strong stochasticity axiom for any system with a target X containing both primary sources 1 and 2.*

Proof. Consider a target X and two sources 1 and 2. The work in [35] defined the measure of redundancy based on information projections in the space of probability distributions. The projection of $p(X|2)$ in the space of distributions of $p(X|1)$, named $p_{2\setminus 1}(X|2)$, is defined as the distribution

$$q(X) = \sum_1 \alpha(1) p(X|1) \tag{A14}$$

with $\alpha(1)$ being a probability distribution optimized such that $p_{2\setminus 1}(X|2) = \arg\min_q \mathrm{KL}(p(X|2); q(X))$. That is, $p_{2\setminus 1}(X|2)$ minimizes the Kullback-Leibler divergence with $p(X|2)$ in the space of the probability distributions defined by Equation (A14). The corresponding projected information of 2 onto 1 with respect to X is defined as:

$$I_{2\searrow1}(X;2) = \sum_{x,2} p(x,2) \log \frac{p_{2\searrow1}(x|2)}{p(x)}. \tag{A15}$$

Redundancy is then defined as $I(X;1.2) = \min\{I_{2\searrow1}(X;2), I_{1\searrow2}(X;1)\}$.

We now examine $I_{2\searrow1}(X;2)$ for the case in which 1 is part of X. In particular define $X = \{Y,1\}$. For this case the distributions $q(X)$ correspond to:

$$q(X) = \sum_{1'} \alpha(1')p(Y,1|1') = \alpha(1)p(Y|1). \tag{A16}$$

The KL-divergence is then:

$$\begin{aligned}
KL(p(Y,1|2);q(Y,1)) &= \sum_{y,1} p(y,1|2) \log \frac{p(y,1|2)}{\alpha(1)p(y|1)} \\
&= \sum_{1} p(1|2) \sum_{y} p(y|12) \log \frac{p(y|12)}{p(y|1)} + \sum_{1} p(1|2) \log \frac{p(1|2)}{\alpha(1)}.
\end{aligned} \tag{A17}$$

The first summand does not depend on $\alpha(1)$ and is thus constant in the minimization space. The second summand is zero if $\alpha(1) = p(1|2)$, which minimizes the divergence. Accordingly, with this value of $\alpha(1)$ into Equation (A16), $p_{2\searrow1}(X|2) = p(1|2)p(Y|1)$. Plugging this distribution into Equation (A15) for $X = \{Y,1\}$, we have:

$$I_{2\searrow1}(Y,1;2) = \sum_{y,1,2} p(y,1,2) \log \frac{p(1|2)p(y|1)}{p(y,1)} = \sum_{1,2} p(1,2) \log \frac{p(1|2)}{p(1)} = I(1;2). \tag{A18}$$

Now, for the case considered in this proposition, both 1 and 2 are part of X. We can thus repeat the derivation above to find that $I_{2\searrow1}(X;2) = I_{1\searrow2}(X;1) = I(1;2)$, and hence $I(X;1.2) = I(1;2)$. This proves that for this case the PID is equal to the one obtained with the strong axiom. $\square$

Note that the proof presented above for I_{red} only contemplates the case in which both primary sources are part of the target, a more specific case than for the proofs presented for SI and I_{dep} in Section 4.4, which show that those measures comply with the strong axiom whenever one of the primary sources is part of the target. This is because, when 1 is part of X, the information projection $I_{2\searrow1}(X;2) = I(1;2)$, but the information projection $I_{1\searrow2}(X;1)$ is in general not equal to $I(1;2)$ unless 2 is also part of X. Despite of its more reduced scope, the proposition for I_{red} encompasses relevant systems as the ones involved in the reduction of target 123 to 12 when 3 is a deterministic function of the primary sources 1 and 2.

Appendix D. The Relation between the Constraints of Equations (11) and (12) for SI, I_{dep}, and I_{red}

In Section 4.3, we have shown that SI and I_{dep} fulfill Equation (12) when $X \cap 12 \neq \varnothing$, and in Appendix C that I_{red} fulfills Equation (12) when $X \cap 12 = 12$. We now show that in these cases Equation (11) is consistently fulfilled. We start with SI:

Proposition A3. *In the case of $X \cap 12 \neq \varnothing$, the PID associated with the redundancy measure SI fulfills Equation (11) consistently with Equation (12).*

Proof. It suffices that $X \cap 12 \neq \varnothing$ so that the joint distribution $p(X,1,2)$ is preserved within the family of distributions $\Delta(p)$. Given Equation (A13), this implies that the synergy $CI(X;1,2)$ vanishes, in agreement with Equation (12). $\square$

We now prove this consistency for I_{dep}:

Proposition A4. *In the case of $X \cap 12 \neq \emptyset$, the PID associated with the redundancy measure I_{dep} fulfills Equation (11) consistently with Equation (12).*

Proof. Consider $X' = X \cap 12$ like in Equation (11) for the case that $X' = S'$. If $X \cap 12 \neq \emptyset$, following the definition of unique information of [43] described in Section 4.3 (according to Appendix B of [43]), the unique information of 2 with respect to 1 in the PID decomposing $I(X; 12|X')$ can be expressed as:

$$I(X; 2\backslash 1|X') = \min\{I_{X2,1}(X; 2|X'), I_{X2,12}(X; 2|X'), I_{X1,X2}(X; 2|X'), I_{X1,X2,12}(X; 2|X')\}, \quad \text{(A19)}$$

where $I_{X1,X2}(X; 2|X')$ indicates the conditional mutual information for the maximum entropy distribution preserving $p(X, 1)$ and $p(X, 2)$, and analogously for the other mutual informations. If $1 \in X'$, $p(X, 1, 2)$ is preserved for all the maximum entropy distributions. If $2 \in X'$, all the mutual informations vanish. Accordingly, it suffices $X \cap 12 \neq \emptyset$ so that $I(X; 2\backslash 1|X') = I(X; 2|X')$. Because, from Equation (4), $I(X; 2|1, X') = I(X; 2\backslash 1|X') + I(X; 12\backslash 1, 2|X')$, this implies that $I(X; 12\backslash 1, 2|X') = 0$ in agreement with Equation (12). $\square$

We now prove this consistency for I_{red}:

Proposition A5. *In the case of $X \cap 12 = 12$, the PID associated with the redundancy measure I_{red} fulfills Equation (11) consistently with Equation (12).*

Proof. Given Equation (6), the synergistic term of the PID of $I(X; 12|X')$ can be expressed as:

$$I(X; 12\backslash 1, 2|X') = I(X; 1|2, X') - I(X; 1|X') + I(X; 1.2|X'). \quad \text{(A20)}$$

For $X' = X \cap 12 = 12$, both $I(X; 1|2, X')$ and $I(X; 1|X')$ vanish. From the definition of the projected information of 2 onto 1 with respect to X conditioned on X', analogous to Equation (A15), $I_{2\searrow 1}(X; 2|X') = 0$ for $X' = 12$. Similarly, also $I_{1\searrow 2}(X; 1|X')$ vanishes. Accordingly, also $I(X; 1.2|X') = 0$, and hence $I(X; 12\backslash 1, 2|X')$ vanishes in agreement with Equation (12). $\square$

Appendix E. Derivations of the Trivariate Decomposition with Target-Source Overlap

We here derive in more detail the trivariate deterministic PID components. We start with the derivations following the weak stochasticity axiom. If we consider the unique information of one primary source with respect to the other two, for example $I(X; 3\backslash 12)$, we have that:

$$I(X; 3\backslash 12) = \Delta(X; 3)$$
$$= I(X; 3|12) - [\Delta(X; 123) + \Delta(X; 13) + \Delta(X; 23) + \Delta(X; 13.23)]. \quad \text{(A21)}$$

The weak axiom imposes for the trivariate case that synergy deterministic components upper than the single source nodes have to be zero (Equation (27)). Accordingly, any deterministic component of $I(X; 3|12)$ has to be contained in $\Delta(X; 3)$. Decomposing this conditional mutual information with the partitioning order that considers first the dependencies with the non-overlapping target variables:

$$I(X; 3\backslash 12) = I(X - 123; 3\backslash 12) + H(X \cap 3|12, X - 123), \quad \text{(A22)}$$

and thus in general:

$$\Delta_d(X; i) = H(X \cap i|jk, X - ijk). \quad \text{(A23)}$$

We now consider the conditional information of two primary sources given the third, for example:

$$I(X; 23|1) = I(X - 123; 23|1) + H(X \cap 23|1, X - 123). \quad \text{(A24)}$$

The deterministic part $H(X \cap 23|1, X - 123)$ again can only be contained in the PID terms contributing to $I(X; 23|1)$ that are lower than the single source nodes. This means that it has to be contained in the terms:

$$\Delta_d(X; 2) + \Delta_d(X; 3) + \Delta_d(X; 2.3) + \Delta_d(X; 3.12) + \Delta_d(X; 2.13). \tag{A25}$$

Furthermore, this conditional entropy can be decomposed considering explicitly the part of the uncertainty associated with conditional entropies of the form of Equation (A23):

$$
\begin{aligned}
H(X \cap 23|1, X - 123) &= H(X \cap 3|1, X - 123) + H(X \cap 2|1, X \cap 3, X - 123) \\
&= H(X \cap 3|12, X - 123) + I(2; X \cap 3)|1, X - 123) + H(X \cap 2|1, X \cap 3, X - 123).
\end{aligned} \tag{A26}
$$

Accordingly, using the definition of the terms $\Delta_d(X; i)$ in Equation (A23) and combining Equations (A24) and (A25), we get the following equalities. First,

$$\Delta_d(X; i) + \Delta_d(X; i.j) + \Delta_d(X; i.jk) + \Delta_d(X; j.ik) = H(i|k, X - ijk) \text{ if } X \cap i \neq \emptyset, \tag{A27}$$

and second:

$$\Delta_d(X; i.j) + \Delta_d(X; i.jk) + \Delta_d(X; j.ik) = I(i; j|k, X - ijk) \text{ if } X \cap i \neq \emptyset. \tag{A28}$$

Like in the expressions of the deterministic PID components in the tables of Sections 4.1 and 5.1, we here for simplicity indicate the equalities that hold when the primary source i overlaps with the target. The symmetries of each $\Delta_d(X; \beta)$ term indicate when it can be nonzero. For example, $\Delta_d(X; i.j)$ is constrained by an equality of the form of Equation (A28) both if i or j overlap with the target.

Finally, we consider also how an unconditional mutual information is decomposed in PID terms. For example, again using the partitioning order that considers first stochastic target-source dependencies, we have:

$$
\begin{aligned}
I(X; 3) &= I(X - 123; 3) + I(X \cap 123; 3|X - 123) \\
&= I(X - 123; 3) + H(X \cap 3|X - 123) \text{ if } X \cap 3 \neq \emptyset.
\end{aligned} \tag{A29}
$$

When 3 is part of the target the deterministic part of this information has to be contained in the nodes reached descending from 3, and thus in general:

$$\sum_{\beta \in \downarrow i} \Delta_d(X; \beta) = H(i|X - ijk) \text{ if } X \cap i \neq \emptyset. \tag{A30}$$

Combining Equation (A30) with Equation (A27), we get that:

$$
\begin{aligned}
\Delta_d(X; i.j) + \Delta_d(X; i.j.k) - \Delta_d(X; k.ij) &= I(i; j|X - 123) \text{ if } X \cap i \neq \emptyset \\
&= H(i|X - 123) - H(i|j, X - 123).
\end{aligned} \tag{A31}
$$

Altogether, from Equations (A23), (A27), (A28), (A30), and (A31), we can proceed to obtain expressions of the PID terms as a function of mutual informations and entropies. Doing so, the rest of PID terms remain as a function also of the terms $\Delta_d(X; i.jk)$. These terms can be understood by comparing the trivariate decomposition and a bivariate decomposition with only sources j and k. For the latter, if i is part of the target, $I(i; jk \backslash j, k)$ quantifies a stochastic synergistic contribution, because i is not a source. Conversely, in the trivariate decomposition i is a source and this information is now redundant with the information provided by variable i itself. This means that we can identify $\Delta_d(X; i.jk)$ by comparing synergy between these two decompositions. For example, for the bivariate decomposition of $I(X; 12)$, 3 is not a source and according to the weak axiom synergy can provide information about the non-overlapping part of the target, which can comprise 3. Moving to the trivariate case by adding 3 as a source this synergy stochastic component becomes redundant to information source 3 has about itself, and thus:

$$I(X; 12\backslash 1, 2) = I(X - 12; 12\backslash 1, 2)$$
$$= I(X - 123; 12\backslash 1, 2) + [I(X - 12; 12\backslash 1, 2) - I(X - 123; 12\backslash 1, 2)]. \tag{A32}$$

In general, this means that these types of PID terms can be quantified as:

$$\Delta_d(X; i.jk) = I(X - jk; jk\backslash j, k) - I(X - ijk; jk\backslash j, k). \tag{A33}$$

These terms are non-negatively defined, because according to the axiom adding a new source can only reduce synergy. After calculating these terms we can obtain all the expressions collected in Table 3.

For the strong stochasticity axiom, instead of repeating all the derivations we proceed by arguing about what has to change with respect to the decomposition obtained for the weak axiom. Changes originate from the difference in the constraints that both versions of the axiom impose on the existence of synergistic components and from the alternative mutual information partitioning order that leads to an additive separation of stochastic and deterministic PID components depending on the axiom. With the strong axiom this additive separation is reached using the partitioning order that first considers deterministic target-source dependencies. This means that the conditioning of entropies and mutual informations on $X - ijk$ will in this case not be present. Moreover, since the strong axiom restricts also synergy with the non-overlapping target variables, even a stochastic component of $I(X; 12\backslash 1, 2)$ can only be nonzero if 3, but neither 1 or 2, overlap with the target. Since once further adding 3 to the sources any synergistic component should be zero, the expression of the terms $\Delta_d(X; i.jk)$ in Equation (A33) is reduced to $I(i; jk\backslash j, k)$ when only i overlaps with X, and to zero otherwise. Implementing these two modifications, the expressions of Table 4 are obtained from the ones of Table 3.

Appendix F. The Counterexample of Nonnegativity of Bertschinger et al. (2012), Rauh et al. (2014), and Rauh (2017)

We here describe in more detail the previous proofs of existence of negative terms for the trivariate PID of $I(123; 1, 2, 3)$, with 3 being the output of the XOR systems and 1 and 2 the inputs. These proofs show that, if preserving the original axioms of [34], it suffices to add the identity axiom, or even just the independent identity property, to originate negative terms. In these previous studies the terms $I(123; i.j)$ were calculated assuming that redundancy is invariant to the reduction of 123 to 12, that is, that $I(123; i.j) = I(ij; i.j)$. This invariance was not motivated in terms of what is expected from the notion of redundancy, but was accepted as inherited from the fact that it holds for the mutual information. We have now indicated that, given its implications regarding the assignment of information identity, it should be considered as an assumption at the same level of the ones more explicitly discussed in the proofs.

Given the equality $I(123; i.j) = I(ij; i.j)$, by the independent identity property $I(ij; i.j) = I(i; j) = 0$ because all three variables of the XOR system are independent. Furthermore, the terms $I(123; i.jk)$ were determined by the monotonicity axiom [41] as $I(123; i.ijk) = I(123; i) = 1$, or by the identity axiom [44] as $I(i, jk; i.jk) = I(i; jk) = 1$. This means that the minimal extra assumptions, apart from the original axioms, are the independent identity property, as pointed out by [41], and the invariance under the reduction of 123 to 12.

Accordingly, from the nodes reachable descending from $I(123; 12.13.23)$, all redundancies are determined except $I(X; 1.2.3)$ and $I(123; 12.13.23)$ itself. Because the objective of the proof was only to show that a negative term has to exist, and not to actually calculate each PID term, [41,42,44] considered that $I(X; 1.2.3)$ vanishes, which preserves the monotonicity of $I(X; 1.2.3) \leq I(X; i.j) = 0$. Given that, they determined a bound $I(123; 12.13.23) \leq -1$. In fact, one may argue that $I(X; 1.2.3)$ can be negative, which still preserves the monotonicity of $I(X; 1.2.3) \leq I(X; i.j) = 0$, and already would lead to the existence of a negative term. In particular, $I(X; 1.2.3)$ can be negative if it is compensated by the terms $\Delta(X; i.j)$ being positive and of equal magnitude, so that $I(X; i.j) = 0$. This is actually what we find in Figure 4B with the strong axiom, and this decomposition is completely compatible with the values $I(X; i, j) = 0$ and $I(X; i.jk) = 1$ determined in these proofs using the axioms as described above. Taking this into account, we see that these proofs work as proofs of existence of at least one negative term, but this negative term can either be $\Delta(X; 1.2.3)$ or $\Delta(123; 12.13.23)$.

References

1. Amari, S. Information geometry on hierarchy of probability distributions. *IEEE Trans. Inf. Theory* **2001**, *47*, 1701–1711.
2. Schneidman, E.; Still, S.; Berry, M.J.; Bialek, W. Network information and connected correlations. *Phys. Rev. Lett.* **2003**, *91*, 238701.
3. Ince, R.A.A.; Senatore, R.; Arabzadeh, E.; Montani, F.; Diamond, M.E.; Panzeri, S. Information-theoretic methods for studying population codes. *Neural Netw.* **2010**, *23*, 713–727.
4. Panzeri, S.; Schultz, S.; Treves, A.; Rolls, E.T. Correlations and the encoding of information in the nervous system. *Proc. R. Soc. Lond. B Biol. Sci.* **1999**, *266*, 1001–1012.
5. Chicharro, D. A Causal Perspective on the Analysis of Signal and Noise Correlations and Their Role in Population Coding. *Neural Comput.* **2014**, *26*, 999–1054.
6. Timme, N.; Alford, W.; Flecker, B.; Beggs, J.M. Synergy, redundancy, and multivariate information measures: An experimentalist's perspective. *J. Comput. Neurosci.* **2014**, *36*, 119–140.
7. Watkinson, J.; Liang, K.C.; Wang, X.; Zheng, T.; Anastassiou, D. Inference of regulatory gene interactions from expression data using three-way mutual information. *Ann. N. Y. Acad. Sci.* **2009**, *1158*, 302–313.
8. Erwin, D.H.; Davidson, E.H. The evolution of hierarchical gene regulatory networks. *Nat. Rev. Genet.* **2009**, *10*, 141–148.
9. Chatterjee, P.; Pal, N.R. Construction of synergy networks from gene expression data related to disease. *Gene* **2016**, *590*, 250–262.
10. Panzeri, S.; Magri, C.; Logothetis, N.K. On the use of information theory for the analysis of the relationship between neural and imaging signals. *Magn. Reson. Imaging* **2008**, *26*, 1015–1025.
11. Marre, O.; El Boustani, S.; Fregnac, Y.; Destexhe, A. Prediction of Spatiotemporal Patterns of Neural Activity from Pairwise Correlations. *Phys. Rev. Lett.* **2009**, *102*, 138101.
12. Faes, L.; Marinazzo, D.; Nollo, G.; Porta, A. An Information-Theoretic Framework to Map the Spatiotemporal Dynamics of the Scalp Electroencephalogram. *IEEE Trans. Biomed. Eng.* **2016**, *63*, 2488–2496.
13. Katz, Y.; Tunstrøm, K.; Ioannou, C.C.; Huepe, C.; Couzin, I.D. Inferring the structure and dynamics of interactions in schooling fish. *Proc. Natl. Acad. Sci. USA* **2011**, *108*, 18720–18725.
14. Flack, J.C. Multiple time-scales and the developmental dynamics of social systems. *Philos. Trans. R. Soc. B Biol. Sci.* **2012**, *367*, 1802–1810.
15. Ay, N.; Der, R.; Prokopenko, M. Information-driven self-organization: The dynamical system approach to autonomous robot behavior. *Theory Biosci.* **2012**, *131*, 125–127.
16. Latham, P.E.; Nirenberg, S. Synergy, Redundancy, and Independence in Population Codes, Revisited. *J. Neurosci.* **2005**, *25*, 5195–5206.
17. Rauh, J.; Ay, N. Robustness, canalyzing functions and systems design. *Theory Biosci.* **2014**, *133*, 63–78.
18. Tishby, N.; Pereira, F.C.; Bialek, W. The Information Bottleneck Method. In Proceedings of the 37th Annual Allerton Conference on Communication, Control, and Computing, Monticello, IL, USA, 22–24 September 1999; pp. 368–377.
19. Averbeck, B.B.; Latham, P.E.; Pouget, A. Neural correlations, population coding and computation. *Nat. Rev. Neurosci.* **2006**, *7*, 358–366.
20. Panzeri, S.; Harvey, C.D.; Piasini, E.; Latham, P.E.; Fellin, T. Cracking the neural code for sensory perception by combining statistics, intervention and behavior. *Neuron* **2017**, *93*, 491–507.
21. Wibral, M.; Vicente, R.; Lizier, J.T. *Directed Information Measures in Neuroscience*; Springer: Berlin/Heidelberg, Germany, 2014.

22. Timme, N.M.; Ito, S.; Myroshnychenko, M.; Nigam, S.; Shimono, M.; Yeh, F.C.; Hottowy, P.; Litke, A.M.; Beggs, J.M. High-Degree Neurons Feed Cortical Computations. *PLoS Comput. Biol.* **2016**, *12*, e1004858.

23. Panzeri, S.; Brunel, N.; Logothetis, N.K.; Kayser, C. Sensory neural codes using multiplexed temporal scales. *Trends Neurosci.* **2010**, *33*, 111–120.

24. Panzeri, S.; Macke, J.H.; Gross, J.; Kayser, C. Neural population coding: Combining insights from microscopic and mass signals. *Trends Cogn. Sci.* **2015**, *19*, 162–172.

25. Valdes-Sosa, P.A.; Roebroeck, A.; Daunizeau, J.; Friston, K. Effective connectivity: Influence, causality and biophysical modeling. *Neuroimage* **2011**, *58*, 339–361.

26. Vicente, R.; Wibral, M.; Lindner, M.; Pipa, G. Transfer entropy: A model-free measure of effective connectivity for the neurosciences. *J. Comput. Neurosci.* **2011**, *30*, 45–67.

27. Ince, R.A.A.; van Rijsbergen, N.J.; Thut, G.; Rousselet, G.A.; Gross, J.; Panzeri, S.; Schyns, P.G. Tracing the Flow of Perceptual Features in an Algorithmic Brain Network. *Sci. Rep.* **2015**, *5*, 17681.

28. Deco, G.; Tononi, G.; Boly, M.; Kringelbach, M.L. Rethinking segregation and integration: Contributions of whole-brain modelling. *Nat. Rev. Neurosci.* **2015**, *16*, 430–439.

29. McGill, W.J. Multivariate information transmission. *Psychometrika* **1954**, *19*, 97–116.

30. Bell, A.J. The co-information lattice. In Proceedings of the 4th International Symposium Independent Component Analysis and Blind Source Separation, Nara, Japan, 1–4 April 2003; pp. 921–926.

31. Olbrich, E.; Bertschinger, N.; Rauh, J. Information decomposition and synergy. *Entropy* **2015**, *17*, 3501–3517.

32. Perrone, P.; Ay, N. Hierarchical quantification of synergy in channels. *Front. Robot. AI* **2016**, *2*, 35.

33. Williams, P.L.; Beer, R.D. Nonnegative Decomposition of Multivariate Information. *arXiv* **2010**, arXiv:1004.2515.

34. Williams, P.L. Information Dynamics: Its Theory and Application to Embodied Cognitive Systems. PhD. Thesis, Indiana University, Bloomington, IN, USA, 2011.

35. Harder, M.; Salge, C.; Polani, D. Bivariate measure of redundant information. *Phys. Rev. E* **2013**, *87*, 012130.

36. Bertschinger, N.; Rauh, J.; Olbrich, E.; Jost, J.; Ay, N. Quantifying unique information. *Entropy* **2014**, *16*, 2161–2183.

37. Griffith, V.; Koch, C. Quantifying synergistic mutual information. *arXiv* **2013**, arXiv:1205.4265v6.

38. Ince, R.A.A. Measuring multivariate redundant information with pointwise common change in surprisal. *Entropy* **2017**, *19*, 318.

39. Rauh, J.; Banerjee, P.K.; Olbrich, E.; Jost, J.; Bertschinger, N. On Extractable Shared Information. *Entropy* **2017**, *19*, 328.

40. Chicharro, D. Quantifying multivariate redundancy with maximum entropy decompositions of mutual information. *arXiv* **2017**, arXiv:1708.03845v1.

41. Rauh, J. Secret Sharing and shared information. *Entropy* **2017**, *19*, 601.

42. Bertschinger, N.; Rauh, J.; Olbrich, E.; Jost, J. Shared Information—New Insights and Problems in Decomposing Information in Complex Systems. In *Proceedings of the European Conference on Complex Systems 2012*; Gilbert, T., Kirkilionis, M., Nicolis, G., Eds.; Springer: Cham, Switzerland, 2012; pp. 251–269.

43. James, R.G.; Emenheiser, J.; Crutchfield, J.P. Unique Information via Dependency Constraints. *arXiv* **2017**, arXiv:1709.06653v1.

44. Rauh, J.; Bertschinger, N.; Olbrich, E.; Jost, J. Reconsidering unique information: Towards a multivariate information decomposition. In Proceedings of the 2014 IEEE International Symposium on Information Theory (ISIT), Honolulu, HI, USA, 29 June–4 July 2014; pp. 2232–2236.

45. Cover, T.M.; Thomas, J.A. *Elements of Information Theory*, 2nd ed.; John Wiley and Sons: New York, NY, USA, 2006.

46. Chicharro, D.; Panzeri, S. Synergy and Redundancy in Dual Decompositions of Mutual Information Gain and Information Loss. *Entropy* **2017**, *19*, 71.

47. Pica, G.; Piasini, E.; Chicharro, D.; Panzeri, S. Invariant components of synergy, redundancy, and unique information among three variables. *Entropy* **2017**, *19*, 451.

48. Griffith, V.; Chong, E.K.P.; James, R.G.; Ellison, C.J.; Crutchfield, J.P. Intersection Information based on Common Randomness. *Entropy* **2014**, *16*, 1985–2000.

49. Banerjee, P.K.; Griffith, V. Synergy, redundancy, and common information. *arXiv* **2015**, arXiv:1509.03706v1.

50. Barrett, A.B. Exploration of synergistic and redundant information sharing in static and dynamical Gaussian systems. *Phys. Rev. E* **2015**, *91*, 052802.

51. Faes, L.; Marinazzo, D.; Stramaglia, S. Multiscale Information Decomposition: Exact Computation for Multivariate Gaussian Processes. *Entropy* **2017**, *19*, 408.

52. James, R.G.; Crutchfield, J.P. Multivariate Dependence Beyond Shannon Information. *Entropy* **2017**, *19*, 531.

53. Shannon, C.E. A mathematical theory of communication. *Bell. Syst. Tech. J.* **1948**, *27*, 379–423, 623–656.
54. Kullback, S. *Information Theory and Statistics*; Dover: Mineola, NY, USA, 1959.
55. Wibral, M.; Lizier, J.T.; Priesemann, V. Bits from brains for biologically inspired computing. *Front. Robot. AI* **2015**, *2*, 5.
56. Thomson, E.E.; Kristan, W.B. Quantifying Stimulus Discriminability: A Comparison of Information Theory and Ideal Observer Analysis. *Neural Comput.* **2005**, *17*, 741–778.
57. Pica, G.; Piasini, E.; Safaai, H.; Runyan, C.A.; Diamond, M.E.; Fellin, T.; Kayser, C.; Harvey, C.D.; Panzeri, S. Quantifying how much sensory information in a neural code is relevant for behavior. In Proceedings of the 31st Conference on Neural Information Processing Systems, Long Beach, CA, USA, 4 December 2017.
58. Granger, C.W.J. Investigating Causal Relations by Econometric Models and Cross-Spectral Methods. *Econometrica* **1969**, *37*, 424–38.
59. Stramaglia, S.; Cortes, J.M.; Marinazzo, D. Synergy and redundancy in the Granger causal analysis of dynamical networks. *New J. Phys.* **2014**, *16*, 105003.
60. Stramaglia, S.; Angelini, L.; Wu, G.; Cortes, J.M.; Faes, L.; Marinazzo, D. Synergetic and redundant information flow detected by unnormalized Granger causality: Application to resting state fMRI. *IEEE Trans. Biomed. Eng.* **2016**, *63*, 2518–2524.
61. Williams, P.L.; Beer, R.D. Generalized Measures of Information Transfer. *arXiv* **2011**, arXiv:1102.1507v1.
62. Marko, H. Bidirectional communication theory—Generalization of information-theory. *IEEE Trans. Commun.* **1973**, *12*, 1345–1351.
63. Schreiber, T. Measuring information transfer. *Phys. Rev. Lett.* **2000**, *85*, 461–464.
64. Beer, R.D.; Williams, P.L. Information Processing and Dynamics in Minimally Cognitive Agents. *Cogn. Sci.* **2015**, *39*, 1–39.
65. Chicharro, D.; Panzeri, S. Algorithms of causal inference for the analysis of effective connectivity among brain regions. *Front. Neuroinform.* **2014**, *8*, 64.
66. O'Connor, D.H.; Hires, S.A.; Guo, Z.V.; Li, N.; Yu, J.; Sun, Q.Q.; Huber, D.; Svoboda, K. Neural coding during active somatosensation revealed using illusory touch. *Nat. Neurosci.* **2013**, *16*, 958–965.
67. Otchy, T.M.; Wolff, S.B.E.; Rhee, J.Y.; Pehlevan, C.; Kawai, R.; Kempf, A.; Gobes, S.M.H.; Olveczky, B.P. Acute off-target effects of neural circuit manipulations. *Nature* **2015**, *528*, 358–363.

entropy

MDPI

Article

Assessing the Relevance of Specific Response Features in the Neural Code

Hugo Gabriel Eyherabide [1,*] and Inés Samengo [2]

[1] Department of Computer Science and Helsinki Institute for Information Technology,
    University of Helsinki Gustaf Hällströmin katu 2b, FI00560 Helsinki, Finland
[2] Department of Medical Physics, Centro Atómico Bariloche and Instituto Balseiro,
    8400 San Carlos de Bariloche, Argentina; samengo@cab.cnea.gov.ar
* Correspondence: neuralinfo@eyherabidehg.com; Tel.: +358-050-368-1758

Received: 1 October 2018; Accepted: 13 November 2018; Published: 15 November 2018

check for updates

Abstract: The study of the neural code aims at deciphering how the nervous system maps external stimuli into neural activity—the encoding phase—and subsequently transforms such activity into adequate responses to the original stimuli—the decoding phase. Several information-theoretical methods have been proposed to assess the relevance of individual response features, as for example, the spike count of a given neuron, or the amount of correlation in the activity of two cells. These methods work under the premise that the relevance of a feature is reflected in the information loss that is induced by eliminating the feature from the response. The alternative methods differ in the procedure by which the tested feature is removed, and the algorithm with which the lost information is calculated. Here we compare these methods, and show that more often than not, each method assigns a different relevance to the tested feature. We demonstrate that the differences are both quantitative and qualitative, and connect them with the method employed to remove the tested feature, as well as the procedure to calculate the lost information. By studying a collection of carefully designed examples, and working on analytic derivations, we identify the conditions under which the relevance of features diagnosed by different methods can be ranked, or sometimes even equated. The condition for equality involves both the amount and the type of information contributed by the tested feature. We conclude that the quest for relevant response features is more delicate than previously thought, and may yield to multiple answers depending on methodological subtleties.

Keywords: neural code; representation; decoding; spike-time precision; discrimination; noise correlations; information theory; mismatched decoding

---

1. Introduction

Understanding the neural code involves, among other things, identifying the relevant response features that participate in the representation of information. Different studies have proposed several candidates, for example, the spiking rate [1,2], the response latency [3], the temporal organisation of spikes [4], the amount of synchrony in a given brain area [5], the amount of correlation between the activity of different neurons [6], or the phase of the local field potential at the time of spiking [7], to cite a few. One way of evaluating the relevance of each candidate feature is to assess how much information is lost by ignoring that feature. This strategy involves the comparison of the mutual information between the stimulus and the so-called *full response* (a collection of response features including the tested one) and the same information calculated with a *reduced response*, obtained by dropping the tested feature from the full response. If the tested feature is relevant, the information encoded by the reduced response should be smaller than that of the full response.

The procedure is fairly straightforward when the response features are defined in terms of variables that take definite values in each stimulus presentation, as for example, the spike count C fired in a fixed time window, or the latency L between the stimulus and the first spike. The full response in this case is a two-component vector $[C, L]$, the value of which is uniquely defined for each stimulus presentation—let us assume that in this example, C is never equal to 0, so L is always well defined. The reduced response is a one-component vector, either C or L, depending whether we are evaluating the relevance of the latency or the spike count, respectively. If the latency or the spike count are relevant, then the information encoded by C or L, respectively, should be smaller than that of the pair $[C, L]$. Throughout this paper, we often use C and L as examples of response features that take a precise value in each trial, to contrast with other features that are only defined in the whole collection of trials, as discussed below.

The method becomes more controversial when applied to response properties that can only be defined in multiple stimulus presentations, as for example, the amount of correlation in the activity of two or more neurons, or the temporal precision of the elicited spikes. These properties cannot be calculated from single responses, so more sophisticated methods are required to delete the tested feature. There are several alternative procedures to perform such deletion, and several are also the ways in which the lost information can be calculated. Interestingly, the lost information depends markedly on the chosen method, implying that the so-called *relevance* of a given feature is a subtle concept, that needs to be specified precisely. When assessing the relevance of noise correlations, two different sets of strategies have been proposed by the seminal works of Nirenberg et al. [8] and Schneidman et al. [9]. The first proposal evaluated the role of noise correlations in *decoding* the information represented in neural activity, whereas the second, in the amount of *encoded* information. Quite surprisingly, the contribution of correlations to the decoded information was shown to sometimes exceed the amount of encoded information [9], seemingly contradicting the intuitive idea that the encoded information constitutes an upper bound to the decoded information. The apparent inconsistency between the two measures has not been observed in later extensions of the technique, where the relevance of other response aspects was evaluated, such as spike-time precision, spike-counts or spike-onsets. Moreover, it has even been argued that the inconsistency was exclusively observed when assessing the role of noise correlations [10–13].

In this paper, for the first time, the different methods used in the literature to delete a given response feature are distinguished, and the implications of each method are discussed and compared. We show that the data processing inequality, stating that the decoded information cannot surpass the encoded information, can only be invoked with some - and not all - deletion procedures. The distinction between such procedures allows us to identify the conditions in which the decoded information can exceed the encoded information, and to demonstrate that there was no logical inconsistency in previous studies. We also show explicit examples where the decoded information surpasses the encoded information also when assessing the role of other response aspects different from noise correlations. In order to explain why such behaviours have not been identified until now, we scrutinise the arguments given in the literature to claim that only noise correlations could exhibit such syndrome. We conclude that although the measures employed to assess the relevance of individual response features initially distinguished clearly between the relevance for encoding and the relevance for decoding, this distinction was eventually lost in later modifications of the measures. By diagnosing the confusion, we prove that indeed, the response features for which the decoded information can surpass the encoded information are not restricted to noise correlations.

More generally, we discuss a wide collection of strategies employed to assess the relevance of individual response features, ranging from those encoded-oriented to those decoded-oriented. This distinction is related to the way the tested feature contributes to the performance of decoders, which can be mismatched or not. The relevance of the tested feature obtained with some of the measures is always bounded by the relevance of another measure. Yet, not all measures can be ordered hierarchically. There are examples where the relevance of a feature obtained with one method may surpass or be surpassed by the relevance of another, depending on the specific values taken by the prior stimulus probability and the conditional response probabilities. We analyse a collection of carefully chosen examples to identify the cases where this is so. In certain restricted conditions, however, the hierarchy, or even the equality, can be ensured. Here we establish these conditions by means of analytic reasoning, and discuss their implications in terms of the amount and type of information encoded by the tested feature.

We also present examples in which the measures to assess the relevance of a given feature can be used to extract qualitative knowledge about the type of information encoded by the feature. In other words, we assess not only *how much* information is encoded by an individual feature, but also *what kind* of information is provided, with respect to individual stimulus attributes. Again, we prove that the type of encoded information depends on the method employed to assess it.

Finally, given that one important property of measures of relevance hinges on whether they represent the operation of matched or mismatched decoders, we also explore the consequences of operating mismatched decoders on noisy responses, instead of real responses. We conclude that it may be possible to improve the performance of a mismatched decoder by adding noise. From the theoretical point of view, this observation underscores the fact that the conditions for optimality for matched decoders need not hold for mismatched decoders. From the practical perspective, our results open new opportunities for potentially simpler, more efficient and more resilient decoding algorithms.

In Section 2.1, we establish the notation, and we introduce some of the key concepts that will be used throughout the paper. These concepts are employed in Section 2.2 to determine the cases where the data-processing inequality can be ensured. In Section 2.3 we introduce 9 measures of feature relevance that were previously defined in the literature, and briefly discuss their meaning, similarities and discrepancies. A numeric exploration of a set of carefully chosen examples is employed in Section 2.4 to detect the pairs of measures for which no general hierarchical order exists. In Section 2.5 we discuss the consequences of employing measures that are conceptually linked to matched or mismatched decoders. Later, in Section 2.6, we explore the way in which different measures of feature relevance arrogate different qualitative meaning to the type of information encoded by the tested feature. In Section 2.7 we discuss the conditions under which encoding-oriented measures provide the same amount of information as their decoding-oriented counterparts, and also the conditions under which the equality extends also to the content of that information. Then, in Section 2.8, we observe that sometimes, mismatched decoders may improve their performance when operating upon noisy responses. We discuss some relations of our work with other approaches and to the limiting sampling problem in Section 3, and we close with a summary of the main results of the paper in Section 4.

2. Results

2.1. Definitions

2.1.1. Statistical Notation

When no risk of ambiguity arises, we here employ the standard abbreviated notation of statistical inference [14], denoting random variables with letters in upper case, and their values, in lower case. For example, the symbol $P(x|y)$ always denotes the conditional probability of the random variable X taking the value x given that the random variable Y takes the value y. This notation may lead to confusion or be inappropriate, for example, when the random variable X takes the value u given that the random variable Y takes the value v. In those cases, we explicitly indicate the random variables and their values, as for example $P(X = u|Y = v)$.

In the study of the neural code, the relevant random variables are the stimulus S and the response $\mathbf{R}$ generated by the nervous system. In this paper, we discuss the statistics of the true responses observed experimentally, and compare them with a theoretical model that describes how responses would be, if the encoding strategy were different. To differentiate these two situations, we employ the variable $\mathbf{R}_{ex}$ for the experimental responses (the real ones), and $\mathbf{R}_{su}$ for the surrogate responses (the fictitious ones). The associated conditional probability distributions are $P_{ex}(\mathbf{R}_{ex} = \mathbf{r}|S = s)$ and $P_{su}(\mathbf{R}_{su} = \mathbf{r}|S = s)$, which are often abbreviated as $P_{ex}(\mathbf{r}|s)$ and $P_{su}(\mathbf{r}|s)$, respectively. Once these distributions are known, and given the prior stimulus probabilities $P(s)$, the joint probabilities $P_{ex}(\mathbf{r}, s)$ and $P_{su}(\mathbf{r}, s)$ can be deduced, as well as the marginals $P_{ex}(\mathbf{r})$ and $P_{su}(\mathbf{r})$. When interpreting the abbreviated notation, readers should keep in mind that P_{ex} governs the variable $\mathbf{R}_{ex}$, and P_{su}, $\mathbf{R}_{su}$. If a statement is made about a distribution P or a response variable $\mathbf{R}$ that has no sub-index, the argument is intended for both the real and surrogate distributions or variables.

2.1.2. Encoding

The process of converting stimuli S into neural responses $\mathbf{R}$ (e.g., spike-trains, local-field potentials, electroencephalographic or other brain signals, etc.) is called "encoding" [9,15]. The encoding process is typically noisy, in the sense that repeated presentations of the same stimulus may yield different neural responses, and is characterised by the joint probability distribution $P(s, \mathbf{r})$. The associated marginal probabilities are

$$
\begin{aligned}
P(s) &= \sum_{\mathbf{r}} P(s, \mathbf{r}), \\
P(\mathbf{r}) &= \sum_{s} P(s, \mathbf{r}),
\end{aligned}
$$

from which the conditional response probability $P(\mathbf{r}|s) = P(s, \mathbf{r})/P(s)$, and the posterior stimulus probability $P(s|\mathbf{r}) = P(s, \mathbf{r})/P(\mathbf{r})$ can be defined.

The mutual information that $\mathbf{R}$ contains about S is

$$
I(S; \mathbf{R}) = \sum_{s, \mathbf{r}} P(s, \mathbf{r}) \log_2 \frac{P(s|\mathbf{r})}{P(s)}. \tag{1}
$$

More generally, the mutual information $I(S; X)$ about S contained in any random variable X, including but not limited to $\mathbf{R}$, can be computed using the above formula with $\mathbf{R}$ replaced by X. For compactness, we denote $I(S; X)$ as I_X unless ambiguity arises.

2.1.3. Data Processing Inequalities

When the response $\mathbf{R}_2$ is a post-processed version of the response $\mathbf{R}_1$, the joint probability distribution $P(s, \mathbf{r}_1, \mathbf{r}_2)$ can be written as $P(s, \mathbf{r}_1) \, P(\mathbf{r}_2|\mathbf{r}_1)$. This decomposition implies that $\mathbf{R}_2$ is conditionally independent of S. In these circumstances, the information about S contained in $\mathbf{R}_2$ cannot exceed the information about S contained in $\mathbf{R}_1$ [16]. In addition, the accuracy of the optimal decoder operating on $\mathbf{R}_2$ cannot exceed the accuracy of the optimal decoder operating on $\mathbf{R}_1$ [17]. These results constitute the data processing inequalities.

2.1.4. Decoding

The process of transforming responses $\mathbf{r}$ into estimated stimuli $\hat{s}$ is called "decoding" [9,15]. More precisely, a decoder is a mapping $\mathbf{r} \rightarrow \hat{s}$ defined by a function $\hat{s} = D(\mathbf{r})$. The inverse of this function is D^{-1}, and when D is not injective, D^{-1} is a multi-valued mapping. The joint probability $P(s, \hat{s})$ of the presented and estimated stimuli, also called "confusion matrix" [12], is

$$
P(s, \hat{s}) = \sum_{\mathbf{r} \in D^{-1}(\hat{s})} P(s, \mathbf{r}), \tag{2}
$$

where the sum runs over all responses **r** that are mapped onto $\hat{s}$ by D. The information that $\hat{S}$ preserves about S is $I_{\hat{S}}$, and can be calculated from the confusion matrix of Equation (2). The decoding accuracy above chance level is here defined as

$$A = \sum_s P(S{=}s, \hat{S}{=}s) - \max_s P(s). \qquad (3)$$

2.1.5. Optimal Decoding

Although all mappings D are formally admissible as decoders, not all are useful. The aim of a decoder is to make a good guess of the external stimulus S from the neural response **R**. It is therefore important to be able to construct decoders that make good guesses, or at least, as good as the mapping from stimuli to responses allows. Optimal decoders (also called *Bayesian* or *maximum-a-posteriori* decoders, as well as *ideal homunculus*, or *observer*, among other names) are defined as [18,19]

$$\hat{s} = D_{\mathrm{opt}}(\mathbf{r}) = \arg\max_s P(s|\mathbf{r}) = \arg\max_s P(s, \mathbf{r}). \qquad (4)$$

This mapping selects, for each response **r**, the stimulus $\hat{s}$ that most likely generated **r**. It is optimal in the sense that any other decoding algorithm yields a confusion matrix with lower decoding accuracy. Equation (4) depends on $P(s, \mathbf{r})$, so the decoder cannot be defined before knowing the functional shape of the joint probability distribution between stimuli and responses. The process of estimating $P(s, \mathbf{r})$ from real data, and the subsequent insertion of the obtained distribution in Equation (4) is called the *training* of the decoder. The word "training" makes reference to a gradual process, originally stemming from a computational strategy employed to estimate the distribution progressively, while the data was being gathered. However, in this paper we do not discuss estimation strategies from limited samples, so for us, "training a decoder" is equivalent to constructing a decoder from Equation (4).

2.1.6. Extensions of Optimal Decoding

The study of Ince et al. [20] introduced the concept of ranked decoding, in which each response **r** is mapped onto a list of K stimuli $\hat{\mathbf{s}} = (\hat{s}_1, \ldots, \hat{s}_K)$ ordered according to their posterior probabilities so that $P(\hat{s}_k|\mathbf{r}) \geq P(\hat{s}_{k+1}|\mathbf{r})$ (with $1 \leq k < K$, and $K \leq$ the total number of stimuli in the experiment). Ranked decoding can provide useful models for intermediate stages in the decision pathway, and the information loss induced by ranked decoding was computed recently [17]. The joint probability associated with ranked decoding is

$$P(s, \hat{\mathbf{s}}) = \sum_{\mathbf{r} \in D^{-1}(\hat{\mathbf{s}})} P(s, \mathbf{r}), \qquad (5)$$

where the sum runs over all response vectors **r** that produce the same ranking $\hat{\mathbf{s}}$. Although $P(s, \hat{\mathbf{s}})$ can be used to compute the information $I_{\hat{S}}$ between S and $\hat{\mathbf{S}}$, it cannot be used to compute the decoding accuracy above chance level because the support of $\hat{\mathbf{S}}$ (i.e., the set of stimulus lists) is not contained in the support of S (i.e., the set of stimuli).

2.1.7. Approximations to Optimal Decoding

For given probabilities $P(\mathbf{r}|s)$ and $P(s)$, Equation (4) defines a mapping between each response **r** and a candidate stimulus $\hat{s}$. In the study of the neural code, scientists often wonder what would happen if responses were not governed by the experimentally recorded distribution $P_{\mathrm{ex}}(\mathbf{r}|s)$, but by some other surrogate distribution $P_{\mathrm{su}}(\mathbf{r}|s)$. If we replace $P_{\mathrm{ex}}(\mathbf{r}|s)$ by $P_{\mathrm{su}}(\mathbf{r}|s)$ in Equation (4), we define a new decoding algorithm

$$\hat{s} = D_{\mathrm{su}}(\mathbf{r}) = \arg\max_s P_{\mathrm{su}}(s|\mathbf{r}) = \arg\max_s P_{\mathrm{su}}(s, \mathbf{r}). \qquad (6)$$

which, as discussed below, may or may not be optimal, depending on how the decoder is used.

2.1.8. Two Different Decoding Strategies

One alternative, here referred to as "decoding method α" is that, for each response **r** obtained experimentally, one decodifies a stimulus $\hat{s}$ using the new mapping of Equation (6). In this case, the chain $s \rightarrow \mathbf{r} \rightarrow \hat{s}$ gives rise to the confusion matrix

$$P^{\alpha}(s, \hat{s}) = \sum_{\mathbf{r} \in D_{su}^{-1}(\hat{s})} P_{ex}(s, \mathbf{r}), \tag{7}$$

where the sum runs over all response vectors **r** that are mapped onto $\hat{s}$ by the new decoding algorithm D_{su}, and the probability $P_{ex}(\mathbf{r}, s)$ appearing in the right-hand side is the real one, since responses **r** are generated experimentally. It is easy to see that in this case, the decoding accuracy of the new algorithm is suboptimal, since responses **r** are generated with the original distribution $P_{ex}(\mathbf{r}|s)$, and for that distribution, the optimal decoder is given by Equation (4) with $P = P_{ex}$. In the literature, training a decoder with a probability $P_{su}(\mathbf{r}|s)$ and then operating it on variables that are generated with $P_{ex}(\mathbf{r}|s)$ is called *mismatched decoding*. In what follows, information values calculated from the distribution of Equation (7) are noted as $I_{\hat{s}}^{\alpha}$.

A second alternative, "decoding method β," is that, for each stimulus s, a surrogate response $\mathbf{R}_{su}$ is drawn using the new distribution $P_{su}(\mathbf{r}|s)$. If the sampled value is $\mathbf{R}_{su} = \mathbf{r}$, the stimulus $\hat{s} = D_{su}(\mathbf{r})$ is decoded. In this case, the confusion matrix is

$$P^{\beta}(s, \hat{s}) = \sum_{\mathbf{r} \in D_{su}^{-1}(\hat{s})} P_{su}(s, \mathbf{r}), \tag{8}$$

where as before, the sum runs over all response vectors **r** that are mapped onto $\hat{s}$ by the decoding algorithm $D_{su}(\mathbf{r})$, but now the probability $P_{su}(\mathbf{r}, s)$ appearing in the right-hand side is the surrogate one, since responses $\mathbf{R}_{su}$ are not generated experimentally. In this case, there is no mismatch between the construction and operation of the decoder, and D_{su} is optimal, in the sense that no other algorithm decodes $\mathbf{R}_{su}$ with higher decoding accuracy. One should bear in mind, however, that the surrogate responses are not the responses observed experimentally, that they may well take values in a response set that does not coincide with the set of real responses, and that $\mathbf{R}_{su}$ is not necessarily obtained by transforming the real response $\mathbf{R}_{ex}$ with a stimulus-independent mapping (see below). In what follows, information values calculated from the distribution of Equation (8) are noted as $I_{\hat{s}}^{\beta}$. Methods α and β can be easily extended to encompass also ranked decoding, *mutatis mutandis*.

The two alternative decoding methods yield two different decoding accuracies. To distinguish them, we use the notation $A_{\mathbf{R}_1}^{\mathbf{R}_2}$. The superscript indicates the variable whose probability distribution is used to construct the decoder in Equation (4), and consequently, determines the set of $\mathbf{r} \in D_{su}^{-1}(\hat{s})$ that contribute to the sums of Equations (7) and (8). The subscript indicates the variable upon which the decoder is applied, and its probability distribution is summed in the right-hand side of Equations (7) and (8). That is, $A_{\mathbf{R}_1}^{\mathbf{R}_2}$ is computed through Equation (3) with

$$P_{\mathbf{R}_1}^{\mathbf{R}_2}(s, \hat{s}) = \sum_{\mathbf{r} \in D_{\mathbf{R}_2}^{-1}(\hat{s})} P(S = s, \mathbf{R}_1 = \mathbf{r}), \tag{9}$$

so that $P^{\alpha}(s, \hat{s}) = P_{\mathbf{R}_{ex}}^{\mathbf{R}_{su}}(s, \hat{s})$ and $P^{\beta}(s, \hat{s}) = P_{\mathbf{R}_{su}}^{\mathbf{R}_{su}}(s, \hat{s})$.

2.2. The Applicability of the Data-Processing Inequality

Assessing the relevance of a response feature typically involves a subtraction $\Delta I = I - I'$, where I and I' represent the mutual information between stimuli and a set of response features containing or not containing the tested feature, respectively. The magnitude of ΔI is often interpreted as the information provided by the tested feature. This interpretation requires ΔI to be positive, since intuitively, one would imagine that removing a response feature cannot increase the encoded information.

As shown below, a formal proof of this intuition may or may not be possible invoking the data processing inequality (see Section 2.1.3 and reference [16]), depending on the method used to eliminate the tested feature. As a consequence, there are cases in which ΔI is indeed negative (see below). In these cases, the tested feature is detrimental to information encoding [9].

2.2.1. Reduced Representations

There are several procedures by which the tested feature can be removed from the response. The validity of the data-processing inequalities (see definition in Section 2.1.3) depends on the chosen procedure. In order to specify the conditions in which the inequalities hold, we here introduce the concept of *reduced representations*. When the response feature under evaluation is removed from $\mathbf{R}_{ex}$ by a deterministic mapping $\mathbf{R}_{su} = f(\mathbf{R}_{ex})$, we call the obtained variable $\mathbf{R}_{su}$ a *reduced representation* of $\mathbf{R}_{ex}$. A required condition for a mapping to be a reduced representation is that the function f be stimulus-independent, that is, that the value of $\mathbf{R}_{su}$ be conditionally independent from s. Mathematically, this means that $P(\mathbf{r}_{su}, s | \mathbf{r}_{ex}) = P(\mathbf{r}_{su} | \mathbf{r}_{ex}) \, P(s | \mathbf{r}_{ex})$. If the mapping f and the conditional response distribution $P_{ex}(\mathbf{r}|s)$ are known, the distribution $P_{su}(\mathbf{r}|s)$ can be derived using standard methods. The data processing inequality ensures that for all reduced representations, $I_{\mathbf{R}_{ex}} \geq I_{\mathbf{R}_{su}}$.

Reduced representations are usually employed when the response feature whose relevance is to be assessed takes a definite value in each trial, as happens for example, with the number of spikes in a fixed time window, the latency of the firing response, or the activity of a specific neuron in a larger population of neurons. In these cases it is easy to construct $\mathbf{R}_{su}$ simply by dropping from $\mathbf{R}_{ex}$ the tested feature, or by fixing its value with some deterministic rule.

Reduced representations can also be used in other cases, for example, when the relevance of the feature *response accuracy* is assessed. This feature does not take a specific value in each trial; only by comparing multiple trials can the response accuracy be determined. A widely-used strategy is to represent spike trains with temporal bins of increasing duration, and to evaluate how the amount of information decreases as the representation becomes coarser. A sequence of surrogate responses is thereby defined, by progressively disregarding the fine temporal precision with which spike trains were recorded (Figure 1).

Several studies have reported an information $I_{\mathbf{R}_{su}}$ that decreases monotonically with the duration δt of the time bin (for example [21–23]). If there is a specific temporal scale in which spike-time precision is relevant—the alleged argument goes—a sudden drop in $I_{\mathbf{R}_{su}}(\delta t)$ appears at the relevant scale. It should be noted, however, that the data processing inequality does not ensure that $I_{\mathbf{R}_{su}}(\delta t)$ be a monotonically decreasing function of δt. In the example of Figure 1, representations $\mathbf{R}_{su}^1$ and $\mathbf{R}_{su}^2$ are defined with long temporal bins, the durations of which are integer multiples of the bin used for $\mathbf{R}_{ex}$. Hence, $\mathbf{R}_{su}^1$ and $\mathbf{R}_{su}^2$ are reduced representations of $\mathbf{R}_{ex}$, and the data processing inequality does indeed guarantee that $I_{\mathbf{R}_{ex}} \geq I_{\mathbf{R}_{su}^1}$ and $I_{\mathbf{R}_{ex}} \geq I_{\mathbf{R}_{su}^2}$. However, $\mathbf{R}_{su}^2$ is not a reduced representation of $\mathbf{R}_{su}^1$, so there is no reason why $I_{\mathbf{R}_{su}^2}$ should be smaller than $I_{\mathbf{R}_{su}^1}$, and indeed, Figure 1b shows an example where it is not. The representation constructed with bins of intermediate duration, namely 10 ms, does not distinguish between the two stimuli, whereas those of shorter and longer duration, 5 and 15 ms, do. A similar effect can be observed in the experimental data (freely available online) of Lefebvre et al. [24], when analysed with bins of sizes 5, 10 and 15 ms in windows of total duration 60 ms. Although these examples are rare, they demonstrate that there is no theoretical substantiation to the expectation of $I_{\mathbf{R}_{su}}$ to drop monotonically with increasing δt.

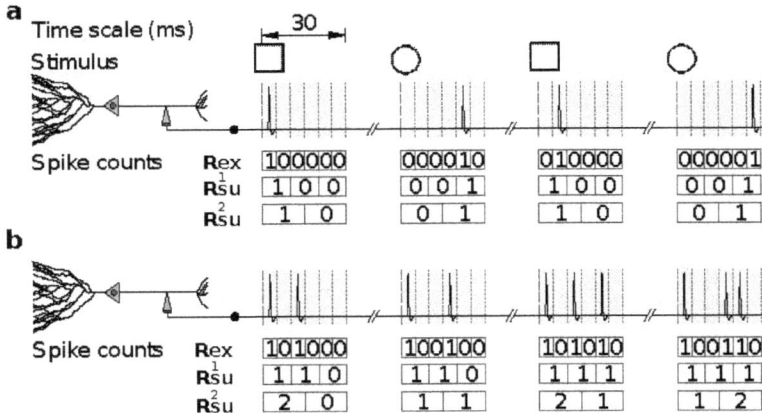

Figure 1. Assessing the relevance of response accuracy by varying the duration of the temporal bin. (a) Hypothetical intracellular recording of the spike patterns elicited by a single neuron after presenting in alternation two visual stimuli, □ and ○, each of which triggers two possible responses displayed in columns 1 and 3 for □, and 2 and 4 for ○. Stimulus probabilities and conditional response probabilities are arbitrary. Time is discretized in bins of 5 ms. The responses are recorded within 30 ms time-windows after stimulus onset. Spikes are fired with latencies that are uniformly distributed between 0 and 10 ms after the onset of □, and between 20 and 30 ms after the onset of ○. Responses are represented by counting the number of spikes within consecutive time-bins of size 5, 10 and 15 ms starting from stimulus onset, thereby yielding discrete-time sequences $\mathbf{R}_{ex}$, $\mathbf{R}_{su}^1$ and $\mathbf{R}_{su}^2$, respectively; (b) Same as a, but with stimuli producing two different types of response patterns composed of 2 or 3 spikes.

2.2.2. Stochastically Reduced Representations

When the response feature under evaluation is removed from the response variable $\mathbf{R}_{ex}$ by a stochastic mapping $\mathbf{R}_{ex} \rightarrow \mathbf{R}_{su}$, the obtained variable $\mathbf{R}_{su}$ is called a *stochastically reduced representation* of $\mathbf{R}_{ex}$. A required condition for a mapping to be a stochastically reduced representation is that the probability distribution of each $\mathbf{R}_{su}$ be dependent on $\mathbf{R}_{ex}$, but conditionally independent from s. In these circumstances, the data processing inequality ensures that $I_{\mathbf{R}_{ex}} \geq I_{\mathbf{R}_{su}}$. If the statistical properties of the noisy components of the mapping are known, as well as the conditional response probability distribution $P_{ex}(\mathbf{r}|s)$, the distribution $P_{su}(\mathbf{r}|s)$ can be derived using standard methods. Formally, stochastic representations $\mathbf{R}_{su}$ are obtained through stimulus-independent stochastic functions of the original representation $\mathbf{R}_{ex}$. After observing that $\mathbf{R}_{ex}$ adopted the value $\mathbf{r}_{ex}$, these functions produce a single value $\mathbf{r}_{su}$ for $\mathbf{R}_{su}$ chosen with transition probabilities $Q(\mathbf{r}_{su}|\mathbf{r}_{ex})$ such that

$$P_{su}(\mathbf{r}_{su}|s) = \sum_{\mathbf{r}_{ex}} P_{ex}(\mathbf{r}_{ex}|s) \, Q(\mathbf{r}_{su}|\mathbf{r}_{ex}) \, . \tag{10}$$

To illustrate the utility of stochastically reduced representations, we discuss their role in providing alternative strategies when assessing the relevance of spike-timing precision, not by changing the bin size as in Figure 1, but by randomly manipulating the responses, as illustrated in Figure 2.

Entropy **2018**, *20*, 879

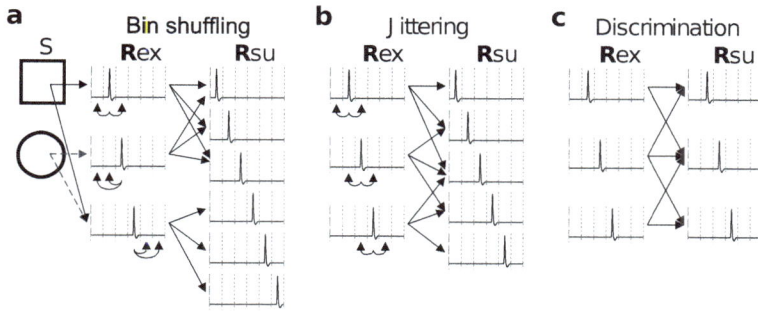

Figure 2. Examples of stochastic codes. Alternative ways of assessing the relevance of spike-timing precision. (**a**) Stochastic function (arrows on the left) modeling the encoding process. The elicited response r_{ex} is turned into a surrogate response r_{su} with a transition probability $Q(r_{su}|r_{ex})$ given by Equation (11). This function turns $\mathbf{R}_{ex}$ into a stochastic representation $\mathbf{R}_{su}$ by shuffling spikes and silences within bins of 15 ms starting from stimulus onset; (**b**) Responses r_{ex} in panel (a) are transformed by a stochastic function with $Q(r_{su}|r_{ex})$ given by Equation (12), which introduces jitter uniformly distributed within 15 ms windows centered at each spike; (**c**) Responses r_{ex} in panel (a) are transformed by a stochastic function with $Q(r_{su}|r_{ex})$ given by Equation (13), which models the inability to distinguish responses with spikes occurring in adjacent bins, or equivalently, with distances $d^{\text{spike}}[q = 1] \leq 1$ or $d^{\text{interval}}[q = 1] \leq 1$ (see [25,26] for further remarks on these distances). Notice that $\mathbf{R}_{su}$ samples the same response set as $\mathbf{R}_{ex}$.

The method of Figure 2a yields the same information $I_{\mathbf{R}_{su}}$ and response accuracy as the method producing $\mathbf{R}^2_{su}$ in Figure 1. Each method yields responses that can be related to the responses of the other method through a stimulus-independent deterministic or stochastic function. Both methods suffer from the same drawback: They treat spikes differently depending on their location within the 15 ms time window. Indeed, both methods preserve the distinction between two spikes located in different windows, but not within the same window, even if the separation between the spikes is the same. The mapping illustrated in Figure 2a has transition probabilities

$$Q(\mathbf{r}_{su}|\mathbf{r}_{ex}) = \frac{1}{3} \begin{bmatrix} 1 & 1 & 1 & 0 & 0 & 0 \\ 1 & 1 & 1 & 0 & 0 & 0 \\ 0 & 0 & 0 & 1 & 1 & 1 \end{bmatrix}, \tag{11}$$

where rows enumerate the elements of the ordered set $\mathcal{R}_{ex}=\{[2],[3],[4]\}$ from where $\mathbf{R}_{ex}$ is sampled, and columns enumerate the elements of the ordered set $\mathcal{R}_{su}=\{[1],[2],[3],[4],[5],[6]\}$ from where $\mathbf{R}_{su}$ is sampled.

A third method, jittering, consists in shuffling the recorded spikes within time windows centered at each spike (Figure 2b). The responses generated by this method need not be obtainable from the responses generated by the mappings of Figure 2a or Figure 1 through stimulus-independent stochastic functions. Still, the method of Figure 2b inherently yields a stochastic code, and, unlike the methods discussed previously, treats all spikes in the same manner. The mapping illustrated in Figure 2b has transition probabilities

$$Q(\mathbf{r}_{su}|\mathbf{r}_{ex}) = \frac{1}{3} \begin{bmatrix} 1 & 1 & 1 & 0 & 0 \\ 0 & 1 & 1 & 1 & 0 \\ 0 & 0 & 1 & 1 & 1 \end{bmatrix}, \tag{12}$$

where rows enumerate the elements of the ordered set $\mathcal{R}_{ex}=\{[2],[3],[4]\}$ from where $\mathbf{R}_{ex}$ is sampled, and columns enumerate the elements of the ordered set $\mathcal{R}_{su}=\{[1],[2],[3],[4],[5]\}$ from where $\mathbf{R}_{su}$ is sampled.

As a fourth example, consider the effect of response discrimination, as studied in the seminal work of Victor and Purpura [25]. There, two responses were considered indistinguishable when some measure of distance between the responses was less than a predefined threshold. However, neural responses were transformed through a method based on cross-validation that is not guaranteed to be stimulus-independent. Depending on the case, hence, this fourth method may or may not be a stochastically reduced representation. The case chosen in Figure 2c is a successful example, and the associated matrix of transition probabilities is

$$Q(\mathbf{r}_{su}|\mathbf{r}_{ex}) = \frac{1}{6} \begin{bmatrix} 3 & 3 & 0 \\ 2 & 2 & 2 \\ 0 & 3 & 3 \end{bmatrix}, \tag{13}$$

where rows and columns enumerate the elements of the ordered set $\mathcal{R}_{ex}=\mathcal{R}_{su}=\{[2],[3],[4]\}$ from where both $\mathbf{R}_{ex}$ and $\mathbf{R}_{su}$ are sampled.

Other methods exist which merge indistinguishable responses, thereby yielding reduced representations. These methods, however, are limited to notions of similarity that are transitive, a condition not fulfilled, for example, by those based on Euclidean distance, edit distance, or by the case of Figure 2c.

Stochastically reduced representations include reduced representations as limiting cases. Indeed, when for each $\mathbf{r}_{ex}$ there is a $\mathbf{r}_{su}$ such that $Q(\mathbf{r}_{su}|\mathbf{r}_{ex}) = 1$, stochastic representations become reduced representations (Figure 3). The possibility to include stochasticity, however, broadens the range of alternatives. Consider for example the hypothetical experiment in Figure 3a, in which the neural responses $\mathbf{R}_{ex}=[L,C]$ can be completely characterized by the first-spike latencies (L) and the spike counts (C). The importance of C can be studied for example by using a reduced code that replaces all C-values with a constant (Figure 3b). In this case,

$$Q(\mathbf{r}_{su}|\mathbf{r}_{ex}) = \begin{bmatrix} 1 & 0 & 0 \\ 0 & 1 & 0 \\ 0 & 1 & 0 \\ 0 & 0 & 1 \end{bmatrix}, \tag{14}$$

where rows enumerate the elements of the ordered set $\mathcal{R}_{ex}=\{[2,1],[3,1],[3,2],[4,2]\}$ from where $\mathbf{R}_{ex}$ is sampled, and columns enumerate the elements of the ordered set $\mathcal{R}_{su}=\{[2,1],[3,1],[4,1]\}$ from where $\mathbf{R}_{su}$ is sampled.

Another alternative is to assess the relevance of C by means of a stochastic code that shuffles the values of C across all responses with the same L (Figure 3c). In this case,

$$Q(\mathbf{r}_{su}|\mathbf{r}_{ex}) = \begin{bmatrix} 1 & 0 & 0 & 0 \\ 0 & a & \bar{a} & 0 \\ 0 & a & \bar{a} & 0 \\ 0 & 0 & 0 & 1 \end{bmatrix} \tag{15}$$

where rows enumerate the elements of the ordered set $\mathcal{R}_{ex}=\{[2,1],[3,1],[3,2],[4,2]\}$ from where $\mathbf{R}_{ex}$ is sampled, and columns enumerate the elements of the ordered set $\mathcal{R}_{su}=\{[2,1],[3,1],[3,2],[4,2]\}$ from where $\mathbf{R}_{su}$ is sampled. The parameter a is arbitrary, as long as $0 < a < 1$. We use the notation $\bar{a} = 1 - a$.

A third option is to use a stochastic code that preserves the original value of L but chooses the value of C from some possibly L−dependent probability distribution (Figure 3d), for which

$$Q(\mathbf{r}_{su}|\mathbf{r}_{ex}) = \begin{bmatrix} b & 0 & 0 & \bar{b} & 0 & 0 \\ 0 & c & 0 & 0 & \bar{c} & 0 \\ 0 & c & 0 & 0 & \bar{c} & 0 \\ 0 & 0 & d & 0 & 0 & \bar{d} \end{bmatrix} \tag{16}$$

where rows enumerate the elements of the ordered set $\mathcal{R}_{ex}=\{[2,1],[3,1],[3,2],[4,2]\}$ from where $\mathbf{R}_{ex}$ is sampled, and columns enumerate the elements of the ordered set $\mathcal{R}_{su}=\{[2,1],[3,1],[4,1],[2,2],[3,2],[4,2]\}$ from where $\mathbf{R}_{su}$ is sampled. The parameters a,b,c and d are arbitrary, as long as $0<a,b,c,d<1$; and we have used the notation $\bar{x}=1-x$ for any number x.

a

Time scale (ms)   ⊢── 30 ──⊣

Stimulus   S

Latency   L   [2]   [3]   [3]   [4]
Spike counts   C   [1]   [1]   [2]   [2]

b   Reduced code   **c**   Stochastic code   **d**   Stochastic code

Rex=[L,C] ──[Ĺ, 1]=Rsu   Rex=[L,C] ──[Ĺ, Ĉ]=Rsu   Rex=[L,C] ──[Ĺ, Ĉ]=Rsu

[2, 1] ── [2, 1]   [2, 1] ── [2, 1]   [2, 1] ⟍ [2,2]

                                         [2,1]

[3, 2] ⟍   [3, 2] ⟍ [3, 2]   [3, 2] ⟍ [3,2]

       ⟩ [3, 1]   ⨯   ⨯

[3, 1] ⟋   [3, 1] ⟋ [3, 1]   [3, 1] ⟋ [3, 1]

                                           [4,2]

[4, 2] ── [4, 1]   [4, 2] ── [4,2]   [4, 2] ⟋ [4, 1]

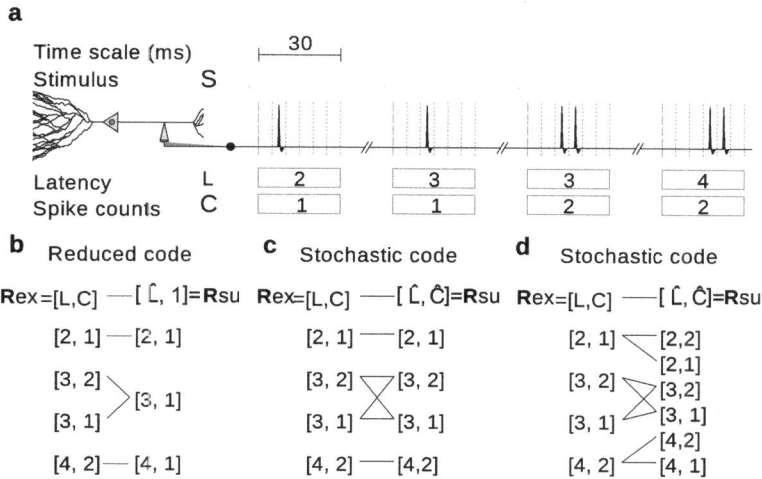

Figure 3. Stochastically reduced representations include and generalize deterministically reduced representations. (**a**) Analogous description to Figure 1a, but with responses characterized using a representation $\mathbf{R}_{ex} = [L, C]$ based on the first-spike latency (L) and the spike-count (C); (**b**) Deterministic transformation (arrows) of $\mathbf{R}_{ex}$ in panel a into a reduced code $\mathbf{R}_{su} = [\hat{L}, 1]$, which ignores the additional information carried in C by considering it constant and equal to unity. This reduced code can also be reinterpreted as a stochastic code with transition probabilities $Q(\mathbf{r}_{su}|\mathbf{r}_{ex})$ defined by Equation (14); (**c**) The additional information carried in C is here ignored by shuffling the values of C across all trails with the same L, thereby turning $\mathbf{R}_{ex}$ in panel a into a stochastic code $\mathbf{R}_{su}=[\hat{L}, \hat{C}]$ with transition probabilities $Q(\mathbf{r}_{su}|\mathbf{r}_{ex})$ defined by Equation (15); (**d**) The additional information carried in C is here ignored by replacing the actual value of C for one chosen with some possibly L-dependent probability distribution (Equation (16)).

2.2.3. Modification of the Conditional Response Probability Distribution

When the response feature under evaluation is removed by altering the real conditional response probability distribution $P_{ex}(\mathbf{r}|s)$, and transforming it into a surrogate distribution $P_{su}(\mathbf{r}|s)$, the obtained response model is here said to implement a *probabilistic removal* of the tested feature. Probabilistic removals are usually employed when assessing the relevance of correlations between neurons in a population, since correlations are not a variable that can be deleted from each individual response. For example, if $\mathbf{R}=(R_1, \ldots, R_n)$ represents the spike count of n different neurons, the real distribution $P_{ex}(r_1, \ldots, r_n|s)$ is replaced by a new distribution $P_{su}(r_1, \ldots, r_n|s)$ in which all neurons are conditionally independent, that is,

$$P_{su}(\mathbf{r}|s) = P_{NI}(\mathbf{r}|s) = \prod_{i=1}^{n} P_{ex}(r_i|s), \tag{17}$$

where, following the notation introduced previously [17], the generic subscript "su" was replaced by "NI" to indicate "noise-independent".

The probabilistic removal of a response feature may or may not be describable in terms of a deterministically or a stochastically reduced representation. In other words, there may or may not exist a mapping $\mathbf{R}_{ex} \rightarrow \mathbf{R}_{su}$, or equivalently, a matrix of transition probabilities $Q(\mathbf{r}_{su}|\mathbf{r}_{ex})$, that captures the replacement of $P_{ex}(\mathbf{r}|s)$ by $P_{su}(\mathbf{r}|s)$. It is important to assess whether such a matrix exists, since the data processing inequality is only guaranteed to hold with reduced representations, stochastic or not. If no reduced representation can capture the effect of a probabilistic removal, the data processing inequality may not hold, and $I_{\mathbf{R}_{su}}$ may well be larger than $I_{\mathbf{R}_{ex}}$.

In order to determine whether a stochastically reduced representation exists, the first step is to discern whether Equation (10) constitutes a compatible or an incompatible linear system for the matrix elements $Q(\mathbf{r}_{su}|\mathbf{r}_{ex})$. If the system is incompatible, there is no solution. In the compatible case, which is often indeterminate, a solution entirely composed of non-negative numbers that sum up to unity in each row is required. Given enough time and computational power, the problem can always be solved in the framework of linear programming [27]. In practical cases, however, the search is often hampered by the curse of dimensionality. To facilitate the labour, here we list a few necessary (though not sufficient) conditions that must be fulfilled for the mapping to exist. If any of the following properties does not hold, Equation (10) has no solution, so there is no need to begin a search.

Property 1. *Let* $\mu(s)$ *be a probability distribution defined in the set of stimuli that may or may not be equal to the actual distribution with which stimuli appear in the experiment under study. For any stimulus s, the inequality* $I_\mu(\mathbf{R}_{su}; S = s) \leq I_\mu(\mathbf{R}_{ex}; S = s)$ *between stimulus-specific informations [28,29] must hold, where*

$$I_\mu(\mathbf{R}; S = s) = \sum_{\mathbf{r}} P(\mathbf{r}|s) \, \log_2 \frac{P(\mathbf{r}|s)}{\sum_{s'} P(\mathbf{r}|s') \, \mu(s')}. \tag{18}$$

Proof. If $Q(\mathbf{r}_{su}|\mathbf{r}_{ex})$ exists, then Equation (10) can be inserted in Equation (18). Using the log-sum inequality [16], Property 1 follows. □

If we multiply both sides of the inequality by $\mu(s')$ and sum over s', we obtain an inequality between the mutual informations $I_\mu(\mathbf{R}_{su}; S) \leq I_\mu(\mathbf{R}_{ex}; S)$. If $\mu(s) = P(s)$, this result reduces to the data-processing inequality $I_{\mathbf{R}_{su}} \leq I_{\mathbf{R}_{ex}}$.

Property 2. *If* $Q(\mathbf{r}_{su}|\mathbf{r}_{ex})$ *exists, then* $Q(\mathbf{r}_{su}|\mathbf{r}_{ex}) = 0$ *whenever* $P_{ex}(s, \mathbf{r}_{ex}) > 0$ *and* $P_{su}(s, \mathbf{r}_{su}) = 0$ *for at least some s.*

Proof. Suppose that $Q(\mathbf{r}_{su}|\mathbf{r}_{ex}) > 0$ when $P_{ex}(s, \mathbf{r}_{ex}) > 0$ for some s. Then, Equation (10) yields $P_{su}(\mathbf{r}_{su}|s) > 0$, contradicting the hypothesis that $P_{su}(\mathbf{r}_{su}|s) = 0$. Hence, $Q(\mathbf{r}_{su}|\mathbf{r}_{ex})$ must vanish. □

For example, in Figure 4a, we decorrelate first-spike latencies (L) and spike counts (C) by replacing the true conditional distribution $P_{ex}(\mathbf{r}|s)$ (left panel) by its noise-independent version $P_{su} = P_{NI}(\mathbf{r}|s)$ defined in Equation (17) (middle panel). Before searching for a mapping $\mathbf{R}_{ex} \rightarrow \mathbf{R}_{su}$, we verify that the condition $I_{\mathbf{R}_{ex}} > I_{\mathbf{R}_{su}}$ holds. Moreover, for several choices of $\mu(\bigcirc)$ and $\mu(\square)$, one may confirm that $I_\mu(\mathbf{R}_{ex}; S = \bigcirc) > I_\mu(\mathbf{R}_{su}; S = \bigcirc)$, as well as $I_\mu(\mathbf{R}_{ex}; S = \square) > I_\mu(\mathbf{R}_{su}; S = \square)$. These results motivate the search for a solution of Equation (10) for $Q(\mathbf{r}_{su}|\mathbf{r}_{ex})$. The transition probability must be zero at least whenever $\mathbf{R}_{su} \in \{[1,3]; [2,3]; [3,3]; [3,2]; [3,1]\}$ and $\mathbf{R}_{ex} \in \{[1,2]; [2,1]\}$ (Property 2). One possible solution is

$$Q(\mathbf{r}_{su}|\mathbf{r}_{ex}) = \frac{1}{2} \begin{bmatrix} 2b & \bar{b}c & \bar{c}\bar{b} & \bar{b}c & 0 & 0 & \bar{c}\bar{b} & 0 & 0 \\ \bar{a} & 2a & 0 & 0 & \bar{a} & 0 & 0 & 0 & 0 \\ a & 0 & 0 & 2\bar{a} & a & 0 & 0 & 0 & 0 \\ 0 & b & 0 & b & 2\bar{b}c & \bar{c}\bar{b} & 0 & \bar{c}\bar{b} & 0 \\ 0 & 0 & b & 0 & 0 & \bar{b}c & b & \bar{b}c & 2\bar{c}\bar{b} \end{bmatrix}. \tag{19}$$

where each response is defined by a vector $[L, C]$, and rows and columns enumerate the elements of the ordered sets $\mathcal{R}_{ex} = \{[1,1], [1,2], [2,1], [2,2], [3,3]\}$ and $\mathcal{R}_{su} = \{[1,1], [1,2], [1,3], [2,1], [2,2], [2,3], [3,1], [3,2], [3,3]\}$ from where $\mathbf{R}_{ex}$ and $\mathbf{R}_{su}$ are sampled, respectively. In Equation (19), $a = P_{ex}([1,2]|\square)$; $b = P_{ex}([1,1]|\bigcirc)$; and $c = P_{ex}([2,2]|\bigcirc)/\bar{b}$.

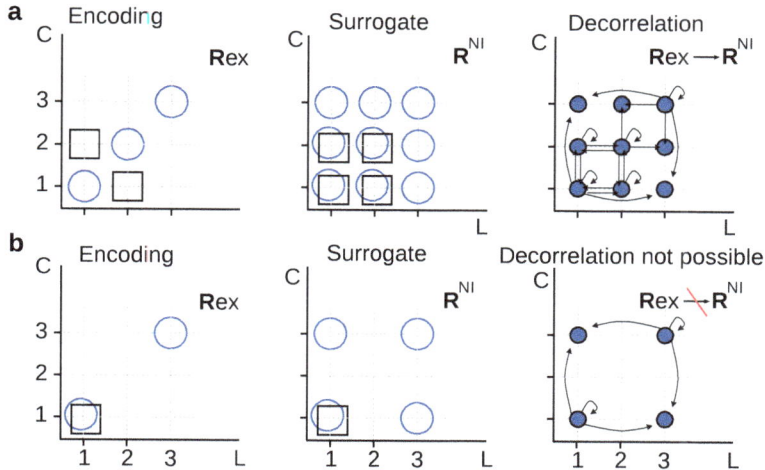

Figure 4. Relation between probabilistic removal and stochastic codes. (**a**) Cartesian coordinates depicting: on the left, responses $\mathbf{R}_{ex}$ of a neuron for which L and C are positively correlated when elicited by $\bigcirc$, and negatively correlated when elicited by $\square$; in the middle, the surrogate responses $\mathbf{R}_{su} = \mathbf{R}_{NI}$ that would occur should L and C be noise independent (middle); and on the right, a stimulus-independent stochastic function that turns $\mathbf{R}_{ex}$ into $\mathbf{R}_{su}$ with $Q(\mathbf{r}_{su}|\mathbf{r}_{ex})$ given by Equation (19); (**b**) Same description as in (**a**), but with L and C noise independent given $\square$, and with the stochastic function depicted on the right turning $\mathbf{R}_{ex}$ into $\mathbf{R}_{NI}$ given $\bigcirc$ but not $\square$.

However, stochastically reduced representations are not always guaranteed to exist. For example, in Figure 4b, it is easy to verify that the condition $I_\mu(\mathbf{R}_{ex}; S = \square) < I_\mu(\mathbf{R}_{su}; S = \square)$ holds for any $\mu(\bigcirc) \neq 0$. Therefore, no stochastic mapping can transform $\mathbf{R}_{ex}$ into $\mathbf{R}_{su}$ in such a way that $P_{ex}(\mathbf{r}|s)$ is converted into $P_{su}(\mathbf{r}|s)$. Schneidman et al. [9] employed an analogous example, but involving different neurons instead of response aspects. The two examples of Figure 4 motivate the following theorem:

Theorem 1. *No deterministic mapping $\mathbf{R}_{ex} \rightarrow \mathbf{R}_{su}$ exists transforming the conditional probability $P_{ex}(\mathbf{r}|s)$ into its noise-independent version $P_{su} = P_{NI}(\mathbf{r}|s)$ defined in Equation (17). Stochastic mappings $\mathbf{R}_{ex} \rightarrow \mathbf{R}_{su}$ may or may not exist, depending on the conditional probability $P_{ex}(\mathbf{r}|s)$.*

Proof. See Appendix B.2. □

In addition, when a stochastic mapping $\mathbf{R}_{ex} \rightarrow \mathbf{R}_{su}$ exists, the values of the probabilities $Q(\mathbf{r}_{su}|\mathbf{r}_{ex})$ may well depend on the discarded response aspect, as well as on the preserved response aspects. We mention this fact, because when assessing the relevance of noise correlations, the marginals $P_{ex}(r_i|s)$ suffice for us to write down the surrogate distribution $P_{su}(\mathbf{r}|s) = P_{NI}(\mathbf{r}|s)$, with no need to know the full distribution $P_{ex}(\mathbf{r}|s)$ containing the noise correlations. One could have hoped that perhaps also the mapping $\mathbf{R}_{ex} \rightarrow \mathbf{R}_{su}$ (assuming that such a mapping exists) could be calculated with no knowledge of the noise correlations. This is, however, not always true, as stated in the theorem below. Two experiments with the same marginals and different amounts of noise correlations may require different mappings to eliminate noise correlations, as illustrated in the the example of Figure 5. More formally:

Theorem 2. *The transition probabilities $Q(\mathbf{r}_{su}|\mathbf{r}_{ex})$ of stochastic codes that ignore noise correlations may depend both on the marginal likelihoods (preserved at the output of the mapping), and on the noise correlations (eliminated at the output of the mapping).*

Proof. See Appendix B.3. □

The solution of Equation (10) for the example of Figure 5 is

$$
Q(\mathbf{r}_{su}|\mathbf{r}_{ex}) = \frac{1}{2}
\begin{bmatrix}
\bar{a} & 2a & 0 & \bar{a} & 0 & 0 & 0 \\
a & 0 & 2\bar{a} & a & 0 & 0 & 0 \\
0 & 0 & 0 & b & 2\bar{b} & 0 & b \\
0 & 0 & 0 & \bar{b} & 0 & 2b & \bar{b}
\end{bmatrix},
\tag{20}
$$

where each response is defined by a vector $[L, C]$, and rows and columns enumerate the elements of the ordered sets $\mathcal{R}_{ex} = \{[1,2],[2,1],[2,3],[3,2]\}$ and $\mathcal{R}_{su} = \{[1,1],[1,2],[2,1],[2,2],[2,3],[3,2],[3,3]\}$ from where $\mathbf{R}_{ex}$ and $\mathbf{R}_{su}$ are sampled, respectively. In Equation (20), $a = P(\mathbf{R}_{ex} = [1,2]|S = \square)$; and $b = P(\mathbf{R}_{ex} = [3,2]|S = \bigcirc)$. The fact that the matrix in Equation (20) bears an explicit dependence on these parameters–and not only on $P_{ex}(L|S)$ and $P_{ex}(C|S)$–implies that the transformation between $\mathbf{R}_{ex}$ and $\mathbf{R}_{su}$ depends on the amount of noise correlations in $\mathbf{R}_{ex}$.

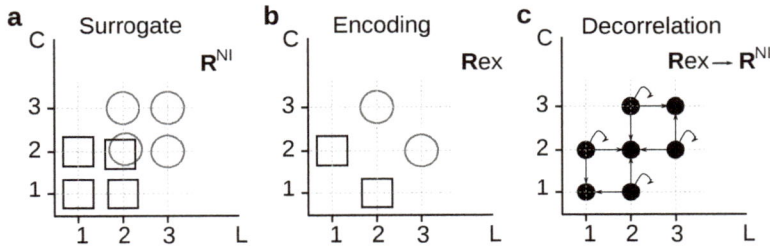

Figure 5. Stochastically reduced representations that ignore noise correlations may depend on them.(**a**) Cartesian coordinates representing a hypothetical experiment in which two different stimuli, $\square$ and $\bigcirc$, elicit single neuron responses ($\mathbf{R}_{su} = \mathbf{R}_{NI}$) that are completely characterized by their first-spike latency (L) and spike counts (C). Both L and C are noise independent; (**b**) Cartesian coordinates representing a hypothetical experiment with the same marginal probabilities $P_{ex}(l|s)$ and $P_{ex}(c|s)$ as in panel (a), with one among many possible types of noise correlations between L and C; (**c**) Stimulus-independent stochastic function transforming the noise-correlated responses $\mathbf{R}_{ex}$ of panel (b) into the noise-independent responses $\mathbf{R}_{su} = \mathbf{R}_{NI}$ of panel (a). The transition probabilities $Q(\mathbf{r}_{su}|\mathbf{r}_{ex})$ are given in Equation 20, and they bear an explicit dependence on the amount of noise correlations.

2.3. Multiple Measures to Assess the Relevance of a Specific Response Feature

The importance of a specific response feature has been previously quantified in many ways (see [17,30] and references therein), which have oftentimes led to heated debates about their merits and drawbacks [9,11,12,17,31–33]. Here we consider several measures, to underscore the diversity of the meanings with which the relevance of a given feature has been assessed so far. They are mathematically defined as

$$\Delta I_{\mathbf{R}_{su}} = I_{\mathbf{R}_{ex}} - I_{\mathbf{R}_{su}} \tag{21}$$

$$\Delta I_{\hat{\mathsf{S}}} = I_{\mathbf{R}_{ex}} - I_{\hat{\mathsf{S}}}^{\beta} \tag{22}$$

$$\Delta I_{\hat{\varsigma}} = I_{\mathbf{R}_{ex}} - I_{\hat{\varsigma}}^{\beta} \tag{23}$$

$$\Delta A_{\mathbf{R}_{su}} = A_{\mathbf{R}_{ex}}^{\mathbf{R}_{ex}} - A_{\mathbf{R}_{su}}^{\mathbf{R}_{su}} \tag{24}$$

$$\Delta I^{D} = \sum_{s,\mathbf{r}} P_{ex}(s,\mathbf{r}) \ln \frac{P_{ex}(s|\mathbf{r})}{P_{su}(s|\mathbf{r})} \tag{25}$$

$$\Delta I^{DL} = \min_{\theta} \sum_{s,\mathbf{r}} P_{ex}(s,\mathbf{r}) \ln \frac{P_{ex}(s|\mathbf{r})}{P_{su}(s|\mathbf{r},\theta)} \tag{26}$$

$$\Delta I^{LS} = I_{\mathbf{R}_{ex}} - I_{\hat{\mathsf{S}}}^{\alpha} \tag{27}$$

$$\Delta I^{B} = I_{\mathbf{R}_{ex}} - I_{\hat{\varsigma}}^{\alpha} \tag{28}$$

$$\Delta A^{B} = A_{\mathbf{R}_{ex}}^{\mathbf{R}_{ex}} - A_{\mathbf{R}_{ex}}^{\mathbf{R}_{su}} \tag{29}$$

Equations (22)–(24) are based on matched decoders, that is, decoders operating on responses governed by the same probability distribution involved in their construction (method β). Instead, Equations (25)–(28) are based on the operation of mismatched decoders (method α). Each measure of Equations (21)–(24) has one or two homologous measures in Equations (25)–(29), as illustrated in Figure 6.

Figure 6. Relations between the measures defined in Equations (21)–(29). The four measures on the left are either encoding-oriented ($\Delta I_{\mathbf{R}_{su}}$, on a pink background), or half-way between encoding- and decoding-oriented (the last three, gray background). The five measures on the right are all decoding-oriented (light-blue background). Each measure on the left has a conceptually related measure on the right on the same line, except for $\Delta I_{\mathbf{R}_{su}}$, which has two associated decoding-oriented measures: ΔI^{D} and ΔI^{LD}. The distinction between the measures on pink and on gray background relies on the fact that $\Delta I_{\mathbf{R}_{su}}$ does not involve a decoding process. Instead, $\Delta I_{\hat{\mathsf{S}}}, \Delta I_{\hat{\varsigma}}$ and $\Delta A_{\mathbf{R}_{su}}$ decode a stimulus (or rank the stimuli) with decoding method β. This decoding is not meant to be applicable to real experiments, since (as opposed to the truly decoding-oriented measures on the right, that operate with method α) the decoding is applied to the surrogate responses $\mathbf{R}_{su}$, not the real ones $\mathbf{R}_{ex}$.

We here describe the measures briefly, and refer the interested reader to the original papers.

In Equation (21), $I_{\mathbf{R}_{ex}}$ and $I_{\mathbf{R}_{su}}$ are the mutual informations between the set of stimuli and a set of responses governed by the distributions $P_{ex}(\mathbf{r}|s)$ and $P_{su}(\mathbf{r}|s)$, respectively. Thus, $\Delta I_{\mathbf{R}_{su}}$ is the simplest way in which the information encoded by the true responses can be compared with that of the surrogate responses. This comparison has been employed for more than six decades in neuroscience [34,35] to study, for example, the encoding of different stimulus features in spike counts, in synchronous spikes, and in other forms of spike patterns, both in single neurons and populations (see [30] and references therein).

The measure ΔI^{D} defined in Equation (25) was introduced by Nirenberg et al. [8] to study the role of noise correlations, and was later extended to arbitrary deterministic mappings [10,12,13]. Here we use the supra-script D to indicate that the measure is the "divergence" (in the Kullback-Leibler sense) between the posterior stimulus distributions calculated with the real and the surrogate responses, respectively. In [10], Nirenberg and Latham argued that the important feature of ΔI^D is that it represents the information loss of a mismatched decoder trained with $P_{su}(\mathbf{r}|s)$ but operated on the real responses, sampled from $P_{ex}(\mathbf{r}|s)$. Not before long, Schneidman et al. [9] noticed that ΔI^D can exceed $I_{\mathbf{R}_{ex}}$. The interpretation of ΔI^D as a measure of information loss would imply that decoders trained with surrogate responses can lose more information than the one encoded by the real response. In fact, ΔI^D tends to infinity if $P_{su}(s|\mathbf{r}) \to 0$ when $P(s|\mathbf{r}) > 0$ for some s. In the limit, ΔI^D becomes undefined when $P_{su}(\mathbf{r}) = 0$ and $P_{ex}(\mathbf{r}) > 0$. To avoid this peculiar behavior, Latham and Nirenberg generalized the theoretical framework used to derive ΔI^D [11], giving rise to the measure ΔI^{DL} of Equation (26). Here, the supra-script DL makes reference to "Divergence Lowest", since the measure was presented as the lowest possible information loss of a decoder trained with $P_{su}(\mathbf{r}|s)$. In the definition of ΔI^{LD}, the parameter θ is a real scalar. The distribution $P_{su}(s|\mathbf{r}, \theta)$ was defined by Latham and Nirenberg [11] as proportional to $P(s) P_{su}(\mathbf{r}|s)^\theta$. This definition has several problems, as discussed in [11,17,36–39]. In Appendix B.1 we demonstrate a theorem that resolves the issues appearing in previous definitions, and justifies the use of

$$P_{su}(s|\mathbf{r}, \theta) \propto \begin{cases} P(s) & \text{if } \exists \hat{s}, \hat{\mathbf{r}} \text{ such that } P_{ex}(\hat{\mathbf{r}}|\hat{s}) > P_{su}(\hat{\mathbf{r}}|\hat{s}) = 0 \\ 0 & \text{if } P_{su}(\mathbf{r}|s) = P_{ex}(\mathbf{r}|s) = 0 \text{ for some but not all } s \\ P(s) P_{su}(\mathbf{r}|s)^\theta & \text{otherwise} \end{cases} \quad (30)$$

From the conceptual point of view, ΔI^{DL} represents the information loss of a mismatched decoder trained with $P_{su}(\mathbf{r}|s)$ and operated on $\mathbf{R}_{ex}$. Latham and Nirenberg [11] showed that, unlike ΔI^D, it is possible to demonstrate that $\Delta I^{DL} \leq I_{\mathbf{R}_{ex}}$. Hence, ΔI^{DL} never yields a tested feature encoding more information than the full response. The proof in [11] ignored a few specific cases that we discuss in the Theorem A1 of Appendix B.1. Still, even in those additional cases, the inequality $\Delta I^{DL} \leq I_{\mathbf{R}_{ex}}$ holds.

In Equations (22) and (23), $\hat{\mathbf{S}}$ and $\hat{s}$ denote a sorted stimulus list and the most-likely stimulus, respectively, both decoded by evaluating Equation (6) (or its ranked version) on a response $\mathbf{r}$ sampled from the surrogate distribution $P_{su}(\mathbf{r}|s)$ (method β). Estimating mutual informations using decoders can be traced back at least to Gochin et al. [40], and comparing the estimations of two decoders that take different response features into account, at least to Warland et al. [41].

The measures $\Delta I_{\hat{\mathbf{S}}}$ and $\Delta I_{\hat{s}}$ are paired with ΔI^{LS} and ΔI^B, respectively, since the latter are obtained from the former when replacing the decoding method from β to α. The measure ΔI^{LS} was introduced by Ince et al. [20], and quantifies the difference between the information in $\mathbf{R}_{ex}$, and the one in the output of decoders that, after observing a variable $\mathbf{r}$ sampled with distribution $P_{ex}(\mathbf{r}|s)$ (method α), produce a stimulus list sorted according to $P_{su}(s|\mathbf{r})$. The supra-script LS indicates "List of Stimuli". Similarly, ΔI^B, quantifies the difference between the information encoded in $\mathbf{R}_{ex}$ and that encoded in the output of a decoder trained by inserting $P_{su}(s|\mathbf{r})$ into Equation (6), and operated on $\mathbf{r}$ sampled with distribution $P_{ex}(\mathbf{r}|s)$ (method α). The supra-script B stands for the "Bayesian" nature of the involved decoder. The use of these measures can be traced back at least to Nirenberg et al. [8], although in that case, decoders were restricted to be linear. The measure $\Delta I_{\hat{\mathbf{S}}}$ of Equation (22) is new, and we have introduced it here as the homologous of ΔI^{LS}. When the number of stimuli is two, $\Delta I_{\hat{\mathbf{S}}} = \Delta I_{\hat{s}}$, since selecting the optimal stimulus is (as a computation) in one-to-one correspondence with ranking the two candidate stimuli.

The accuracy loss $\Delta A_{\mathbf{R}_{su}}$ defined in Equation (24) entails the comparison between the performance of two decoders, one trained with and applied on $\mathbf{R}_{ex}$, and one trained with and applied on $\mathbf{R}_{su}$. Such comparisons have also a long history in neuroscience [42,43] (see [9,12] for further discussion). The accuracy loss ΔA^B also compares two decoders. The first, is the same as for $\Delta A_{\mathbf{R}_{su}}$, but the second is trained with $\mathbf{R}_{su}$ and applied on $\mathbf{R}_{ex}$.

The measures ΔI^{LS}, ΔI^B, and ΔA^B are undefined if the actual responses $\mathbf{R}_{ex}$ are not contained in the set of surrogate responses $\mathbf{R}_{su}$. In other words, a decoder constructed with $P_{su}(\mathbf{r}|s)$ does not know what output to produce when evaluated in a response $\mathbf{r}$ for which $P_{su}(\mathbf{r}) = 0$. This situation never happens when evaluating the relevance of noise correlations with $P_{su} = P_{NI}$, but it may well be encountered in more general situations, as for example, in Figure 3B.

2.4. Relating the Values Obtained with Different Measures

If a mapping $\mathbf{R}_{ex} \rightarrow \mathbf{R}_{su}$ exists transforming $P_{ex}(\mathbf{r}|s)$ into $P_{su}(\mathbf{r}|s)$, we may use the decoding procedure of Equation (6) to construct the transformation chain $\mathbf{R}_{ex} \rightarrow \mathbf{R}_{su} \rightarrow \hat{\mathbf{S}} \rightarrow \hat{S}$ [17,44]. Consequently, $\Delta I_{\mathbf{R}_{su}}$, $\Delta I_{\hat{\mathbf{S}}}$ and $\Delta I_{\hat{S}}$ can be interpreted as accumulated information losses after the first, second and third transformations, respectively, and $\Delta A_{\mathbf{R}_{su}}$, as the accuracy loss after the first transformation. The data processing theorems (Section 2.1.3) ensure that these measures are never negative. This property, however, cannot be guaranteed in the absence of a reduced transformation $\mathbf{R}_{ex} \rightarrow \mathbf{R}_{su}$, stochastic or deterministic. Indeed, in the example of Figure 4b, if both stimuli are equiprobable, and both responses $\mathbf{R}_{ex}$ associated with $\bigcirc$ are equiprobable, then $\Delta I_{\mathbf{R}_{su}} = \Delta I_{\hat{\mathbf{S}}} = \Delta I \hat{S} \approx -79\,\%$ of $I_{\mathbf{R}_{ex}} \approx 0.31$ bits, implying that the surrogate responses encode more information about the stimulus than the original, experimental responses. Removing the correlations between spike count and latency, hence, increases the information, so correlations can be concluded to be detrimental to information encoding.

Irrespective of whether a (deterministic or stochastic) mapping $\mathbf{R}_{ex} \rightarrow \mathbf{R}_{su}$ exists, the data processing inequality guarantees that $\Delta I_{\mathbf{R}_{su}} \leq \Delta I_{\hat{\mathbf{S}}} \leq \Delta I_{\hat{S}}$, since $\hat{\mathbf{S}}$ is a deterministic function of $\mathbf{R}_{su}$, and $\hat{S}$ is a deterministic function of $\hat{\mathbf{S}}$. The inequality holds irrespective of the sign of each measure.

All decoder-oriented measured are guaranteed to be non-negative. The very definitions of ΔI^D and of ΔI^{DL} imply they cannot be negative, since they are both Kullback-Leibler divergences between two probability distributions. The sequence of reduced transformations $\mathbf{R}_{ex} \rightarrow \hat{S} \rightarrow S$, in turn, guarantees the non-negativity of ΔI^{LS}, ΔI^B and ΔA^B, through the Data Processing Inequalities.

In order to assess whether decoding-oriented measures are always larger or smaller than their encoding (or gray) counterparts, we performed a numerical exploration comparing each encoding/gray-oriented measure with its decoding-oriented homologue. The exploration was conducted by calculating the values of these measures for a large collection of possible stimulus prior probabilities $P(s)$, and response conditional probabilities $P_{ex}(\mathbf{r}|s)$ in the examples of Figures 2–4 and 7. The details of the numerical exploration are in Appendix A. The measures in the first group were sometimes greater and sometimes smaller than those of the second group, depending on the case and the probabilities (Table 1). Consequently, our results demonstrate that there is no general rule by which measures of one type bound the measures of the other type.

The exploration also included the example of Figure 7a. In panel (a), the transition probabilities are

$$Q(\mathbf{r}_{su}|\mathbf{r}_{ex}) = \begin{bmatrix} 0 & 1 & 0 & 0 \\ 0 & 0 & 1 & 0 \\ 0 & 0 & 0 & 1 \\ 1 & 0 & 0 & 0 \end{bmatrix}, \tag{31}$$

where rows and columns enumerate the elements of the ordered sets $\mathcal{R}_{ex} = \mathcal{R}_{su} = \{[1],[2],[3],[4]\}$ from where both $\mathbf{R}_{ex}$ and $\mathbf{R}_{su}$ are sampled. For panel b,

$$Q(\mathbf{r}_{su}|\mathbf{r}_{ex}) = \frac{1}{2} \begin{bmatrix} 2\bar{a} & a & a & 0 & 0 & 0 \\ 0 & b & b & 2\bar{b} & 0 & 0 \\ 0 & 0 & 0 & 2\bar{b} & b & b \\ 2\bar{a} & 0 & 0 & 0 & a & a \end{bmatrix}, \tag{32}$$

with $0 < a, b < 1$, rows enumerating the elements of $\mathcal{R}_{ex} = \{[2],[3],[5],[6]\}$, and columns those of $\mathcal{R}_{su} = \{[1],[2],[3],[4],[5],[6]\}$.

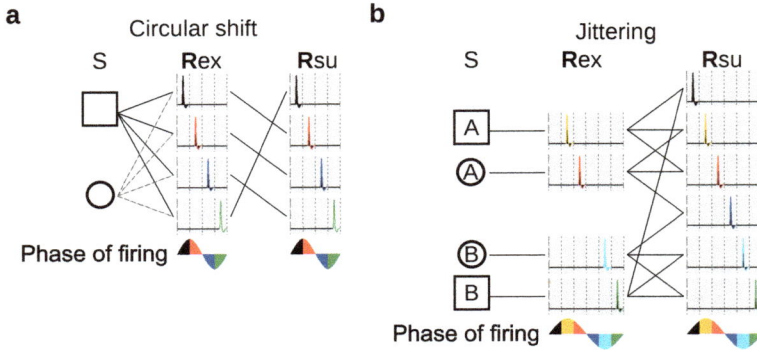

Figure 7. Stochastic codes may play different roles in encoding and decoding. (**a**) Hypothetical experiment with two stimuli □ and ○, which are transformed (solid and dashed lines) into neural responses containing a single spike ($C = 1$) fired at different phases (Φ) with respect to a cycle of 20 ms period starting at stimulus onset. The phases have been discretized in intervals of size $\pi/2$ and wrapped to the interval $[0, 2\pi)$. The encoding process is followed by a circular phase-shift that transforms $\mathbf{R}_{ex} = \Phi$ into another code $\mathbf{R}_{su} = \hat{\Phi}$ with transition probabilities $Q(\mathbf{r}_{su}|\mathbf{r}_{ex})$ defined by Equation (31). The set of all $\mathbf{R}_{su}$ coincides with the set of all $\mathbf{R}_{ex}$; (**b**) Same as (**a**), except that stimuli are four (Ⓐ, Ⓐ, Ⓑ, and Ⓑ), and phases are measured with respect to a cycle of 30 ms period and discretized in intervals of size $\pi/3$. The encoding process is followed by a stochastic transformation (lines on the right) that introduces jitter, thereby transforming $\mathbf{R}_{ex} = \Phi$ into another code $\mathbf{R}_{su} = \hat{\Phi}$ with transition probabilities $Q(\mathbf{r}_{su}|\mathbf{r}_{ex})$ defined by Equation (32).

Table 1. Numerical exploration of the maximum and minimum differences between several measures of information and accuracy losses. The values are expressed as percentages of $I_{\mathbf{R}_{ex}}$ (the information encoded in $\mathbf{R}_{ex}$) or $A_{\mathbf{R}_{ex}}^{\mathbf{R}_{ex}}$ (the maximum accuracy above chance level when decoders operate on $\mathbf{R}_{ex}$). All examples involve two stimuli, so $\Delta I_{\hat{S}} = \Delta I_{\hat{S}}$ and $\Delta I^{LS} = \Delta I^B$. The absolute value of $\Delta A_{\mathbf{R}_{su}} - \Delta A^B$ can become extremely large when $A_{\mathbf{R}_{su}}^{\mathbf{R}_{su}} \approx 0$. Dashes represent cases in which decoding-oriented measures are undefined, as explained in Section 2.4.

Cases		Figure 4a	Figure 2b	Figure 2c	Figure 3d	Figure 2a	Figure 3b	Figure 7a
$\Delta I_{\hat{R}} - \Delta I^D$	min	−79	−51	−34	0	0	—	≤999
	max	26	32	51	0	0	—	−20
$\Delta I_{\hat{R}} - \Delta I^{DL}$	min	−34	−32	−16	0	0	−100	−100
	max	59	41	98	0	0	0	0
$\Delta I_{\hat{R}} - \Delta I^B$	min	−67	−62	−46	−63	−87	—	−100
	max	57	81	96	0	0	—	0
$\Delta I_{\hat{S}} - \Delta I^D$	min	−79	−48	−34	0	0	—	≤999
	max	67	92	93	63	87	—	70
$\Delta I_{\hat{S}} - \Delta I^{DL}$	min	−34	−27	−16	0	0	-100	−100
	max	91	92	99	63	87	0	97
$\Delta I_{\hat{S}} - \Delta I^B$	min	−51	−31	−17	0	0	—	−100
	max	59	91	98	0	0	—	100
$\Delta A_{\hat{R}} - \Delta A^B$	min	−386	−200	−150	0	0	—	≤999
	max	95	67	100	0	0	—	0

An important issue is to identify the situations in which $\Delta I_{\mathbf{R}_{su}}$ gives exactly the same result as either ΔI^D or ΔI^{DL}. It is not easy to determine the conditions for the equality between $\Delta I_{\mathbf{R}_{su}}$ and ΔI^{DL}. Yet, for the equality between $\Delta I_{\mathbf{R}_{su}}$ and ΔI^D, and in the specific case in which $P_{su}(\mathbf{r}|s) = P_{NI}(\mathbf{r}|s)$ as given by Equation (17), the following theorem holds.

Theorem 3. *When assessing the relevance of noise correlations, $\Delta I^D = \Delta I_{\mathbf{R}_{su}}$ if and only if*

$$\lambda = \sum_{\mathbf{r}}[P_{ex}(\mathbf{r}) - P_{su}(\mathbf{r})]\,\log_2[P_{su}(\mathbf{r})] = 0. \tag{33}$$

Moreover, $\lambda \lessgtr 0$ implies that $\Delta I^D \lessgtr \Delta I_{\mathbf{R}_{su}}$.

Proof. See Appendix B.4. □

Equation (33) implies that neither the prior stimulus probabilities $P(s)$ nor the conditional response probabilities $P_{ex}(\mathbf{r}|s)$ intervene in the condition for the equality, beyond the effect they have in fixing the value of $P_{ex}(\mathbf{r})$ and $P_{su}(\mathbf{r})$. Each response $\mathbf{r}$ makes a contribution to the value of λ, which favours ΔI^D whenever $P_{su}(\mathbf{r}) > P_{ex}(\mathbf{r})$, and $I_{\mathbf{R}_{ex}}$ in the opposite case. As pointed out by [10], all responses $\mathbf{r}$ for which $P_{ex}(\mathbf{r}) = 0$ and $P_{su}(\mathbf{r}) > 0$ give a null contribution to ΔI^D, and a negative contribution to $I_{\mathbf{R}_{ex}}$, implying that correlations in such responses are irrelevant for decoding, and detrimental to encoding.

The fact that encoding-oriented measures neither bound nor are bounded by decoding-oriented measures is a daunting result. If, when working in a specific example, one gets a positive value with one measure and a negative value with another, the interpretation must carefully distinguish between the two paradigms. One may wonder, however, if such distinction is also required when correlations are absolutely essential for one of the measures, in that they capture the whole of the encoded information. Could the other measure conclude that they are irrelevant? Or that they are only mildly relevant? Luckily, in this case, the answer is negative. In other words, when the tested feature is fundamental, then ΔI^D and $\Delta I_{\mathbf{R}_{su}}$ coincide, and no conflict arises between encoding and decoding, as proven by the following theorem:

Theorem 4. *$\Delta I^{DL} = I_{\mathbf{R}_{ex}}$ if and only if $\Delta I_{\mathbf{R}_{su}} = I_{\mathbf{R}_{ex}}$, regardless of whether stochastic codes exist that map the actual responses $\mathbf{R}_{ex}$ into the surrogate responses $\mathbf{R}_{su} = \mathbf{R}_{NI}$ generated assuming noise independence.*

Proof. See Appendix B.5. □

The conclusion is that if a given feature is 100% relevant for encoding, then it is also 100% relevant for decoding, and vice versa. Hence, although $\Delta I_{\mathbf{R}_{su}}$ and ΔI^{DL} often differ in the relevance they ascribe to a given feature, the discrepancy is only encountered when the tested feature is not the only informative feature in play. When the removal of the feature is catastrophic (in the sense that it brings about a complete information loss), then both $\Delta I_{\mathbf{R}_{su}}$ and ΔI^{DL} diagnose the situation equally.

2.5. Relation between Measures Based on Decoding Strategies α and β

The results of Table 1 may seem puzzling because decoding happens after encoding. Therefore—one may naively reason—the data processing theorems should have forbidden both $\Delta I_{\mathbf{R}_{su}}$ to surpass ΔI^D, ΔI^{DL}, or ΔI^B, as well as $\Delta A_{\mathbf{R}_{su}}$ to surpass ΔA^B. However, even though decoding indeed happens after encoding, the data processing theorem is not violated. The theorem certainly ensures that $\Delta I_{\mathbf{R}_{su}}$ and $\Delta A_{\mathbf{R}_{su}}$ constitute lower bounds for measures related to decoders that operate on responses generated by $P_{su}(\mathbf{r}|s)$, but not for measures related to decoders that operate on responses generated by $P_{ex}(\mathbf{r}|s)$, such as happens with ΔI^D, ΔI^{DL}, ΔI^B, and ΔA^B.

This observation about the validity of the data processing inequality is different from the one discussed in Section 2.2. There, we discussed the conditions under which $\Delta I_{\mathbf{R}_{su}}$ could be guaranteed to be non-negative, the crucial factor being the existence of a stochastic mapping $\mathbf{R}_{ex} \rightarrow \mathbf{R}_{su}$. Now we are discussing a different aspect, regarding whether decoding-related measures can or cannot be bounded by encoding-oriented measures. The conclusion is that in general terms, the answer is negative, because decoding-related measures operate with decoding strategy α, a strategy never addressed by the encoding measures. The surrogate variable $\mathbf{R}_{su}$ participating in the encoding measure $\Delta I_{\mathbf{R}_{su}}$ is *not* the response decoded by the measures of Equations (25)–(28), so the data processing inequalities need not hold. That being said, there are specific instances in which both types of measures coincide, two of them discussed in Theorems 3 and 4 and a third case later in Theorem 5.

Other explanations have been given in the literature for the fact that sometimes, decoding oriented measures surpass their encoding counterparts. For example, it has been alleged [10] that when $\Delta I^D, \Delta I^{DL}$ or ΔI^B are smaller than $\Delta I_{\mathbf{R}_{su}}$, this is either due to (a) the impossibility to define a stimulus-independent reduction $\mathbf{R}_{ex} \rightarrow \mathbf{R}_{su}$ that yields $P_{ex}(\mathbf{r}|s) \rightarrow P_{su}(\mathbf{r}|s)$ (and therefore the data-processing inequality is not guaranteed to hold), or due to (b) the fact that surrogate responses often sample values of response space that are never reached by real responses (and therefore, the losses of matched decoders may be larger than the ones of mismatched ones). However, Figure 2c constitutes a counterexample of both arguments, since there, the stimulus-independent stochastic reduction exists, and the response set of $\mathbf{R}_{ex}$ and $\mathbf{R}_{su}$ coincide.

One could also wonder whether the discrepancy between the values obtained with encoding-oriented measures and decoding-oriented measures only occurs in examples where a stochastic reduction $\mathbf{R}_{ex} \rightarrow \mathbf{R}_{su}$ exists, and the involved transition matrix $Q(\mathbf{r}_{su}|\mathbf{r}_{ex})$ depends on the joint probabilities $P_{ex}(\mathbf{r}, s)$, and not only on the marginals, as discussed in Theorem 2. However, Figure 2b,c provide examples in which $Q(\mathbf{r}_{su}|\mathbf{r}_{ex})$ does not depend on $P(\mathbf{r}, s)$, and yet, the discrepancies are still observed.

The distinction between decoding strategies α and β is also crucial when using the measure ΔI^D. This measure was introduced by Nirenberg et al. [8] for the specific case in which the tested feature is the amount of noise correlations, that is, when $P_{su}(s|\mathbf{r}) = P_{NI}(s|\mathbf{r})$. The measure was later extended to arbitrary deterministic mappings $\mathbf{R}_{su} = f(\mathbf{R}_{ex})$ [10,12,13], with the instruction to use an expression like Equation (25), but with $P_{su}(s|\mathbf{r})$ replaced by $P(s|\mathbf{R}_{su} = f(\mathbf{r})) = P_{su}(s|f(\mathbf{r}))$. It should be noted, however, that as soon as this replacement is made, ΔI^D becomes exactly equal to $\Delta I_{\mathbf{R}_{su}}$. Specifically, the measure ΔI^D now describes the information loss of a decoder that operates on a response variable generated with the surrogate distribution $P_{su}(\mathbf{r}|s)$ (decoding method β). If we want to keep the original spirit, and associate ΔI^D with a decoder that operates on a response variable generated with the real distribution $P_{ex}(\mathbf{r}|s)$ (decoding method α), in Equation 25, $P_{su}(s|\mathbf{r})$ should not be modified. Only the evaluation of the surrogate variable $\mathbf{R}_{su}$ in the experimentally observed value $\mathbf{R}_{ex} = \mathbf{r}$ describes a mismatched decoder constructed with $P_{su}(\mathbf{r}|s)$ and operated on $\mathbf{R}_{ex}$ (mathematical details in Appendix C).

2.6. Assessing the Type of Information Encoded by Individual Response Features

When the stimulus contains several attributes (as shape, color, sound, etc.), by removing a specific response feature it is possible to assess not only *how much* information is encoded by the feature, but also, *what type* of information. Identifiying the type of encoded information implies determining the stimulus feature represented by the tested response feature. As shown in this section, the type of encoded information is as dependent on the method of removal as is the amount. In other words, the different measures defined in Equations (21)–(29) sometimes associate a feature with the encoding of different stimulus attributes.

In the example of Figure 8, we use four compound stimuli $S = [S_F, S_L]$, generated by choosing independently a frame ($S_F = \square$ or $\bigcirc$) and a letter (S_L = A or B), thereby yielding $\boxed{A}$, $\textcircled{A}$, $\boxed{B}$, and $\textcircled{B}$. Stimuli are transformed into neural responses $\mathbf{R} = [L, C]$ with different number of spikes ($1 \leq C \leq 5$) fired at different first-spike latencies ($1 \leq L \leq 4$; time has been discretized in 5 ms bins). Latencies are only sensitive to frames whereas spikes counts are only sensitive to letters, thereby constituting independent-information streams: $P(s, \mathbf{r}) = P(s_F, l) P(s_L, c)$ [33]. The equality in the numerical value of two measures does not imply that both measures assign the same meaning to the information encoded by the tested response feature. Indeed, the two measures may sometimes report the tested response feature to encode two different aspects of the set of stimuli. Consider a decoder that is trained using the noisy data $\mathbf{R}_{su}$ shown in Figure 8a, but it is asked to operate on either the same noisy data with which it was trained (strategy β), or with the quality data $\mathbf{R}_{ex}$ of Figure 8b (strategy α). The information losses $\Delta I_{\mathbf{R}_{su}}$, ΔI^D, and ΔI^{DL} are all equal to 50 % of $I(S, \mathbf{R}_{ex}) = 2$ bits. Therefore, the information loss is independent of whether, in the operation phase, the decoder is fed with responses generated with $P_{su}(\mathbf{r}|s)$ or with $P_{ex}(\mathbf{r}|s)$.

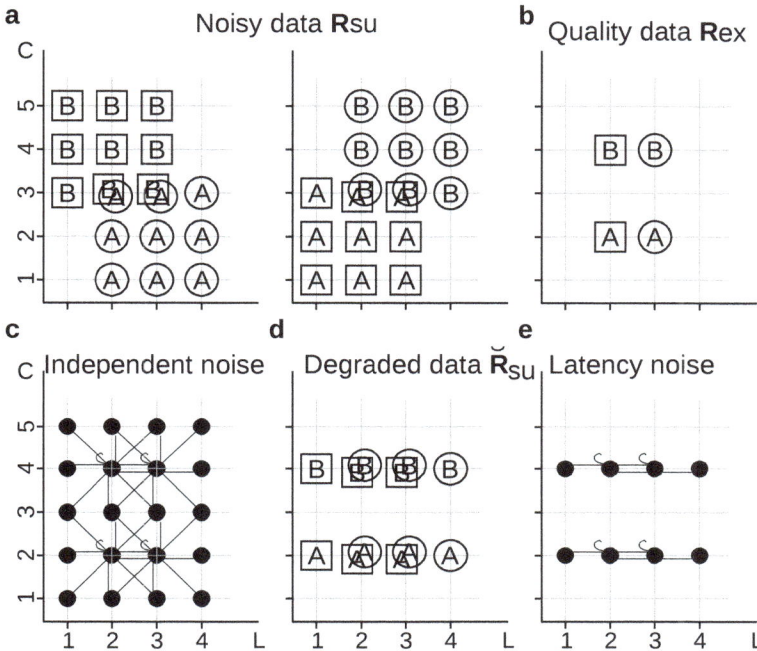

Figure 8. Assessing the amount and type of information encoded by . (**a**) Noisy data $\mathbf{R}_{su} = [L, C]$ recorded in response of the compound stimulus $S = [S_F, S_L]$; (**b**) Quality data $\mathbf{R}_{ex} = [L, C]$ recorded in the case of panel (a), but without noise; (**c**) Stimulus-independent stochastic transformation with transition probabilities $Q(\mathbf{r}_{su}|\mathbf{r}_{ex})$ given by Equation (34), that introduces independent noise both in the latencies and in the spike counts, thereby transforming $\mathbf{R}_{ex}$ into $\mathbf{R}_{su}$ and rendering $\mathbf{R}_{su}$ a stochastic code; (**d**) Degraded data $\check{\mathbf{R}}$ obtained by adding latency noise to the quality data; (**e**) Representation of the stimulus-independent stochastic transformation $\mathbf{R}_{ex} \rightarrow \check{\mathbf{R}}$ with transition probabilities $Q(\check{\mathbf{r}}|\mathbf{r}_{ex})$ given by Equation (35) that adds latency noise in panel (d).

The transformation $Q(\mathbf{r}_{su}|\mathbf{r}_{ex})$ causes some responses $\mathbf{R}_{su}$ to occur for all stimuli, so when decoding with method β, some information about frames is lost (that is, $I(S_F, \mathbf{R}_{su}) \approx 33\%$ of $I(S_F, \mathbf{R}_{ex}) = 1$ bit), as well as some information about letters (that is, $I(S_L, \mathbf{R}_{su}) \approx 67\%$ of $I(S_L, \mathbf{R}_{ex}) = 1$ bit). In other words, decoding $\mathbf{R}_{su}$ causes a partial information loss $\Delta I_{\mathbf{R}_{su}}$ that is composed of both frame and letter information. Instead, when decoding $\mathbf{R}_{ex}$ with method α, there is no information loss about letters: For the responses $\mathbf{R}_{ex}$ that actually occur, the decoder trained with $\mathbf{R}_{su}$ can perfectly identify the letters, because $P_{su}(C = 2|S_L = A) = P_{su}(C = 4|S_L = B) = 1$. The information about frames, on the other hand, is completely lost, since $P_{su}(l|\square) = P_{su}(l|\bigcirc)$ whenever l adopts a value that actually occurs in $\mathbf{R}_{ex}$, namely 2 or 3. This example shows that the fact that two decoding procedures give the same numerical loss does not mean that they draw the same conclusions regarding the role of the tested feature in the neural code. Ananalogous computations yield analogous results for the hypothetical experiment shown in Figure 7b.

If responses r_{ex} and r_{su} are written as vectors $[L, C]$, and the values of $Q(r_{su}|r_{ex})$ are arranged in a rectangular structure, in Figure 8c the transition probabilities are

$$Q(r_{su}|r_{ex}) = \frac{1}{9} \begin{bmatrix} 1 & 1 & 1 & 0 & 1 & 1 & 1 & 0 & 1 & 1 & 0 & 0 & 0 & 0 & 0 & 0 & 0 & 0 \\ 0 & 1 & 1 & 1 & 0 & 1 & 1 & 1 & 0 & 1 & 1 & 1 & 0 & 0 & 0 & 0 & 0 & 0 \\ 0 & 0 & 0 & 0 & 0 & 0 & 0 & 1 & 1 & 1 & 0 & 1 & 1 & 1 & 0 & 1 & 1 & 0 \\ 0 & 0 & 0 & 0 & 0 & 0 & 0 & 0 & 1 & 1 & 1 & 0 & 1 & 1 & 1 & 0 & 1 & 1 \end{bmatrix}, \tag{34}$$

where rows and columns indicate the ordered sets $\mathcal{R}_{ex} = \{[2, 2], [3, 2], [2, 4], [3, 4]\}$ and $\mathcal{R}_{su} = \{1, 2, 3, 4\} \times \{1, 2, 3, 4, 5\}$, where $\times$ denotes the Cartesian product with colexicographical order, that is, ordered as $[1, 1], [2, 1], [3, 1], [4, 1], [1, 2]$, etc. In Figure 8e

$$Q(\check{r}|r_{ex}) = \frac{1}{3} \begin{bmatrix} 0 & 0 & 0 & 0 & 1 & 1 & 1 & 0 & 0 & 0 & 0 & 0 & 0 & 0 & 0 & 0 & 0 & 0 \\ 0 & 0 & 0 & 0 & 0 & 1 & 1 & 1 & 0 & 0 & 0 & 0 & 0 & 0 & 0 & 0 & 0 & 0 \\ 0 & 0 & 0 & 0 & 0 & 0 & 0 & 0 & 0 & 0 & 0 & 0 & 1 & 1 & 1 & 0 & 0 & 0 \\ 0 & 0 & 0 & 0 & 0 & 0 & 0 & 0 & 0 & 0 & 0 & 0 & 0 & 1 & 1 & 1 & 0 & 0 & 0 \end{bmatrix}, \tag{35}$$

with rows and columns with the same convention as in Equation (34).

Finally, the noisy data (Figure 8a) can be obtained by transforming the degraded data (Figure 8d) with the transition matrix

$$Q(r_{su}|\check{r}) = \frac{1}{3} \begin{bmatrix} 1 & 0 & 0 & 0 & 1 & 0 & 0 & 0 & 1 & 0 & 0 & 0 & 0 & 0 & 0 & 0 & 0 & 0 & 0 & 0 \\ 0 & 1 & 0 & 0 & 0 & 1 & 0 & 0 & 0 & 1 & 0 & 0 & 0 & 0 & 0 & 0 & 0 & 0 & 0 & 0 \\ 0 & 0 & 1 & 0 & 0 & 0 & 1 & 0 & 0 & 0 & 1 & 0 & 0 & 0 & 0 & 0 & 0 & 0 & 0 & 0 \\ 0 & 0 & 0 & 1 & 0 & 0 & 0 & 1 & 0 & 0 & 0 & 1 & 0 & 0 & 0 & 0 & 0 & 0 & 0 & 0 \\ 0 & 0 & 0 & 0 & 0 & 0 & 0 & 1 & 0 & 0 & 0 & 1 & 0 & 0 & 0 & 1 & 0 & 0 & 0 \\ 0 & 0 & 0 & 0 & 0 & 0 & 0 & 0 & 1 & 0 & 0 & 0 & 1 & 0 & 0 & 0 & 1 & 0 & 0 \\ 0 & 0 & 0 & 0 & 0 & 0 & 0 & 0 & 0 & 1 & 0 & 0 & 0 & 1 & 0 & 0 & 0 & 1 & 0 \\ 0 & 0 & 0 & 0 & 0 & 0 & 0 & 0 & 0 & 0 & 1 & 0 & 0 & 0 & 1 & 0 & 0 & 0 & 1 \end{bmatrix}. \tag{36}$$

with rows and columns indicating the ordered sets $\{[1, 2, 3, 4] \times [2, 4]\}$ and $\{1, 2, 3, 4\} \times \{1, 2, 3, 4, 5\}$, respectively, where $\times$ denotes the Cartesian product with colexicographical order.

2.7. Conditions for Equality of the Amount and Type of Information Loss Reported by Different Measures

We now derive the conditions under which encoding/gray-oriented measures coincide with their decoding-oriented counterparts, as observed in Figures 2a and 3d. That is, we derive the conditions under which the following equalities hold:

$$\Delta I_{R_{su}} = \Delta I^D = \Delta I^{DL}, \tag{37}$$

$$\Delta I_{\hat{S}} = \Delta I^{LS}, \tag{38}$$

$$\Delta I_{\check{S}} = \Delta I^B, \tag{39}$$

$$\Delta A_{R_{su}} = \Delta A^B. \tag{40}$$

The example in Figure 7a showed that the existence of deterministic mappings does not suffice for a qualitative and quantitative equivalence of different measures. Furthermore, the example of Figure 3b showed that the equalities require the space of R_{su} to include the space of R_{ex}, or else the decoding method α may be undefined. We demonstrate that the Equations (37)–(40) arise, and moreover, that there is no discrepancy in the type of information assessed by these different measures, whenever the mapping from R_{ex} into R_{su} can be described using positive-diagonal idempotent stochastic matrices [45]. Specifically, we prove the following theorem:

Theorem 5. *Consider a stimulus-independent stochastic function f from a representation R_{ex} into another representation R_{su}, such that the range $\mathcal{R}$ of R_{su} includes that of R_{ex}, and with transition probabilities $Q(r_{su}|r_{ex})$ that can be written as positive-diagonal idempotent right stochastic matrices with row and column indices that enumerate the elements of $\mathcal{R}$ in the same order. Then, Equations (37)–(40) hold.*

Proof. See Appendix B.6. □

The theorem states that the equalities of Equations (37)–(40) can be guaranteed whenever the removal of the tested response feature involves a (deterministic or) stochastic mapping $\mathbf{R}_{ex} \rightarrow \mathbf{R}_{su}$ that induces a partition within the set of real responses $\mathbf{R}_{ex}$, and $\mathbf{R}_{su}$ is obtained by rendering all responses inside each partition indistinguishable (but not across partitions). To sample $\mathbf{R}_{su}$, the probabilities of individual responses inside each partition are re-assigned, rendering their distinction uninformative [30].

This theorem provides sufficient but not necessary conditions for the equalities to hold. The important aspect, however, is that it ensures that the equalities hold not only in numerical value, but also, in the type of information that different measures ascribe to the tested feature. Two different methods preserve or lose information of different type if, when decoding a stimulus, the trials with decoding errors tend to confound different attributes of the stimulus, as in the example of Figure 8. The conditions of Theorem 5, however, ensure that the strategies α and β always decode exactly the same stimulus (see Appendix B.6), so there can be no difference in the confounded attributes. Pushing the argument further, one could even argue that responses (real or surrogate) encode more information than the identity of the stimulus that originated them. For a fixed decoded stimulus, the response still contains additional information [46], that refers to (a) the degree of certainty with which the stimulus is decoded, and (b) the rank of the alternative stimuli, in case the decoded stimulus was mistaken [20]. Both meanings are embodied in the whole rank of a posteriori probabilities $P_{su}(s|\mathbf{r})$, not just the maximal one. Yet, under the conditions of the theorem, the entire rankings obtained with methods α and β coincide (see Appendix B.6). Therefore, even within this broader interpretation, there can be no difference in the qualitative aspects of the information preserved or lost by one and the other.

For example, in Figure 7b, we found that all information losses are equal (that is, $\Delta I_{\hat{R}}$, $\Delta I_{\hat{S}}$, $\Delta I_{\check{S}}$, ΔI^D, ΔI^{DL}, ΔI^{LS}, and ΔI^B are all 50%), and both accuracy losses are equal (that is, $\Delta A_{\hat{R}}$ and ΔA^B are both $\approx 67\%$). However, the conditions of Theorem 5 do not hold. The matrix of Equation (32) is not block-diagonal, nor it can be taken to that shape by incorporating new rows (to make it square), and permuting both rows and columns, in such a way that the response vectors are enumerated in the same order by both indices. For this reason, the losses are not guaranteed to be of the same type.

Instead, the transition probabilities of Equations (15) and (16) can be turned into positive-diagonal idempotent right stochastic matrices. Equation (15) is already in the required format. To take Equation (16) to the conditions of Theorem 5, two new rows need to be incorporated, associated to the responses $[4, 1]$ and $[2, 2]$, that do not occur experimentally. Those rows can contain arbitrary values, since the condition $P_{ex}([4,1]|S) = P_{ex}([2,2]|S) = 0$, $\forall S$ renders them irrelevant. Arranging the columns so that both rows and columns enumerate the same list of responses, Equation (16) can be written as

$$Q(\mathbf{r}_{su}|\mathbf{r}_{ex}) = \begin{bmatrix} b & \bar{b} & 0 & 0 & 0 & 0 \\ b & \bar{b} & 0 & 0 & 0 & 0 \\ 0 & 0 & c & \bar{c} & 0 & 0 \\ 0 & 0 & c & \bar{c} & 0 & 0 \\ 0 & 0 & 0 & 0 & d & \bar{d} \\ 0 & 0 & 0 & 0 & d & \bar{d} \end{bmatrix}, \tag{41}$$

with $\mathcal{R}_{ex} = \mathcal{R}_{su} = \{[2,1], [2,2], [3,1], [3,2], [4,1], [4,2]\}$. Hence, in these two examples, both the amount and type of information of encoding and decoding-based measures coincide.

2.8. Improving the Performance of Decoders Operating with Strategy α

In a previous paper [17], we demonstrated that neither ΔI^D nor ΔI^{DL} constitute lower bounds on the information loss induced by decoders constructed by disregarding the tested response feature. This means that some decoders may exist, that perform better than $D_{su}(\mathbf{r})$ defined in Equation (6). In this section we discuss one possible way in which some of these improved decoders may be constructed, inspired in the example of Figure 8. Quite remarkably, the construction involves the addition of noise to the real responses, before feeding them to the decoder of Equation (6). Panel (a) shows a decoder constructed with noisy data ($\mathbf{R}_{su}$), and then employed to decode quality data ($\mathbf{R}_{ex}$;

Figure 8b), thereby yielding information losses $\Delta I^D = \Delta I^{DL} = 50\%$. These losses can be decreased by feeding the decoder with a degraded version $\check{\mathbf{R}}$ of the quality data (Figure 8d) generated through a stimulus-independent transformation that adds latency noise (Figure 8e). Decoding $\mathbf{R}_{ex}$ as if it were $\mathbf{R}_{su}$ by first transforming $\mathbf{R}_{ex}$ into $\check{\mathbf{R}}$ results in $\Delta I^D = \Delta I^{DL} \approx 33\%$, thereby recovering 33% of the information previously lost. On the contrary, adding spike-count noise will tend to increase the losses. Thus, adding suitable amounts and type of noise can increase the performance of approximate decoders, and the result is not limited to the case in which the response aspect is the amount of noise correlations. In addition, this result also indicates that, contrary to previously thought [47], decoding algorithms need not match the encoding mechanisms for performing optimally from an information-theoretical standpoint. All these results are a consequence of the fact that decoders operating with strategy α are not optimal, so it is possible to improve their performance by deterministic or stochastic manipulations of the response. In practice, our results open up the possibility of increasing the efficiency of decoders constructed with approximate descriptions of the neural responses, usually called approximate or mismatched decoders, by adding suitable amounts and types of noise to the decoder input.

3. Related Issues

3.1. Relation to Decomposition-Based Methods

Many measures of different types have been developed to assess how different response features of the neural code interact with each other. Some are based on direct comparisons between the information encoded by individual features, or collections of features (see for example [48–50], to cite just a few among many). Others distinguish between two or more potential dynamical models of brain activity [51], for example, by differentiating between conditional and unconditional correlations between neurons in the frequency domain [52]. Yet others, rely on decompositions or projections based on information geometry. In those, the mutual information between stimuli and responses $I_{\mathbf{R}}$ is broken down as $I_{\mathbf{R}} = \sum_i I'_{R_i} + \text{Synergy Terms} + \text{Redundancy Terms}$, where I'_{R_i} represents the information contributed by the individual response feature R_i, and the remaining terms incorporate the synergy or redundancy between them. In the original approaches [53–57], the terms I'_{R_i} represented the information $I(R_i; S)$ encoded in single response aspects irrespective of what be encoded in other aspects. In later studies, [58–62], these terms accounted for the information that is *only* encoded in individual aspects, taking care of excluding whatever be redundant with other aspects. The approach discussed in this paper is in the line of the studies Nirenberg et al. [8] and Schneidman et al. [9] and all their consequences. This line has some similarities and some discrepancies with the decomposition-based studies. We here comment on some of these relations.

-   First, the measure $\Delta I_{\mathbf{R}_{su}}$ quantifies the relevance of a given feature with the difference $I_{\mathbf{R}_{ex}} - I_{\mathbf{R}_{su}}$. When the surrogate response $\mathbf{R}_{su}$ is equal to the original response $\mathbf{R}_{ex}$ with just a single component R_i eliminated, $\Delta I_{\mathbf{R}_{su}}$ is equal to $I(R_i; s|\bar{R}_i)$, where $\bar{R}_i$ is the collection of all response aspects except R_i. In this case, $\Delta I_{\mathbf{R}_{su}}$ coincides with the sum of the unique and the synergistic contributions of the dual decompositions in the newest set of methods [63].
-   Second, when assessing the relevance of a given response feature, we are often inclined to draw conclusions about the cost of ignoring the tested feature when aiming to decode the original stimulus. As shown in this paper, those conclusions depend not only on how stimuli are encoded, but also, on how they are decoded. The decomposition-based methods are mainly focused in the encoding problem, so they are less suited to draw conclusions about decoding.
-   Finally, as discussed in Figure 8, not only the amount of (encoded or decoded) information matters, but also, what type. Decomposition-based methods, although not yet reaching a full consensus in their formulation, provide a valuable attempt to characterize how both the type and the amount of information is structured within the set of analyzed variables, in a way that is complementary to the present approach, specifically in analyzing the structure of the lattices obtained by associating different response features [58,63].

3.2. The Problem of Limited Sampling

Throughout the paper we assumed that the distribution $P_{ex}(s, \mathbf{r})$ is known, or is accessible to the experimenter. In the examples, when we calculated information values, we plugged the true distributions into the formulas, without discussing the fact that such distribution may not be easily estimated with finite amounts of data. Whichever method is used to estimate $P_{ex}(s, \mathbf{r})$, to a larger or lesser degree, the outcome is no more than an approximation. Hence, even $I_{\mathbf{R}_{ex}}$ (which is supposed to be the full information) is estimated approximately. Since $P_{su}(s, \mathbf{r})$ is a modified version of $P_{ex}(s, \mathbf{r})$, also $P_{su}(s, \mathbf{r})$ can only be estimated approximately. Information measures, including Kullback-Leibler divergences, are highly sensitive to variations in the involved probabilities [20,32,64–69], and the latter are unavoidable in high-dimensional response spaces. The assessment of the relevance of a given feature, hence, requires experiments that contain sufficient samples so as to ensure that the correcting methods work. When the response space is large, the measures ΔI_S, ΔI^B and the loss of accuracies are less sensitive to limited sampling than $\Delta I_{\mathbf{R}_{su}}$, ΔI^D and ΔI^{LD}.

In addition, the problem of finite sampling can also be formulated as an attempt to determine the relevance of the feature "Accuracy in the estimation of $P_{ex}(\mathbf{r}|s)$". This feature is not a property of the nervous system, but rather, of our ability to characterise it. Still, the framework developed here can also handle this methodological problem. The estimated distribution can be interpreted as a stochastic modification $P_{su}(\mathbf{r}|s)$ of the true distribution $P_{ex}(\mathbf{r}|s)$. As long as the caveats discussed in this paper are taken into account, the measures of Equations (21)–(29) may serve to evaluate the cost of modeling $P_{ex}(\mathbf{r}|s)$ out of finite amounts of data.

4. Conclusions

Several measures have been proposed in the literature to assess the relevance of specific response features in the neural code. All proposals are based on the idea that by removing the tested feature from the response, the neural code deteriorates, and the lost information is a useful measure of the relevance of the feature. In this paper, we demonstrated that the neural code may or may not deteriorate when removing a response feature, depending on the nature of the tested feature, and on the method of removal, in ways previously unseen. First, we determined the conditions under which the data processing inequality can be invoked. Second, we showed that decoding-oriented measures may result in larger or smaller losses than their encoding (or gray) counterparts, even for response aspects that, unlike noise correlations, can be modeled as stimulus-independent transformations of the full response. Third, we demonstrated that both types of measures coincide under the conditions of Theorem 5. Fourth, we showed that evaluating the role of a response feature in the neural code involves not only an assessment of its contribution to the amount of encoded information, but also, to the meaning of that information. Such meaning is as dependent as the amount on the measure employed to assess it. Finally, our results open up the possibility that simple and cheap decoding strategies, based on the addition of an adequate type and amount of noise, be more efficient and resilient than previously thought. We conclude that the assessment of the relevance of a specific response feature cannot be performed without a careful justification for the selection of a specific method of removal.

Author Contributions: Conceptualization, methodology, software, validation, formal analysis, investigation, writing, visualization: H.G.E. Formal analysis, resources, writing, editing: I.S.

Funding: This work was supported by the Ella and Georg Ehrnrooth Foundation, Consejo Nacional de Investigaciones Científicas y Técnicas of Argentina (06/C444), Universidad Nacional de Cuyo (PIP 0256), and Agencia Nacional de Promoción Científica y Tecnológica (grant PICT Raíces 2016 1004).

Conflicts of Interest: The authors declare no conflict of interest. The founding sponsors had no role in the design of the study; in the collection, analyses, or interpretation of data; in the writing of the manuscript, and in the decision to publish the results.

Appendix A. On the Information and Accuracy Differences

Each value in Table 1 (except for those associated with Figures 3b; see below) was computed using the Nelder-Mead simplex algorithm for optimization, as implemented by the function fminsearch of Matlab 2016. For accuracy reasons, only examples in which $I_{\mathrm{R_{ex}}} \geq 10^{-6}$ bits and $A_{\mathrm{R_{ex}}}^{\mathrm{R_{ex}}} \geq 10^{-6}$ were considered. Furthermore, parameters defining the joint stimulus-response probabilities and the transition matrices were restricted to the interval $[0.05, 0.95]$. Each difference between two measures defined in Equations (21)–(29) was computed repeatedly, with random initial values for the stimulus-response probabilities and the transition matrices, until the value of the difference failed to increase or decrease in 20 consecutive runs.

The values in Table 1 for Figure 3b were computed analytically with $P_{\mathrm{ex}}([3,2]) > 0$ or $P_{\mathrm{ex}}([4,2]) > 0$, but not both. In those cases, the measures ΔI^D, ΔI^B, and ΔA^B are undefined, whereas $\Delta I^{DL} = 100\,\%$, for the reasons given in Section 2.4. However, $\Delta I_{\mathrm{R_{su}}}$ and $\Delta I_{\hat{s}}$ can vary between $0\,\%$ and $100\,\%$, for example, attaining $0\,\%$ when $P_{\mathrm{ex}}([3,1]) \to 0$, and $100\,\%$ when $P_{\mathrm{ex}}([2,1]) \to 0$ and $P_{\mathrm{ex}}([4,2]) \to 0$. The information $I_{\mathrm{R_{ex}}}$ equals the stimulus entropy, regardless of the response probabilities. The values in Table 1 for Figure 3d were computed by setting $b = c = d = 0.5$ in Equation (16). The values in Section 2.4 for Figure 7b were obtained by setting $P_{\mathrm{ex}}(s,r) = 1/4$ for the stimulus-response pairs shown in the figure, and are valid for any transition probability matrix set as in Equation (32) with $b=a$. The values in Section 2.4 for Figure 8 were obtained by setting $P_{\mathrm{ex}}(s,r) = 1/4$ for the stimulus-response pairs shown in the figure.

Appendix B. Proofs

Appendix B.1. Derivation of Equation (30)

The definition of ΔI^{DL} involves the probability $P_{\mathrm{su}}(s|\hat{r},\theta)$ defined in [11,36,38] as proportional to $P(S) \prod_i P^\theta(R_i|S)$, where the exponent θ is chosen so as to maximize ΔI^{DL}. This definition has been recently shown to be invalid when $\exists\ r,s$ such that $P_{\mathrm{su}}(r|s) = 0$ for a stimulus s or a response r for which $P_{\mathrm{ex}}(r|s) \neq 0$ [17]. This problem never appears when evaluating the relevance of noise correlations with $P_{\mathrm{su}}(r|s) = P_{NI}(r|s)$ as stated by Equation (17). Yet, it may well appear in more general cases, including those arising from stochastically reduced codes. To overcome it, we prove the theorem

Theorem A1. *The probability $P(s|r,\theta)$ that appears in the definition of ΔI^{DL} is*

$$P_{\mathrm{su}}(s|r,\theta) \propto \begin{cases} P(s) & \text{if } \exists \hat{s}, \hat{r} \text{ such that } P_{\mathrm{ex}}(\hat{r}|\hat{s}) > P_{\mathrm{su}}(\hat{r}|\hat{s}) = 0 \\ 0 & \text{if } P_{\mathrm{su}}(r|s) = P_{\mathrm{ex}}(r|s) = 0 \text{ for some but not all } s \\ P(s)\, P_{\mathrm{su}}(r|s)^\theta & \text{otherwise} \end{cases}$$

Proof. According to Latham and Nirenberg [11], the probability $P_{\mathrm{su}}(s|r,\theta)$ is the one that minimizes the Kullback-Leibler divergence $D_{\mathrm{KL}}[P^*(r,s)||p(r)p(s)]$ with respect to the distribution $P^*(r,s)$, subject to the constraints

$$\langle \log_2 P_{\mathrm{su}}(r|s) \rangle_{P^*(r,s)} = \langle \log_2 q(r|s) \rangle_{P(s,r)} \tag{A1}$$

$$\sum_s P^*(r,s) = P(r). \tag{A2}$$

The minimization problem can be formulated in terms of an objective function to be minimized, in which the constraints appear with Lagrange multipliers, and θ is the one accompanying Equation (A1). Using the standard conventions that $0 \log 0 = 0$ and $x \log 0 = \infty$ for $x > 0$, Equation (A1) is fulfilled if $\exists\ \hat{r}, \hat{s}$ such that $P(\hat{s}|\hat{r},\theta) > 0$ if $P_{\mathrm{ex}}(\hat{r},\hat{s}) > P_{\mathrm{su}}(\hat{r}|\hat{s}) = 0$. The first part of the theorem immediately follows by solving Equation (B15) in [11] as there indicated with $\beta = 0$. If $\nexists\ \hat{r}, \hat{s}$ such that $P_{\mathrm{ex}}(\hat{r}, \hat{s}) > P_{\mathrm{su}}(\hat{r}|\hat{s}) = 0$, then Equation (A1) is fulfilled only if $P(s,r|\theta) = 0$ when $P_{\mathrm{su}}(r|s) = P_{\mathrm{ex}}(r,s) = 0$. The second and third parts of the theorem immediately follows using Bayes' rule. □

Appendix B.2. Proof of Theorem 1

Proof. The second part is proved by the two examples in Figure 4. The first part was proved in [9], at least for cases in which the set of the surrogate responses $\mathbf{R}_{su} = \mathbf{R}_{NI}$ differ from the set of the real responses $\mathbf{R}_{ex}$. When they both coincide, we can prove the first part by contradiction, assuming that a deterministic mapping exists from $\mathbf{R}_{ex}$ into $\mathbf{R}_{NI}$. If both variables sample the same response space, the deterministic mapping must be one-to-one, otherwise the variable $\mathbf{R}_{NI}$ would sample a smaller set. Therefore, both $\mathbf{R}_{NI}$ and $\mathbf{R}_{ex}$ maximize the conditional entropy given S over the probability distributions with the same marginals, since one-to-one mappings do not modify the entropy, and $\mathbf{R}_{NI}$ is defined as the distribution with maximal conditional entropy with fixed marginals. Because the probability distribution achieving this maximum is unique [16], $P_{su}(\mathbf{r}|s)$ and $P_{ex}(\mathbf{r}|s)$ must be the same, thereby proving the theorem. $\square$

Appendix B.3. Proof of Theorem 2

Proof. We prove the dependency on the marginal likelihoods by computing $Q(\mathbf{r}_{su}|\mathbf{r}_{ex})$ for the hypothetical experiment of Figure 4a, and observing that the result depends on the marginal likelihood $P_{ex}(L|s)$. To that end, we rewrite Equation (10) for $\mathbf{R}_{su} = [1,2]$ as

$$P_{su}([1,2]|\square) = P_{ex}([1,2]|\square)\, Q([1,2]|[1,2]) + P_{ex}([2,1]|\square)\, Q([1,2]|[2,1]) \,.$$

Note that $P_{ex}([1,2]|\square)=1-P_{ex}([2,1]|\square)=P_{ex}(L=1|\square)$ and $P_{su}([1,2]|\square)=P_{ex}(L=1|\square)^2$. Using this and rearranging the terms, we obtain the quadratic equation

$$P_{ex}(L=1|\square)^2 + P_{ex}(L=1|\square)\,\{Q([1,2]|[2,1]) - Q([1,2]|[1,2])\} - Q([1,2]|[2,1]) = 0\,,$$

that is solved by

$$P_{ex}(L=1|\square) = 0.5\left[\delta q + \left(\delta q^2 + 4\,Q([1,2]|[2,1])\right)^{0.5}\right]\,,$$

where $\delta q = Q([1,2]|[1,2]) - Q([1,2]|[2,1])$. Hence, any change in $P_{ex}(L=1|\square)$ must be followed by some change in $Q(\mathbf{r}_{su}|\mathbf{r}_{ex})$, thereby proving the first part.

We prove the dependency on the noise correlations by computing $Q(\mathbf{r}_{su}|\mathbf{r}_{ex})$ for the hypothetical experiment of Figure 5, and observing that the result not only depends on the marginal likelihoods $P_{ex}(L|s)$ and $P_{ex}(C|s)$, but in many cases, it also depends on the joint distributions $P_{ex}(L,C|s)$. Hence, varying the amount of noise correlations, even if keeping the marginals fixed, yields a variation in the mapping $Q(\mathbf{r}_{su}|\mathbf{r}_{ex})$.

We proceed by reductio ad absurdum. If $Q(\mathbf{r}_{su}|\mathbf{r}_{ex})$ does not depend on the amount of noise correlations in $P_{ex}(\mathbf{r}|s)$, we may assume that if we vary $P_{ex}(\mathbf{r}|s)$ but keep the marginals $P_{ex}(r_i|s)$ fixed, the transition probabilities $Q(\mathbf{r}_{su}|\mathbf{r}_{ex})$ remain unchanged. Under this hypothesis, Equation (10) is valid for many choices of $P_{ex}(\mathbf{r}|s)$. In this context, consider the set of all response distributions with the same marginals as $P_{ex}(\mathbf{r}|s)$ that can be turned into $P_{su}(\mathbf{r}|s)$ through $Q(\mathbf{r}_{su}|\mathbf{r}_{ex})$. This set includes $P_{su}(\mathbf{r}|s)$, and therefore, $Q(\mathbf{r}_{su}|\mathbf{r}_{ex})$ should be able to transform $P_{su}(\mathbf{r}|s)$ into itself. In addition, Property 2 requires that $Q(\mathbf{r}_{su}|[2,2]) = 0$ when $\mathbf{r}_{su} \neq [2,2]$ because either $P(\mathbf{r}_{su}|\square) = 0$ or $P(\mathbf{r}_{su}|\bigcirc) = 0$ for those responses. Normalization yields $Q([2,2]|[2,2]) = 1$. Furthermore, computing Equation (10) for $\mathbf{R}_{su} = [2,2]$ yields

$$0 = P_{ex}([1,1]|\square)\, Q([2,2]|[1,1]) + P_{ex}([1,2]|\square)\, Q([2,2]|[1,2]) + P_{ex}([2,1]|\square)\, Q([2,2]|[2,1])\,,$$

which shows that $Q([2,2]|\mathbf{r}_{ex})=0$ when $\mathbf{r}_{ex} \in \{[1,1],[1,2],[2,1]\}$. Consequently, the resulting $Q(\mathbf{r}_{su}|\mathbf{r}_{ex})$ yields through Equation 10 that $P_{su}([2,2]|\square) = P_{ex}([2,2]|\square)$. After noticing that

$$P_{su}([2,2]|\square)=P_{ex}(L=2|\square)\,P_{ex}(C=2|\square)\,,$$

and that

$$P_{ex}([2,2]|\square) = P_{ex}(L=2|\square)-P_{ex}([1,2]|\square) = P_{ex}(C=2|\square)-P_{ex}([2,1]|\square)\,,$$

we can show that, after some straightforward algebra, Equation (10) only holds if $P_{su}(\mathbf{r}|\square) = P_{ex}(\mathbf{r}|\square)$ for all $\mathbf{r}$. Thus, the initial hypothesis yields a transition matrix $Q(\mathbf{r}_{su}|\mathbf{r}_{ex})$ that is unable to transform $\mathbf{R}_{ex}$ into $\mathbf{R}_{su}$ when $\mathbf{R}_{ex}$ is noise correlated, and thus $Q(\mathbf{r}_{su}|\mathbf{r}_{ex})$ necessarily depends on the amount of noise correlations in $\mathbf{R}_{ex}$. $\square$

Appendix B.4. Proof of Theorem 3

Proof. The condition $\Delta I^D = \Delta I_{\mathbf{R}_{su}}$ implies that

$$\sum_{sr} P_{ex}(s,\mathbf{r}) \ \log \left[\frac{P_{ex}(s|\mathbf{r})}{P_{su}(s|\mathbf{r})} \right] = I_{\mathbf{R}_{ex}} - I_{\mathbf{R}_{su}}. \tag{A3}$$

However,

$$\Delta I^D = I_{\mathbf{R}_{ex}} - \sum_{sr} P_{ex}(s,\mathbf{r}) \ \log \left[\frac{P_{su}(\mathbf{r}|s)}{P_{su}(\mathbf{r})} \right].$$

Hence, Equation (A3) becomes

$$-\sum_{sr} P_{ex}(s,\mathbf{r}) \ \log \left[\frac{P_{su}(\mathbf{r}|s)}{P_{su}(\mathbf{r})} \right] = -I_{\mathbf{R}_{su}} \tag{A4}$$

In addition, when evaluating the relevance of noise correlations, $P_{su}(\mathbf{r},s) = P(s) \, P_{NI}(\mathbf{r}|s)$ as established by Equation (17). Hence,

$$-\sum_{sr} P_{ex}(s,\mathbf{r}) \ \log \left[\frac{P_{su}(\mathbf{r}|s)}{P_{su}(\mathbf{r})} \right] = \sum_{j} H(R_j|s) + \sum_{sr} P_{ex}(\mathbf{r},s) \ \log[P_{su}(\mathbf{r})] \tag{A5}$$

$$-I_{\mathbf{R}_{su}} = \sum_{j} H(R_j|s) + \sum_{sr} P_{su}(s,\mathbf{r}) \ \log[P_{su}(\mathbf{r})]. \tag{A6}$$

Replacing Equations (A5) and (A6) in Equation (A4),

$$\sum_{sr} P_{ex}(s,\mathbf{r}) \ \log[P_{su}(\mathbf{r})] = \sum_{sr} P_{su}(s,\mathbf{r}) \ \log[P_{su}(\mathbf{r})].$$

Summing in s, and rearranging,

$$\sum_{\mathbf{r}} [P_{ex}(\mathbf{r}) - P_{su}(\mathbf{r})] \ \log[P_{su}(\mathbf{r})] = 0.$$

If instead of an equality, we start with an inequality, that same inequality can be kept all through the proof. $\square$

Appendix B.5. Proof of Theorem 4

Proof. Consider a neural code $\mathbf{R}_{ex} = [R_1, \ldots, R_N]$ and recall that the range of $\mathbf{R}_{NI}$ includes that of $\mathbf{R}_{ex}$. Therefore, $\Delta I^{DL} = I_{\mathbf{R}_{ex}}$ implies that the minimum in Eqution (26) is attained when $\theta = 0$. In that case, Equation (B13a) in [11] yields

$$\sum_{s,r_n} P(s,r_n) \ \log_2 P(r_n|s) = \sum_{s,r_n} P(s) \ P_{ex}(r_n) \ \log_2 P_{ex}(r_n|s), \quad \forall \ 1 \le n \le N, \ \forall \ n.$$

After some more algebra and recalling that the Kullback-Leibler divergence is never negative, this equation becomes $I_{R_n} = 0$, implying that when read isolatedly, single responses contain no information about the stimulus. Consequently $\Delta I_{\mathbf{R}_N} = I_{\mathbf{R}_{ex}}$, thereby proving the "only if" part. For the "if" part, it is sufficient to notice that the last equality implies that $P_{NI}(\mathbf{r}|s) = P_{NI}(\mathbf{r})$. $\square$

Appendix B.6. Proof of Theorem 5

Proof. The conditions on f and $Q(\mathbf{r}_{su}|\mathbf{r}_{ex})$ ensure that $Q(\mathbf{r}_{su}|\mathbf{r}_{ex})$ can be written as a block-diagonal matrix, each block composed of the same rows with no zeros, and that each block can be associated with a non-overlapping partition $\mathcal{R}_1, \ldots, \mathcal{R}_M$ of the range of f. Under these conditions, $P(\mathbf{r}_{su}|\mathbf{r}_{ex}) = P(\mathbf{r}_{su}|\mathcal{R}_m)$ when $\mathbf{r}_{ex} \in \mathcal{R}_m$. Hence, for $\mathbf{r}_{su} \in \mathcal{R}_m$, $P(\mathbf{r}_{su}|s) = P(\mathbf{r}_{su}|\mathcal{R}_m) P(\mathcal{R}_m|s)$, yielding $P(s|\mathbf{r}_{su}) = P(s|\mathcal{R}_m)$ and $P(s|\mathbf{r}_{su}, \theta) = P(s|\mathcal{R}_m, \theta)$. Recomputing Equations (21)–(29) with these equalities in mind immediately yields the equalities in the theorem.

Even when the amount of information is equal, differences in the type of information may arise because the measures are based on different decoding strategies, here denoted α and β. However, under the conditions of the theorem, decoding strategy α and decoding strategy β are one and the same. Because $P(s|\mathbf{r}_{su}) = P(s|\mathcal{R}_m)$, both decoding strategies choose s only based on the partition $\mathcal{R}$ of $\mathbf{r}_{ex}$ or $\mathbf{r}_{su}$, respectively. Mathematically, both choose s according to

$$\hat{s} = \arg\max_s P(s|\mathcal{R}(\mathbf{r})),$$

where $\mathcal{R}(\mathbf{r})$ denotes the mapping from $\mathbf{r}$ into $\mathcal{R}$, which is the same regardless of whether $\mathbf{r}$ is $\mathbf{r}_{ex}$ or $\mathbf{r}_{su}$. Because $Q(\mathbf{r}_{su}|\mathbf{r}_{ex})$ maps each partition onto itself, the responses within each partition of $\mathbf{r}_{su}$ is completely generated by the responses in each partition of $\mathbf{r}_{ex}$, and thus the decoding strategies are applied to the same set of $\mathbf{r}_{ex}$. Hence, both decoding strategies are defined and operate in the same manner, yielding the same information. $\square$

Appendix C. On the Computation of ΔI^D

The information loss caused by mismatched decoders (decoding strategy α) when $\mathbf{R}_{su} = f(\mathbf{R}_{ex})$ has previously been computed as ΔI^D but with $P_{su}(s|\mathbf{r})$ replaced by $P(s|\mathbf{R}_{su} = f(\mathbf{r})) = P_{su}(s|f(\mathbf{r}))$ [10,12,13]. The latter represents the probability of s given that $\mathbf{R}_{su}$ takes the value $f(\mathbf{r})$, thereby limiting f to deterministic mappings. However, the probabilities $P_{su}(s|\mathbf{r})$ and $P_{su}(s|f(\mathbf{r}))$ are not equivalent, since

$$P_{su}(s|\mathbf{r}) \quad \propto \quad \sum_{\mathbf{r} = f(\hat{\mathbf{r}})} P_{ex}(\hat{\mathbf{r}}, s)$$

$$P_{su}(s|f(\mathbf{r})) \propto \sum_{f(\mathbf{r}) = f(\hat{\mathbf{r}})} P_{ex}(\hat{\mathbf{r}}, s)$$

These two definitions raise the question of which alternative is the appropriate one when computing the information loss caused by mismatched decoders.

To resolve this question, notice that replacing $P_{su}(s|\mathbf{r})$ with $P_{su}(s|f(\mathbf{r}))$ in Equation (6) yields the decoding algorithm

$$\hat{s} = \arg\max_s P_{su}(s|f(\mathbf{r})).$$

This algorithm entails first transforming the observed $\mathbf{r}$ into $\mathbf{r}_{su} = f(\mathbf{r})$, and then choosing the stimulus $\hat{s} = D_{su}(\hat{\mathbf{r}})$ with a matched probability. Hence, its operation is analogous to the decoding algorithm β, and not, as originally intended, to the decoding algorithm α.

To illustrate the difference, recall the experiment in Figure 7a and suppose that the observed response is $\mathbf{r} = 0.25\pi$. The decoding algorithm α reads this value, computes $P_{su}(s|0.25\pi)$, and decodes $\hat{s} = \square$. Instead, the decoding algorithm proposed in [10,12,13], first transforms the value of $\mathbf{r} = 0.25\pi$ into $f(\mathbf{r}) = 0.75\pi$, then computes $P_{su}(s|0.75\pi)$, and finally decodes $\hat{s} = \bigcirc$. This mode of operation corresponds to the decoding algorithm β.

The above discrepancy can also be seen from the change in the operational meaning of ΔI^D caused by the replacement. To that end, recall that ΔI^D was first introduced as a comparison between the average number of binary questions required to identify s after observing $\mathbf{r}$ when using two optimal question-asking strategies, one tailored for $P_{ex}(s|\mathbf{r})$ and the other for $P_{su}(s|\mathbf{r})$ [8]. Mathematically, this difference can be written as

$$\Delta I^D = \sum_{s,\mathbf{r}} P_{ex}(s,\mathbf{r}) \, \log_2 P_{ex}(s|\mathbf{r}) - \sum_{s,\mathbf{r}} P_{ex}(s,\mathbf{r}) \, \log_2 P_{su}(s|\mathbf{r}) \,. \tag{A7}$$

In each term, the argument of the logarithms is determined by the question-asking strategy, whereas the weight of the averages is determined by the probability distribution of the variables on which the strategy is applied [8,10,16]. Equation (A7) describes the decoding strategy α.

Replacing $P_{su}(s|\mathbf{r})$ with $P_{su}(s|f(\mathbf{r}))$ turns Equation (A7) into

$$\begin{aligned}
\Delta \tilde{I} &= \sum_{s,\mathbf{r}} P_{ex}(s,\mathbf{r}) \, \log_2 P_{ex}(s|\mathbf{r}) - \sum_{s,\mathbf{r}} P_{ex}(s,\mathbf{r}) \, \log_2 P(s|f(\mathbf{r})) \\
&= \sum_{s,\mathbf{r}} P_{ex}(s,\mathbf{r}) \, \log_2 P_{ex}(s|\mathbf{r}) - \sum_{s,\mathbf{r}_{su}} P_{su}(s,\mathbf{r}_{su}) \, \log_2 P_{su}(s|\mathbf{r}_{su}) \\
&= \Delta I_{\mathbf{R}_{su}} \,.
\end{aligned}$$

Unlike ΔI^D, this difference compares the average number of binary questions required to identify s after observing $\mathbf{r}$ using a question-asking strategy that is optimal for $P_{ex}(s|\mathbf{r})$, with the average number of binary questions required to identify s after observing $\mathbf{r}_{su}$ using the a question-asking strategy that is optimal for $P_{su}(s|\mathbf{r}_{su})$. This is the way the decoding strategy β operates, not α.

Naively, one may think that a change in $P_{su}(s|\mathbf{r})$, regardless of its size, may turn the measure $\Delta I_{\mathbf{R}_{su}}$, typically regarded as an encoding-oriented measure and here linked to the decoding algorithm β, into the decoding-oriented measure ΔI^D. However, notice that this change cannot occur through the equations above due to the change induced in $P_{su}(s,\mathbf{r}_{su})$. For that to actually occur, one must write $\Delta I_{\mathbf{R}_{su}}$ differently, as for example:

$$\Delta I_{\mathbf{R}_{su}} = \sum_{s,\mathbf{r}} P_{ex}(s,\mathbf{r}) \, \log_2 P_{ex}(s|\mathbf{r}) - \sum_{s,\mathbf{r}_{su}} P_{ex}(s,\mathbf{r}) \, \log_2 P_{su}(s|\mathbf{r}_{su}) \,.$$

In this reformulation, the second term can be interpreted as the average number of binary questions required to identify s after observing $\mathbf{r}$ using a question-asking strategy that is optimal for $P_{su}(s|\mathbf{r}_{su})$, but only after converting $\mathbf{r}$ into $\mathbf{r}_{su}$. Any change in $P_{su}(s|\mathbf{r}_{su})$ immediately renders $P_{su}(s|\mathbf{r}_{su})$ a mismatched probability for $\mathbf{r}_{su}$, and makes the second term represent the average number of binary questions required to identify s after observing $\mathbf{r}$ using the question-asking strategy that is optimal for an altered version of $P_{su}(s|\mathbf{r}_{su})$ but only after converting $\mathbf{r}$ into $\mathbf{r}_{su}$, which need not resemble the meaning of the second term in ΔI^D.

References

1. Adrian, E.D. The impulses produced by sensory nerve endings. *J. Physiol.* **1926**, *61*, 49–72. [CrossRef] [PubMed]
2. Hubel, D.H.; Wiesel, T.N. Receptive fields of single neurones in the cat's striate cortex. *J. Physiol.* **1959**, *148*, 173–180. [CrossRef]
3. Thorpe, S.; Fize, D.; Marlot, C. Speed of processing in the human visual system. *Nature* **1996**, *6582*, 520–522. [CrossRef] [PubMed]
4. Abeles, M. *Corticonix: Neural Circuits of the Cerebral Cortex*; Cambridge University Press: Cambridge, UK, 1991.
5. Gray, C.M.; König, P.; Engel, A.K.; Singer, W. Oscillatory responses in cat visual cortex exhibit inter-columnar synchronization which reflects global stimulus properties. *Nature* **1989**, *6213*, 334–337. [CrossRef] [PubMed]
6. Franke, F.; Fiscella, M.; Sevelev, M.; Roska, B.; Hierlemann, A.; da Silveira, R.A. Structures of Neural Correlation and How They Favor Coding. *Neuron* **2016**, *89*, 409–422. [CrossRef] [PubMed]
7. O'Keefe, J. Hippocampues, theta, and spatial memory. *Curr. Opin. Neurobiol.* **1993**, *6*, 917–924. [CrossRef]
8. Nirenberg, S.; Carcieri, S.M.; Jacobs, A.L.; Latham, P.E. Retinal ganglion cells act largely as independent encoders. *Nature* **2001**, *411*, 698–701. [CrossRef] [PubMed]
9. Schneidman, E.; Bialek, W.; Berry, M.J. Synergy, redundancy, and independence in population codes. *J. Neurosci.* **2003**, *23*, 11539–11553. [CrossRef] [PubMed]

10. Nirenberg, S.; Latham, P.E. Decoding neuronal spike trains: How important are correlations? *Proc. Natl. Acad. Sci. USA* **2003**, *100*, 7348–7353. [CrossRef] [PubMed]

11. Latham, P.E.; Nirenberg, S. Synergy, redundancy, and independence in population codes, revisited. *J. Neurosci.* **2005**, *25*, 5195–5206. [CrossRef] [PubMed]

12. Quiroga, R.Q.; Panzeri, S. Extracting information from neuronal populations: Information theory and decoding approaches. *Nat. Rev. Neurosci.* **2009**, *10*, 173–185. [CrossRef] [PubMed]

13. Latham, P.E.; Roudi, Y. Role of correlations in population coding. In *Principles of Neural Coding*; Panzeri, S., Quian Quiroga, R., Eds.; CRC Press: Boca Raton, FL, USA, 2013; Chapter 7, pp. 121–138.

14. Casella, G.; Berger, R.L. *Statistical Inference*, 2nd ed.; Duxbury Press: Duxbury, MA, USA, 2002.

15. Panzeri, S.; Brunel, N.; Logothetis, N.K.; Kayser, C. Sensory neural codes using multiplexed temporal scales. *Trends Neurosci.* **2010**, *33*, 111–120. [CrossRef] [PubMed]

16. Cover, T.M.; Thomas, J.A. *Elements of Information Theory*, 2nd ed.; Wiley-Interscience: New York, NY, USA, 2006.

17. Eyherabide, H.G.; Samengo, I. When and why noise correlations are important in neural decoding. *J. Neurosci.* **2013**, *33*, 17921–17936. [CrossRef] [PubMed]

18. Knill, D.C.; Pouget, A. The Bayesian brain: The role of uncertainty in neural coding and computation. *Trends Neurosci.* **2004**, *27*, 712–719. [CrossRef] [PubMed]

19. van Bergen, R.S.; Ma, W.J.; Pratte, M.S.; Jehee, J.F.M. Sensory uncertainty decoded from visual cortex predicts behavior. *Nat. Neurosci.* **2015**, *18*, 1728–1730. [CrossRef] [PubMed]

20. Ince, R.A.A.; Senatore, R.; Arabzadeh, E.; Montani, F.; Diamond, M.E.; Panzeri, S. Information-theoretic methods for studying population codes. *Neural Netw.* **2010**, *23*, 713–727. [CrossRef] [PubMed]

21. Reinagel, P.; Reid, R.C. Temporal coding of visual information in the thalamus. *J. Neurosci.* **2000**, *20*, 5392–5400. [CrossRef] [PubMed]

22. Panzeri, S.; Petersen, R.S.; Schultz, S.R.; Lebedev, M.; Diamond, M.E. The Role of Spike Timing in the Coding of Stimulus Location in Rat Somatosensory Cortex. *Neuron* **2001**, *29*, 769–777. [CrossRef]

23. Rokem, A.; Watzl, S.; Gollisch, T.; Stemmler, M.; Herz, A.V.M.; Samengo, I. Spike-timing precision underlies the coding efficiency of auditory receptor neurons. *J. Neurophysiol.* **2006**, *95*, 2541–2552. [CrossRef] [PubMed]

24. Lefebvre, J.L.; Zhang, Y.; Meister, M.; Wang, X.; Sanes, J.R. γ-Protocadherins regulate neuronal survival but are dispensable for circuit formation in retina. *Development* **2008**, *135*, 4141–4151. [CrossRef] [PubMed]

25. Victor, J.D.; Purpura, K.P. Nature and precision of temporal coding in visual cortex: A metric-space analysis. *J. Neurophysiol.* **1996**, *76*, 1310–1326. [CrossRef] [PubMed]

26. Victor, J.D. Spike train metrics. *Curr. Opin. Neurobiol.* **2005**, *15*, 585–592. [CrossRef] [PubMed]

27. Boyd, S.; Vandenberghe, L. *Convex Optimization*; Cambridge University Press: Cambridge, UK, 2004.

28. Fano, R.M. *Transmission of Information*; The MIT Press: Cambridge, MA, USA, 1961.

29. DeWeese, M.R.; Meister, M. How to measure the information gained from one symbol. *Netw. Comput. Neural Syst.* **1999**, *10*, 325–340. [CrossRef]

30. Eyherabide, H.G.; Samengo, I. Time and category information in pattern-based codes. *Front. Comput. Neurosci.* **2010**, *4*, 145. [CrossRef] [PubMed]

31. Eckhorn, R.; Pöpel, B. Rigorous and extended application of information theory to the afferent visual system of the cat. I. Basic concepts. *Kybernetik* **1974**, *16*, 191–200. [CrossRef] [PubMed]

32. Panzeri, S.; Treves, A. Analytical estimates of limited sampling biases in different information measures. *Network* **1996**, *7*, 87–107. [CrossRef] [PubMed]

33. Eyherabide, H.G. Disambiguating the role of noise correlations when decoding neural populations together. *arXiv* **2016**, arXiv:1608.05501.

34. MacKay, D.M.; McCulloch, W.S. The limiting information capacity of a neuronal link. *Bull. Math. Biophys.* **1952**, *14*, 127–135. [CrossRef]

35. Fitzhugh, R. The statistical detection of threshold signals in the retina. *J. Gen. Physiol.* **1957**, *40*, 925–948. [CrossRef] [PubMed]

36. Merhav, N.; Kaplan, G.; Lapidoth, A.; Shamai Shitz, S. On information rates for mismatched decoders. *IEEE Trans. Inf. Theory* **1994**, *40*, 1953–1967. [CrossRef]

37. Oizumi, M.; Ishii, T.; Ishibashi, K.; Hosoya, T.; Okada, M. A general framework for investigating how far the decoding process in the brain can be simplified. In *Advances in Neural Information Processing Systems*; The MIT Press: Cambridge, MA, USA, 2009; pp. 1225–1232.

38. Oizumi, M.; Ishii, T.; Ishibashi, K.; Hosoya, T.; Okada, M. Mismatched decoding in the brain. *J. Neurosci.* **2010**, *30*, 4815–4826. [CrossRef] [PubMed]

39. Oizumi, M.; Amari, S.I.; Yanagawa, T.; Fujii, N.; Tsuchiya, N. Measuring Integrated Information from the Decoding Perspective. *PLoS Comput. Biol.* **2016**, *12*, e1004654. [CrossRef] [PubMed]

40. Gochin, P.M.; Colombo, M.; Dorfman, G.A.; Gerstein, G.L.; Gross, C.G. Neural ensemble coding in inferior temporal cortex. *J. Neurophysiol.* **1994**, *71*, 2325–2337. [CrossRef] [PubMed]

41. Warland, D.K.; Reinagel, P.; Meister, M. Decoding visual information from a population of retinal ganglion cells. *J. Neurophysiol.* **1997**, *78*, 2336–2350. [CrossRef] [PubMed]

42. Optican, L.M.; Richmond, B.J. Temporal encoding of two-dimensional patterns by single units in primate inferior temporal cortex. III. Information theoretic analysis. *J. Neurophysiol.* **1987**, *57*, 162–178. [CrossRef] [PubMed]

43. Salinas, E.; Abbott, L.F. Transfer of coded information from sensory neurons to motor networks. *J. Neurosci.* **1995**, *10*, 6461–6476. [CrossRef]

44. Geisler, W.S. Sequential ideal-observer analysis of visual discriminations. *Psychol. Rev.* **1989**, *96*, 267–314. [CrossRef] [PubMed]

45. Högnäs, G.; Mukherjea, A. *Probability Measures on Semigroups: Convolution Products, Random Walks and Random Matrices*, 2nd ed.; Springer: New York, NY, USA, 2011.

46. Samengo, I.; Treves, A. The information loss in an optimal maximum likelihood decoding. *Neural Comput.* **2002**, *14*, 771–779. [CrossRef] [PubMed]

47. Shamir, M. Emerging principles of population coding: In search for the neural code. *Curr. Opin. Neurobiol.* **2014**, *25*, 140–148. [CrossRef] [PubMed]

48. Gawne, T.J.; Richmond, B.J. How independent are the messages carried by adjacent inferior temporal cortical neurons? *J. Neurosci.* **1993**, *13*, 2758–2771. [CrossRef] [PubMed]

49. Gollisch, T.; Meister, M. Rapid Neural Coding in the Retina with Relative Spike Latencies. *Science* **2008**, *5866*, 1108–1111. [CrossRef] [PubMed]

50. Reifenstein, E.T.; Kemptner, R.; Schreiber, S.; Stemmler, M.B.; Herz, A.V.M. Grid cells in rat entorhinal cortex encode physical space with independent firing fields and phase precession at the single-trial level. *Proc. Natl. Acad. Sci. USA* **2012**, *109*, 6301–6306. [CrossRef] [PubMed]

51. Park, H.J.; Friston, K. Nonlinear multivariate analysis of neurophysiological signals. *Science* **2013**, *6158*, 1238411. [CrossRef] [PubMed]

52. Dahlhaus, R.; Eichler, M.; Sandkühler, J. Identification of synaptic connections in neural ensembles by graphical models. *J. Neurosci. Methods* **1997**, *77*, 93–107. [CrossRef]

53. Panzeri, S.; Schultz, S.R.; Treves, A.; Rolls, E.T. Correlations and the encoding of information in the nervous system. *Proc. R. Soc. B Biol. Sci.* **1999**, *266*, 1001–1012. [CrossRef] [PubMed]

54. Schultz, S.R.; Panzeri, S. Temporal Correlations and Neural Spike Train Entropy. *Phys. Rev. Lett.* **2001**, *25*, 5823–5826. [CrossRef] [PubMed]

55. Panzeri, S.; Schultz, S.R. A Unified Approach to the Study of Temporal, Correlational, and Rate Coding. *Neural Comput.* **2001**, *13*, 1311–1349. [CrossRef] [PubMed]

56. Pola, G.; Thiele, A.; Hoffmann, K.P.; Panzeri, S. An exact method to quantify the information transmitted by different mechanisms of correlational coding. *Network* **2003**, *14*, 35–60. [CrossRef] [PubMed]

57. Hernández, D.G.; Zanette, D.H.; Samengo, I. Information-theoretical analysis of the statistical dependencies between three variables: Applications to written language. *Phys. Rev. E.* **2015**, *92*, 022813. [CrossRef] [PubMed]

58. Williams, P.L.; Beer, R.D. Nonnegative decomposition of multivariate information. *arXiv* **2010**, arXiv:1004.2515.

59. Harder, M.; Salge, C.; Polani, D. Bivariate Measure of Redundant Information. *Phys. Rev. E.* **2013**, *87*, 012130. [CrossRef] [PubMed]

60. Griffith, V.; Koch, C. Quantifying Synergistic Mutual Information. In *Guided Self-Organization: Inception*; Prokopenko, M., Ed.; Springer: New York, NY, USA, 2014; Chapter 6, pp. 159–190.

61. Bertschinger, N.; Rauh, J.; Olbrich, E.; Jost, J.; Ay, N. Quantifying unique information. *Entropy* **2014**, *16*, 2161–2183. [CrossRef]

62. Ince, R.A.A. Measuring Multivariate Redundant Information with Pointwise Common Change in Surprisal. *Entropy* **2017**, *19*, 318. [CrossRef]

63. Chicharro, D.; Panzeri, S. Synergy and Redundancy in Dual Decompositions of Mutual Information Gain and Information Loss. *Entropy* **2017**, *19*, 71. [CrossRef]

64. Wolpert, D.H.; Wolf, D.R. Estimating functions of probability distributions from a finite set of samples. *Phys. Rev. E.* **1996**, *52*, 6841–6973. [CrossRef]

65. Samengo, I. Estimating probabilities from experimental frequencies. *Phys. Rev. E* **2002**, *65*, 046124. [CrossRef] [PubMed]

66. Nemenman, I.; Bialek, W.; de Ruyter van Steveninck, R. Entropy and information in neural spike trains: Progress on the sampling problem. *Phys. Rev. E.* **2004**, *69*, 056111. [CrossRef] [PubMed]

67. Paninski, L. Estimation of entropy and mutual information. *Neural Comput.* **2003**, *6*, 1191–1253. [CrossRef]

68. Panzeri, S.; Senatore, R.; Montemurro, M.A.; Petersen, R.S. Correcting for the sampling bias problem in spike train information measures. *J. Neurophysiol.* **2007**, *98*, 1064–1072. [CrossRef] [PubMed]

69. Montemurro, M.A.; Senatore, R.; Panzeri, S. Tight data-robust bounds to mutual information combining shuffling and model selection techniques. *Neural Comput.* **2007**, *11*, 2913–2957. [CrossRef] [PubMed]

![entropy logo] *entropy*

MDPI

Article

Information-Theoretical Analysis of the Neural Code in the Rodent Temporal Lobe

Melisa B. Maidana Capitán [1] [ID], Emilio Kropff [2] and Inés Samengo [1],*[ID]

[1] Departament of Medical Physics, Centro Atómico Bariloche and Instituto Balseiro,
    Comisión Nacional de Energía Atómica, Consejo Nacional de Investigaciones Científicas y Técnicas,
    8400 San Carlos de Bariloche, Argentina; melisa.mc89@gmail.com
[2] Fundación Instituto Leloir, Consejo Nacional de Investigaciones Científicas y Técnicas,
    1425 Buenos Aires, Argentina; kropff@gmail.com
*   Correspondence: samengo@cab.cnea.gov.ar; Tel.: +54-294-444-5100

Received: 31 May 2018; Accepted: 25 July 2018; Published: 3 August 2018

Abstract: In the study of the neural code, information-theoretical methods have the advantage of making no assumptions about the probabilistic mapping between stimuli and responses. In the sensory domain, several methods have been developed to quantify the amount of information encoded in neural activity, without necessarily identifying the specific stimulus or response features that instantiate the code. As a proof of concept, here we extend those methods to the encoding of kinematic information in a navigating rodent. We estimate the information encoded in two well-characterized codes, mediated by the firing rate of neurons, and by the phase-of-firing with respect to the theta-filtered local field potential. In addition, we also consider a novel code, mediated by the delta-filtered local field potential. We find that all three codes transmit significant amounts of kinematic information, and informative neurons tend to employ a combination of codes. Cells tend to encode conjunctions of kinematic features, so that most of the informative neurons fall outside the traditional cell types employed to classify spatially-selective units. We conclude that a broad perspective on the candidate stimulus and response features expands the repertoire of strategies with which kinematic information is encoded.

Keywords: mutual information; synergy; redundancy; neural code; hippocampus; entorhinal cortex; navigation

1. Introduction

Hippocampal and entorhinal neurons are selective to the kinematic state of the subject [1]. In the original studies, selectivity was assessed by constructing firing maps of individual neurons, that is, by describing the dependence of the firing rate as a function of the position of the animal [2,3]. These representations can be used to estimate the amount of encoded spatial information [4]. Firing maps constitute the two-dimensional generalizations of the tuning curves employed in sensory domains [5]. In addition to neurons selective to position, other cells responding to head direction [6], velocity [7], environment boundaries [8], direction of motion [9], and combinations of these features [10], have been found. Moreover, not only the firing rate, but also the phase-of-firing with respect to the oscillatory extracellular local field potential (LFP) encodes kinematic information [11,12].

Tuning curves reveal how the mean firing rate, averaged inside a certain window and also across trials, is modulated by the stimulus. However, the neural code may well depend on other statistical properties of the joint distribution between stimuli and responses, beyond the mean [13]. If the variance of the firing rate were also modulated by position, for example, tuning curves would not be capturing part of the information encoded in responses. Shannon's mutual information, defined as the Kullback-Leibler divergence between the joint distribution between stimuli and responses and the product of the marginals, is the most general tool to address all possible encoding mechanisms, since it makes no assumptions about the nature of the probabilistic mapping [14].

A delicate matter, however, is how to choose the set of stimuli and the set of responses between which information is to be calculated. In the sensory domain, this choice has been typically approached with two different strategies. The classical strategy has been to select a certain number of static or stereotyped stimuli using heuristic criteria, and to present them repeatedly to a subject while recording the neural responses in a given brain area. To apply this strategy, scientists need to know from the start the stimuli that are adequate to probe the area under study. For example, when examining early visual areas, gratings of varying contrast, orientation and spatial frequency are usually considered adequate; in the infero-temporal cortex, faces, hands, or tools are often employed; in the temporal lobe of navigating rodents, the position, head direction or direction of motion of the animal are taken as stimuli. Of course, it may well be the case that, by probing a given brain area with the wrong stimuli, one misses entirely the key carriers of information. In fact, the identification of the relevant stimuli has often contained a certain dose of luck, as for example, the discovery of orientation selectivity in V1 [15], of hand or face selective neurons in IT [16], and place cells in CA1 [17]. One may therefore wonder whether still more appropriate stimuli could be found, better than the ones traditionally employed, if only more stimuli were tested.

The second strategy circumvents the arbitrariness in the selection of stimuli by no longer considering them isolated items. The stimulus is seen as a continuous signal that flows into the area under study, as happens in ecological conditions. The underlying assumption is that, if the ensemble from which stimulus are drawn is similar to the natural environment, eventually, the relevant features will come up [18,19]. This approach has allowed scientists to discover many relevant stimulus features that had not been noticed using the classical strategy, as for example, multiple subfields in macaque V1 [20], or the correlated deflection of several whiskers in barrel cortex [21]. The information-theoretical formulation of this strategy was provided by Strong et al. in 1998 [22]. Stimuli were considered to be stochastic processes unfolding in time, and the code was characterized by the information rate, that is, the mutual information per unit time between a set of responses and the set of all stimulus histories contained in the input signal. From the operational point of view, each point in time can be associated with a specific stimulus history. The mutual information between stimuli and responses, hence, quantifies the difference between the response distributions conditional to a given point in time, and the response distributions marginalized over time. To calculate this information, the animal must be exposed to exactly the same stimulus sequence in repeated trials. The appeal of this method is that it makes no assumptions about the stimulus feature the response is selective to, nor whether selectivity is encoded in the mean or the variance of the response distribution. The drawback, is that selectivity is quantified, but the key informative features remain unknown.

In an attempt to develop data analysis techniques that are as open minded as possible with respect to the candidate stimulus features that may be relevant, as well as the candidate response features instantiating the code, here we adapt this strategy to the specific case of the encoding of kinematic variables in the temporal lobe. This is a challenging system, because exposing the animal to a rich set of stimuli requires its active engagement in locomotion, exploring different places and states of motion with a repetitive protocol. To this aim, we analyzed physiological data recorded in animals that in repeated trials followed exactly the same trajectory, as fixated by a careful experimental design. The paradigm allowed us to compare the response probability distributions conditional to one point along the trajectory, with the marginal distribution. We thereby assessed whether specific points in time produced distinctive responses, irrespective of whether such points could be linked to well-known kinematic features as "position", "velocity" or "direction of motion". The information thus obtained was then compared to the mutual information between traditional kinematic features and neural responses. We find that, in agreement with another previous open-minded approach developed with different data analysis techniques [23,24], traditional descriptions of the neural code capture only a limited fraction of the whole repertoire of encoding strategies.

To broaden also the candidate response features encoding information, we not only analyzed the traditional information-carriers, as firing rate and phase-of-firing with respect to the theta-filtered LFP, but also the delta-filtered LFP, since both the hippocampus and the entorhinal cortex contain a significant amount of power in the delta band. We also estimated the degree up to which the three candidate codes (rate, theta and delta) tend to coincide in the same cells, or are rather segregated into different populations. In addition, we also estimated the degree of synergy or redundancy between pairs of kinematic features, or pairs of response variables. We conclude that, by broadening both the stimulus and responses variables that can putatively encode information, the neural code in the temporal lobe represents kinematic information employing a wider collection of attributes than revealed by traditional approaches.

2. Results

A Long Evans rat was trained to run along a 4m linear track, inside a bottomless cart, the position and velocity of which was controlled by a computer (Section 4.1 contains the details of the behavioral task). Multiple trials were thus recorded, with identical kinematic conditions. In each trial, the rat ran first to the right, with two different velocities, while the activity of individual neurons in the hippocampus (H) and the entorhinal cortex (EC) was recorded, together with the LFP. When the animal reached the end of the track, it turned around, and the cart moved leftward returning to the starting point, again employing two different velocities. In this paper, several stimulus attributes are probed. The most general attribute is "Time along the trajectory", henceforth shortened as "Time". This attribute has the advantage of tagging whatever feature (among those that are repeated across trials) that may modulate the response. Relevant stimulus features may be traditional *kinematic variables*, as "Position", or "Velocity", but they may also be some event so far unnoticed by us scientists, as for example, a certain distress produced by the cable attached to the head of the animal when a specific motor action is required, or the awareness that the arrival to the end of the track is imminent, where the food reward is delivered. To determine the degree up to which the information encoded in the variable Time corresponded to the traditional kinematic variables, we introduced the variables Position, Velocity, Direction, Signed Position and Signed Direction by adequately grouping temporal bins (Section 4.2 describes how different variables were defined and binned).

2.1. Population Firing Statistics

In Figure 1, the distribution of firing rates and circular variances throughout the population of recorded neurons are displayed. Equations (1) and (2) in Section 4.4 define the relevant circular variances, as well as the criteria to assess their significance. The distribution of hippocampal (left) and entorhinal (right) firing rates is roughly exponential for low-to-medium frequencies, with a long high-frequency tail extending to the right. In the hippocampus, the mean firing rate is 2.2 ± 5.5 Hz, and in the entorhinal cortex, 2.3 ± 4.1 Hz. The two mean values are not significantly different (Student t-Test $p = 0.9$), but the shapes of the distributions are significantly different (Smirnov-Kolomogorov $p < 0.0001$), mainly due to the difference in the two variances. The average firing rate in the two regions is therefore similar, but in the H there is a larger cell-to-cell dispersion.

Section 4.3 describes the way the LFP is filtered in either the theta or the delta bands. The distribution of circular variances of the phase-of-firing with respect to the theta-filtered LFP is fairly broad, in both areas, with some cells with values even below 0.1 (Figure 1(B1,B2)). In H, the mean circular variance was 0.46 ± 0.24, and in the EC, 0.48 ± 0.26 Hz. The two mean values are not significantly different (Student t-test $p = 0.2$). The shape of the distributions may or may not be considered significantly different, depending on the chosen significance threshold (Smirnov-Kolmogorov $p = 0.02$). Hence, the amount of locking to the theta rhythm in both regions is comparable. In H, cells locked to theta preferred to fire at phase π (Figure 1(C1)), that is, when the LFP reaches a minimum. Instead, in EC the distribution of locking angles was bimodal, with peaks near 0 and π, and a preference for the former (Figure 1(C2)).

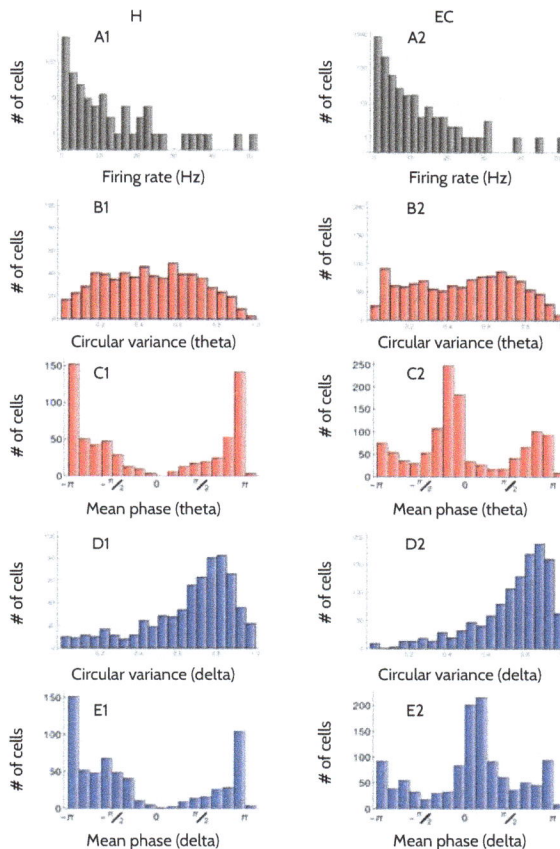

Figure 1. Distribution of response properties in the recorded neurons. Left column: Hippocampus. Right column: Entorhinal cortex. A: Histograms of the firing rate across the population. Vertical axis in logarithmic scale. B: Histograms of the circular variance of the phase-of-firing with respect to the theta-filtered LFP. C: Histograms of the mean phase of firing in the theta-filtered LFP. D: Same as B for the delta-filtered LFP. D: Same as C for the delta-filtered LFP. All the circular variances reported here are significant (Section 4.4).

Neurons are less locked to the delta-filtered LFP than to theta. For delta, the distribution of circular variances peaks at around 0.8 in H, and in 0.9 in the EC, and with means 0.66 ± 0.23 and 0.73 ± 0.20, respectively (Figure 1(D1,D2)). The two distributions are also narrower than in the case of theta. The distributions obtained in H and EC are significantly different (Smirnov-Kolmogorov $p < 0.0001$), and so are the two mean values (Student t-test $p < 0.0001$). Hippocampal cells are more locked to the delta rhythm than entorhinal cells. With respect to delta, hippocampal cells preferred to lock at phase π (Figure 1(E1)), whereas entorhinal cells were divided into two populations, one of them locked at around phase 0, and the other at phase π, the former more numerous (Figure 1(E2)).

Cells that are tightly locked to a rhythm cannot transmit information using a phase code, because the total entropy is small. Cells whose phase-of-firing varies substantially, and the variation is not systematically related to kinematic variables, have no information either, because the noise entropy is large. Hence, neurons can only encode information in the phase of firing if they are locked to a specific phase, and the locking varies with some kinematic condition, either by shifting the preferred firing phase (the mean phase), or the degree of locking (the circular variance).

2.2. Example Neurons with Different Encoding Strategies

A visual inspection of the response properties of individual neurons revealed the existence of firing rate codes (Figure 2A) and phase-of-firing codes, the latter defined in terms of both the theta (Figure 2B) and delta phase (Figure 2C) of the LFP.

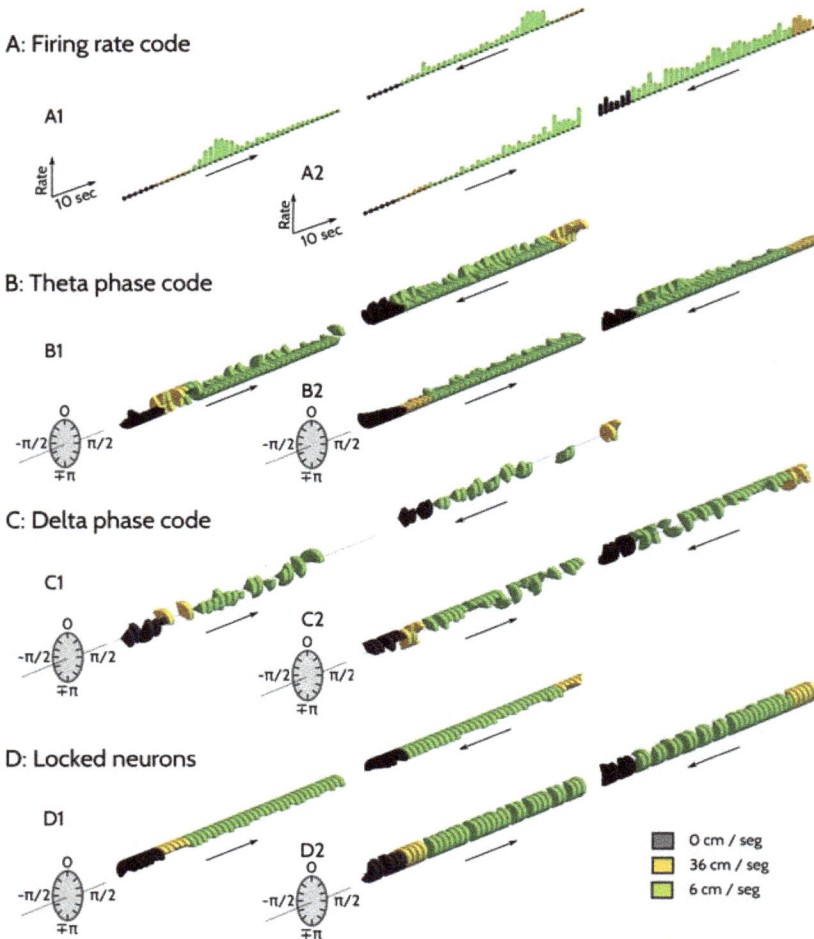

Figure 2. Example codes. Mean firing rate (**A1,A2**), and phase-of-firing with respect to the theta- (**B1,B2,D1**), and delta-filtered (**C1,C2,D2**) LFP. Horizontal axis is time, flowing from left to right. Right and left runs are displayed separately. Arrows indicate direction of motion. (**A**) Maximal rates: 130 Hz (**A1**) and 95 Hz (**A2**); (**B–D**) Mean and SD of phases-of-firing represented disc sectors covering $\langle \phi \rangle \pm \sqrt{-2\ln(\sigma_\phi^2)}$ (Equations (1) and (2)). H: (**A2,B1,B2,C2,D1**). EC: (**A1,C1,D2**).

The entorhinal neuron of Figure 2(A1) shows selectivity to position. Firing rate is maximal when the rat is located in a region around 220 and 250 cm away from the starting point. The hippocampal neuron of Figure 2(A2) is instead selective to direction of motion. The firing rate is maximal when the rat is running towards the left. The hippocampal neuron of Figure 2(B1) encodes velocity with the mean phase-of-firing with respect to the theta-filtered LFP. The phase is almost always locked at π, except for high velocity, where it shifts to $-\pi/2$. In the hippocampal example of Figure 2(B2), the circular variance is fairly constant throughout the trial, except for the initial interval during slow running in the return journey, where it increases. This is an example of information encoded in a modulation of the response variance. In the entorhinal neuron of Figure 2(C1), the delta-filtered firing phases are mostly locked to $\pi/2$, although they occasionally shift towards the angle 0, most notably, at high velocities, and towards the end of the track during the rightward journey. The delta phases of the hippocampal example of Figure 2(C2) are mostly locked to 0 degrees, except at high velocities, and at a spot located approximately at 1 m from the end of the track . The firing phases of the two neurons of Figure 2D (hippocampal in D1 and entorhinal in D2) remain locked all throughout the trajectory (delta and theta bands, respectively).

2.3. The Encoding of the Variable Time by Individual Neurons

In the behavioral task explored here, position, velocity and direction of motion are all functions of Time. Hence, the data processing theorem ensures that the mutual information between a physiological variable (as firing rate, or the phase with respect to the theta, or the delta-filtered LFP) and Time is at least as large as the information about any other derived kinematic feature. Section 4.5 describes the procedure we used to calculate all mutual informations. Figure 3 displays the dependence of the information about Time on each code.

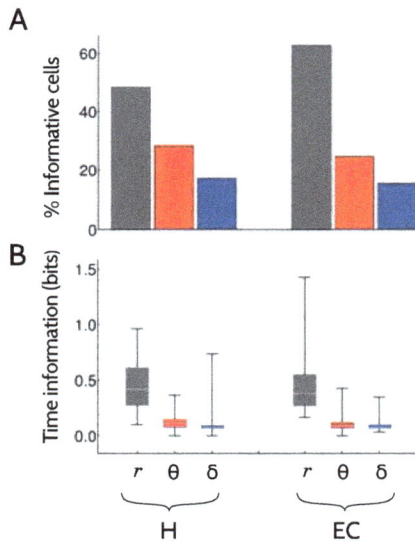

Figure 3. The encoding of the variable Time by individual neurons. (A) Percentage of cells that were informative about Time in the two brain areas (H and EC) and the three tested codes (firing, theta and delta). (B) Box-histograms of the mutual information $I(Y; \text{Time})$ of the cells with significant amounts of information for different choices of the physiological variable Y. The horizontal line represents the median, boxes extend between percentiles 25 and 75, and whiskers reach out from percentile 5 to 95.

In both brain areas, the most informative response feature was the firing rate, both regarding the number of cells encoding significant amounts of information (Figure 3A), and the actual amount of information (Figure 3B). Approximately half (49%) of the hippocampal cells transmitted Time information in their firing rates, and 63% cells in the entorhinal cortex. The difference between these two percentages is significant (Section 4.5 describes the significance criteria). The information encoded in the theta and delta phases, although substantially lower, is not negligible. The fraction of cells encoding significant amounts of information in the theta phase is 29% in H and 25% in EC, and the difference is significant. Fewer cells are involved in the delta code, 17% in H and 16% in EC.

Although the mean information transmitted with the two phase codes is noticeably lower than the mean of the firing rate code (about a quarter), there are a few neurons in the population with high information values (see the tails of the distributions in Figure 3B, indicated by the whiskers extending up to the percentile 95). Table A1 in Appendix B summarizes the mean, standard deviation and maximal information values of Figure 3B. A few outlier cells, hence, encode amounts of information in their firing phases that are comparable to the average information in the rate code.

We then characterized the firing properties of the neurons that encoded significant amounts of information about the variable Time (Figure 4).

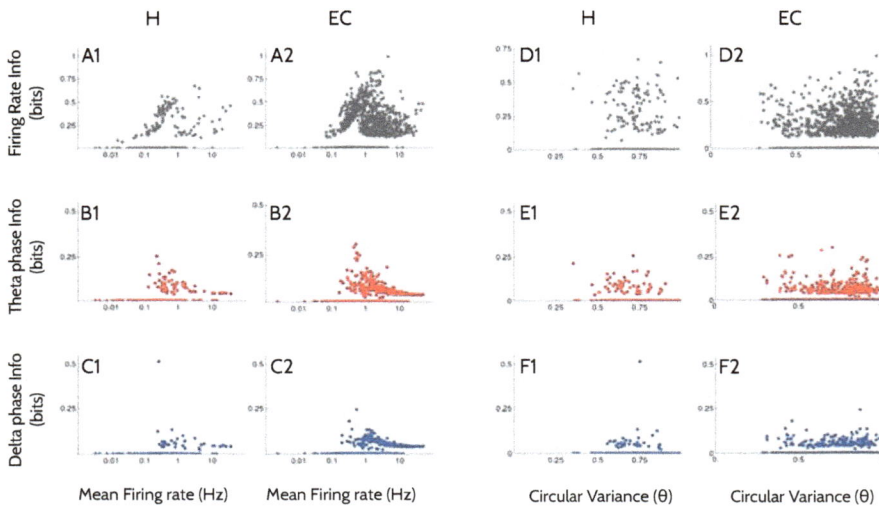

Figure 4. Relation between firing properties and information about Time. In each panel, each data point is a different cell. Cells that encoded an amount of information that was not significant appear clustered on the horizontal axis (the information value was set zero). Pearson correlation coefficients between the values displayed in the horizontal and vertical axes (including only the cells with significant amounts of information) are 0.04 (**A1**); -0.11 * (**A2**); -0.35 * (**B1**); -0.37 * (**B2**); -0.16 * (**C1**); -0.42 * (**C2**); where the asterisk indicates that the value is significantly different from zero at the 0.01 level. All correlation values in panels (**D–F**) are non-significant.

The cells with the lowest mean firing rates encoded amounts of information that were not significant, with all three codes (Figure 4A–C). As the mean firing rate increased beyond 0.1 Hz, the information encoded in the rate code increased linearly with the firing rate, until firing rates of approximately 1 Hz, where the information dropped again (Figure 4A). Instead, the information encoded in the phase-of-firing with respect to the theta and the delta-filtered LFP was a monotonically decreasing function of the mean firing rate (Figure 4B,C).

The degree of locking to the theta rhythm was not related to the information encoded by any of the three codes (Figure 4D–F). Similar results are obtained when the degree of locking is calculated with respect to the delta-filtered LFP (not shown).

An important question regarding the strategy of population signaling is whether the different neural codes are independent from each other. Independent codes would mean that the probability that a neuron encode significant amounts of information with one code (for example, the rate code) does not affect the probability that it employ another code (theta or delta). If codes are not independent, there are two possibilities. On the one hand, neurons may tend to come in disjoint categories, each category specialized in one specific code, for example, rate neurons, theta neurons, or delta neurons. This alternative would be expected if one of the codes, for example, the rate code, required certain biophysical properties that were not compatible with some other code, say, the delta or theta codes. The opposite situation would mean that neurons would tend to be either informative or not informative, and within the informative set, the probability for two or more neural codes to coexist should be higher that in the independent case. This alternative would be expected if the properties that make a neuron suitable for one code could also be exploited by another code.

To assess the independence between codes, we verified if the knowledge that a given cell did or did not encode a significant amount of information with one code could be used to predict whether it also used another code. To this end, we defined the binary variables C_r, C_θ, C_δ representing whether a given cell did or did not encode significant information about the variable Time with the codes rate, theta and delta, respectively. We then estimated from the data the distributions $P_{ij}(c_i, c_j)$, with i, j belonging to $\{r, \theta, \delta\}$, and evaluated whether they coincided or not with the independent hypothesis, which predicts a joint distribution equal to the product of the marginals $P_i(c_i) \, P_j(c_j)$. The evaluation was performed in terms of the mutual information $I^b(C_i; C_j)$ (Section 4.8 displays the mathematical definition of I^b). In both brain areas, and with all three pairs of neural codes, I^b was found to be positive (Figure 5A), implying that the rate, theta and delta codes were not independent. To determine whether pairs of codes tended to coexist or, on the contrary, tended to be segregated into disjoint populations, we calculated the Pearson correlation coefficient $c^b(c_i, c_j)$ (mathematical definition in Section 4.8), obtaining always positive values (Figure 5B). The rate code was only marginally (though significantly) correlated with each of the phase codes. The joint occurrence of the theta and delta codes was more marked, with correlation coefficients of 0.5 in H and 0.4 in EC.

Having determined that the populations of cells using one code or the other have a tendency to coincide, we assessed whether the amounts of encoded information was also correlated (see details in Section 4.8). This second analysis involved real (not binary) variables, representing the pairs of values of mutual information about the variable Time transmitted with any two codes. We only included the cells that transmitted significant amounts of information with the two tested codes (both binary variables equal to 1). The result is displayed in Figure 5C. Except for the comparison between delta and rate codes in hippocampus, all Pearson correlation coefficients were significantly positive. The information values that were most strongly correlated were those involving the theta and delta codes, with Pearson coefficients of 0.6 in H, and almost 0.8 in EC. In both areas, the correlation of the rate information values and any of the phase codes was lower.

In Figure 5D, the slope of a linear fit between the information encoded by each pair of codes is displayed. The data points to be fitted are ordered as pairs (x, y) with components (Info rate, Info theta), (Infor theta, Info delta), (Infor delta, Info rate). The slopes obtained in both areas for the pairs (rate, theta) and (theta, delta) are smaller than unity, which is consistent with the fact that the information values encoded by each individual neuron are typically ordered as Info(rate) $>$ Info(theta) $>$ Info(delta). The large value obtained for the pair (delta, rate) is also consistent with this ordering, since by transitivity, Info(delta) $<$ Info(rate).

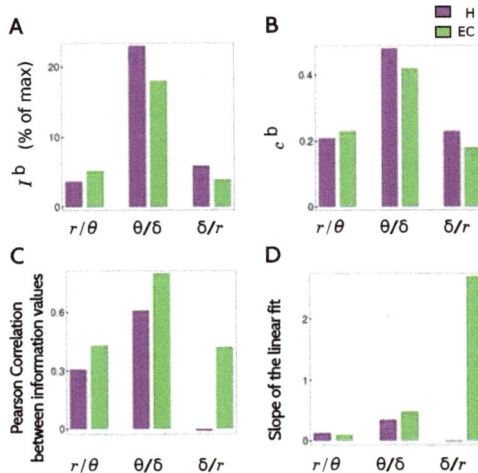

Figure 5. Degree of overlap between pairs of neural codes. (**A**) Mutual information $I^b(C_i; C_j)$ defined in Equation (7). Information values are relative to the maximal possible information, (see Table A3 for the values in bits); (**B**) Pearson correlation coefficients c_{ij}^b defined in Equation (8); (**C**) Pearson correlation between the mutual information $I(i; \text{Time})$ and $I(j; \text{Time})$ with i and j in {firing rate, θ phase, δ phase}; (**D**) Slope of the linear fit linking the information values of each pair of codes. In (**C,D**) only neurons encoding significant amounts of information with both tested codes are considered.

2.4. The Encoding of other Kinematic Variables by Single Neurons

Figure 6 describes the encoding of other kinematic features (Position, Direction, etc.).

In all tested kinematic features, just as it happened with the time variable, the firing rate code was the one that recruited the largest numbers of cells, typically duplicating the number involved in the delta and theta codes (Figure 6A). The latter, however, still represented between 15 and 25% of the population. In the rate code, the kinematic features that recruited the largest number of cells were Position and Signed Position, surpassing even the number of cells that encoded Time. This result derives from the fact that the larger number of bins of the variable Time make its significance analysis stricter than with the other features. In all three codes, Direction was the kinematic feature that recruited the smallest numbers of neurons. In the delta phase code, there was a noticeable difference between the number of cells recruited by the Time variable, and all other kinematic features.

Figure 6B displays the raw information values, so as to determine whether the information encoded in the variable Time can or cannot be interpreted in terms of one of the classical kinematic features. In the firing rate code, the information encoded in the variable Time is explained to a large extent by Signed Positon, in both areas. In the two phase codes, instead, the information in Signed Positon represents a smaller fraction than the information in Time. This result, combined with the fact that Time recruits significantly more cells (panel A), suggests that phase codes (and specifically, the delta code) represent specific conjunctions of position, velocity and direction, that cannot be described by a single tested feature. Indeed, such conjunctions are evident in the examples of Figure 2(C1,C2).

In Figure 6C, the information is normalized with the entropy of the encoded variable, in order to make the values obtained with the different kinematic features comparable. For example, the Time variable contains 82 bins, so in principle, a perfectly informative cell could transmit $\log_2(82) \approx 6.3$ bits. The variable Direction, instead, contains only 2 bins, so the most informative cell could transmit at most 1 bit. When information values are normalized, the Time feature is still the most informative one, implying that its prevalence is not a consequence of the large number of bins it contains. The code indeed contains information about features that are not described by the other variables. The other features, however, appear flatter in panel C than in panel B. In particular, the feature Direction encoded negligible raw information values (panel B), but those values are a non-negligible amount of the maximal attainable values. Hence, the only reason the information about Direction is low, is that it contains only two bits, not because selectivity is poor.

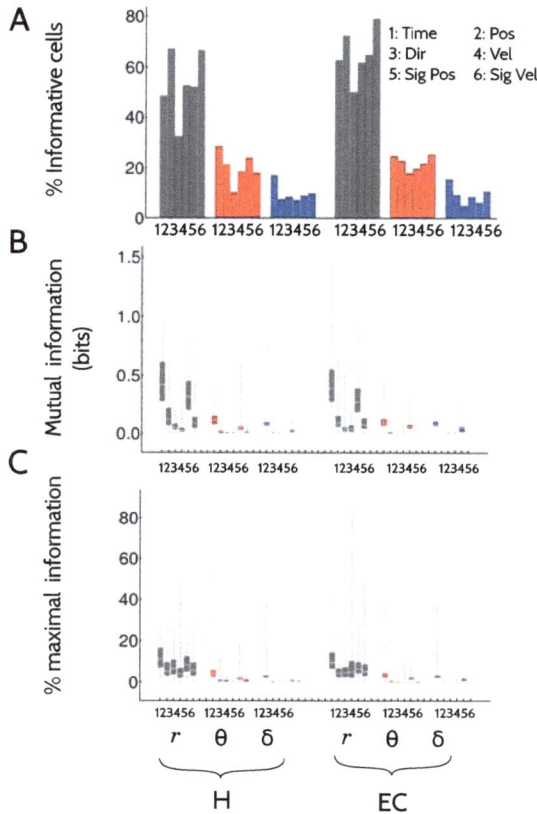

Figure 6. Encoding of multiple kinematic variables by single cells. (A) Percentage of cells that were informative about the different kinematic features in the two brain areas (H and EC) and the three tested codes (firing, theta and delta); **(B)** Box-histograms of the mutual information $I(Y; \text{Feature})$ for different choices of the physiological variable Y and the kinematic feature. Only cells that passed the significance test of Section 4.5 are included; **(C)** Same as **(B)**, but with information values normalized by the maximal attainable information (the entropy of the set of kinematic features). Table A2 in the Appendix summarizes the mean, standard deviation and maximal information values of panel **(C)**. Parameters of box histograms: same as in Figure 3.

To assess the significance of the differences in the values obtained for the two brain areas, the three neural codes, and the six kinematic features, we ran statistical tests for all 64×64 combinations of area + code + feature. The detailed procedure and the results are presented in Appendix A.

When the analysis of Figure 4 was repeated with all kinematic features, the results obtained with the variable Signed Position were similar to those of Time. The mutual information of all other kinematic features were monotonically decreasing functions of the mean firing rate.

When the analysis of Figure 5A,B was repeated using kinematic features different from Time, similar results were obtained in both brain areas, with information and correlation values that either remain unchanged, or decrease slightly. Except for the case of Position and SignedPosition in H, for which both the information and the correlation between the firing and the theta code increased, as compared with Time ($I^b = 15\%$ and 19%, and $c^b = 0.34$ and 0.43, for Position and SignedPosition, respectively). A visual inspection of the firing properties of the neurons that employ both codes reveals place cells that precess in theta.

When the analysis Figure 5A,B was repeated using other kinematic features, the result was approximately the same as with Time, with slightly smaller values for the correlation between the information of the rate code and any of the phase codes, but slightly larger for the correlation between the information values of the two phase codes.

We next checked whether positions and velocities were better encoded in their signed or unsigned versions by each neural code. This point was addressed by calculating the amount of synergy S_k between position and direction, and also between velocity and direction (see Equation (3) in Section 4.6). If the information about a signed variable was significantly higher than the sum of the information of the unsigned variable and the information about direction, the neural code represented conjunctions of kinematic features. In these cases, $S_k > 0$. The opposite case appeared when $S_k < 0$, resulting in redundant encoding.

In Figure 7 we see that the firing rate code represents position and direction synergistically in both areas. The same is true for velocity and direction. With the firing rate code, S_k is never negative, implying that direction and position (or direction and velocity) are never encoded redundantly. From Equation (4) we see that redundancy is only possible if the two tested stimulus attributes are statistically related, that is, if $I(\text{Position}; \text{Direction}) > 0$, or if $I(\text{Velocity}; \text{Direction}) > 0$. These two mutual informations involve only kinematic variables, and are solely determined by the experimental paradigm. In the task analyzed here, $I(\text{Position}; \text{Direction}) = 0$ and $I(\text{Velocity}; \text{Direction}) = 0$, since for each position and each velocity, there are two possible directions of motion, both with equal probability. For this reason, this experiment rules out the possibility of finding redundancy. Some other pairs of attributes, as for example position and velocity, are correlated by the behavioral task. Since such correlation is determined by the experiment, and not by the neural code, the analysis of synergy between attributes is here restricted to position and direction, and to velocity and direction.

The theta code, instead, shows a few examples of redundancy between direction and position. These cases appear because in the phase codes, not all time bins are used to calculate the mutual information. The firing phase of a neuron cannot be calculated if the cell remains silent, so whenever a mutual information is calculated, the silent time bins are discarded from the set of stimuli. If the number of discarded bins occurring during the right run differs from those during the left run, a correlation between position and direction is likely to appear. Redundancy in a phase code, hence, can only be observed in neurons with unequal firing rates in the two running directions. In addition to these few redundancy examples, there is a large number of cells that encode position and direction synergistically. The theta phase, hence, is modulated by conjunctions of location and direction, and not by each attribute separately. Quite notably, no cells were found (neither in H nor in EC) with significant amounts of synergy or redundancy between velocity and running direction. A few neurons in both brain areas encoded position and direction, as well as velocity and direction, synergistically with the delta code. No instances of redundancy were found in the delta code.

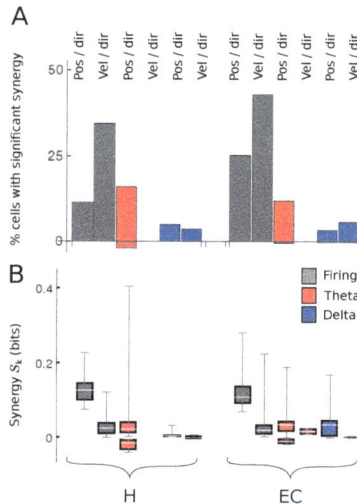

Figure 7. Amount of synergy and redundancy between pairs of kinematic features. (**A**) Percentage of recorded cells for which the encoding of position and direction, or of velocity and direction, is synergistic, in different brain areas or neural codes. Positive (negative) bars account for the cases where $S_k > 0$ ($S_k < 0$); (**B**) Box histograms displaying the amount of synergy between the two kinematic attributes, with positive and negative values displayed separately. Only significant values are included. Parameters of box histograms: same as in Figure 3.

2.5. The Encoding of Kinematic Variables by Pairs of Neurons

Finally, we verified whether pairs of neurons that were recorded simultaneously encoded information synergistically, redundantly or independently. Specifically, we determined whether the information encoded by pairs of neurons was equal (independent), larger (synergistic) or smaller (redundant) than the sum of the informations encoded by each member of the pair [25–27]. This analysis could only be performed in the rate code, since the phase codes are undefined in time bins where no spikes are fired, and those bins varied from neuron to neuron. The rate code, instead, is defined for all temporal bins. We modeled the conditional probability distribution that two neurons respond with firing rates r_1 and r_2 with a bivariate Gaussian probability distribution $P(r_1, r_2|\text{time})$ characterized by two mean values, two variances and a correlation coefficient, all parameters estimated from the experimental data (see Methods, Sections 4.5 and 4.6).

In Figure 8(A1,A2), we see the histogram of the obtained synergy values for H and EC, respectively, for the example of signed velocity. Similar behavior was observed with other kinematic features. Notably, most of the values were not significant. In H a few pairs were significantly redundant, and in EC, a few were significantly synergistic. In B1–B2, we show the percentages of pairs in each area with significant amount of synergy (positive values) or redundancy (negative values) for different kinematic attributes. In H (Figure 8(B1)), pairs were always redundant, except for the encoding of running direction. In EC (Figure 8(B2)), both redundant and synergistic pairs were found, with a preference for synergy. However, in both areas the fraction of cells with significant amounts of synergy (positive or negative) was small. In addition, for all stimulus attributes, the actual values of the synergy were low when compared to the mutual information encoded by the members of each pair. Indeed, Schneidman et al. [27] showed that if R_1 and R_2 are the firing rates of the two cells in the pair and S is a given kinematic feature, then $S_r = \langle I(R_1; R_2|S) \rangle_S - I(R_1; R_2)$. Hence, S_r in a given pair is equal to the distance from the diagonal of the data point representing that pair in Figure 8(C1,C2). The collection of points does not show a noticeable displacement from the diagonal, implying that even for the few pairs with significant synergy, the magnitude of such synergy is a small fraction of the information values.

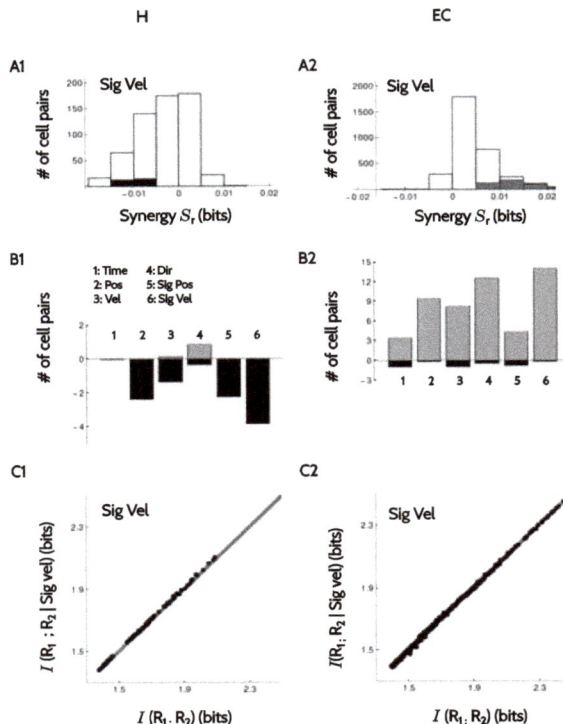

Figure 8. Amount of synergy and redundancy between pairs of neurons. (**A1,A2**) Histograms with the amount of synergy S_r defined in Equation (5) in pairs of simultaneously recorded neurons in H and EC for signed velocity with the firing rate code. Significant values are marked in gray and black, depending on their sign; (**B1,B2**) Percentages of pairs of cells encoding significant amounts of synergy in the 6 tested kinematic attributes. Black and gray: redundant and synergistic encoding, respectively; (**C1,C2**) Scatter plot of $\langle I(R_1; R_2|S) \rangle_S$ vs $I(R_1; R_2)$ for S = sig vel. Each data point is a pair of neurons. The amount of synergy S_r is equal to the displacement from the diagonal.

3. Discussion

Here we explored the neural code with which populations of neurons in H and EC represent kinematic information. We tested three neural codes, one based on the firing rates of cells in 1-second windows, and two based on the phase-of-firing with respect to the theta and delta filtered LFP, respectively. Firing rate codes have been observed ubiquituously in the nervous systems, with or without fine temporal structure [5,14]. Phase codes, instead, are comparatively new, and so far, have only been reported in the theta [11,12] and delta [28,29] bands. Two theoretical studies [30,31], and an experimental one [32] have also identified a possible mechanism in which phase codes can be translated into a pattern-based firing code by bursting neurons.

The firing rate code recruited a larger number of cells and transmitted more information than the other two. The strategy to estimate information values with the rate code was different from the phase codes, since in one case we fitted the experimental data with Gaussian distributions, and in the other, with Von Mises distributions. Moreover, all time bins could be associated with a Gaussian distribution, but not with a Von Mises distribution, since the latter required the bin to contain a minimum number of spikes. Therefore, information estimates performed with the rate code include all temporal bins, but the phase codes typically involved a smaller number of bins. Due to these differences, we believe that comparisons between the information values of the rate and phase codes should be made with caution, since two different estimation methods are in play. Still, the rate code surpasses the phase codes both in number of cells and magnitude in a sufficiently large difference as to make its primacy tenable. Yet, just for precaution, we prefer not to underscore the exact numerical difference between the information encoded by the rate code and the phase codes, also in view of the findings in other studies [33].

The theta and delta codes, instead, were estimated with the same method, the only difference being that recordings were selected on the base of a high-quality LFP, a condition requiring a well-defined theta peak in the LFP, and not a delta peak [34]. The similarity between the two estimation methods, and the knowledge that at most the delta code was in disadvantage, allows us to conclude that the information encoded in the delta code is perhaps smaller, but by no means negligibly smaller, than the one in the well-known theta code. To our knowledge, this is the first study reporting a role of the delta phase in the encoding of kinematic information in the temporal lobe. From the examples and information values obtained in the two brain areas, we conclude that the delta code is mainly involved in representing the Time variable, singling out specific moments during the run, which cannot be straightforwardly associated with a position, a velocity, or a direction.

All three codes tended to be most informative in cells with low mean firing rates (Figure 4A–C). This property suggests that the principal cells are more involved than the interneurons in the encoding of kinematic information. We also found that the circular variance with respect to the theta-filtered LFP (Figure 4D–F) was not a good predictor of the amount of encoded information. The same result was obtained with the circular variance with respect to the delta-filtered LFP (data not shown). Therefore, we conclude that all three codes recruit cells with preferentially low firing rates, but with varying degree of locking to the theta or delta signals.

The rate code recruited approximately half the cells in both areas in the encoding of Time. The delta and theta codes involved approximately 25% and 15% of the cells, respectively. In the case of Position and DirectionalPosition, the percentages with the firing rate code reach as high as 60–80%. These values are larger than the ones usually reported with traditional studies, in which the encoding of kinematic variables is assessed with well-defined attributes as position, head direction or velocity. Moreover, the hippocampal encoding of space is usually associated with place cells, and the entorhinal with grid cells. The fraction of cells devoted to such paradigmatic codes, however, is fairly low, around 20% [7,24]. Yet, as remarkable as regular-firing maps are, position-selective codes need not be restricted to them. At the very least, grid cells have been reported to vary in the degree of periodicity of the location of the fields [35] and the field-to-field peak firing rate [24,36]. The high percentages of cells encoding significant amounts of information about position found in this study, and also in other approaches with a similarly broad perspective [23,24], suggests a wider repertoire of coding strategies, that are not necessarily covered by the traditional categories.

If we assume that these percentages represent the probability to pick a cell that makes use of the rate, theta and delta codes respectively, Figure 5 provides evidence that, for any pair of codes, the probability that a cell employs the two codes is higher than the product of the two marginals. The normalization restrictions imply that the probability that a cell employs neither of the two codes is also higher than the one predicted for independent populations. We conclude that cells are not segregated into a rate population, a theta population and a delta population. They are rather segregated into informative or non informative clusters. This effect is most pronounced for conjunctive encoding with the theta and delta codes, for which also the information values are correlated. The correlation is less clear when comparing cells using the rate code, and one of the phase codes, except for phase-precessing neurons in H, encoding Directional Position. In general terms, the informative population includes a subpopulation of phase neurons, in which delta and theta codes largely overlap.

Cells seem to be apt or inapt to use a phase code, but do not seem to be highly selective to a single frequency band.

The firing rate code is the one in which conjunctions of attributes (as position + direction, or velocity + direction) is most synergistic. Moreover, the amount of synergy between attributes is comparable to the total amount of information encoded by signed variables. We therefore conclude that both in H and in EC, position and direction are mainly encoded in conjunction with running direction with the firing rate code. The same conclusion holds for the phase codes, although the information values are lower.

When assessing the amount of synergy encoded in the firing rates of pairs of neurons recorded simultaneously, small differences in the amount of redundancy and synergy were found in the two brain areas. However, the values were too small to be considered significant. We therefore conclude that pairs of cells tend to encode information independently.

4. Materials and Methods

4.1. Electrophysiology and Behavioral Task

The activity of individual cells, as well as the LFP, was recorded in repeated trials in the hippocampus (CA1 and CA3) and the entorhinal cortex (layers II and III). The experiment was designed to ensure that in all trials, the trajectory of the animal was the same, in order to gather enough statistics in each position, velocity and direction of motion. The animal was placed inside a bottomless cart that moved along a 4 m long linear track. The position, velocity and direction of motion of the cart was controlled by a computer [7]. Since the cart had no floor, the animal had to engage in active locomotion at the experimenter-determined speed in order to reach the end of the track, where a food reward was delivered. At the beginning of each trial, the animal was placed inside the cart. A 6-s beep of increasing pitch indicated the beginning of the next run. After the beep, the cart started moving to the right at a speed of 36 cm/s. Halfway along the track, the cart slowed down to 6 cm/s until it reached the end of the track, where the cart stopped and the rat was rewarded. The animal then turned around to face the opposite end of the cart. The return run was also announced with a 6-sec beep, after which the cart started moving at 6 cm/s, for 30 s. Then, the cart was accelerated to 36 cm/s until it reached the initial position of the track. Figure 9 illustrates the behavioral task.

Typically, a session consisted of 10–25 runs on the linear track, all of them lasting at most 36 min.

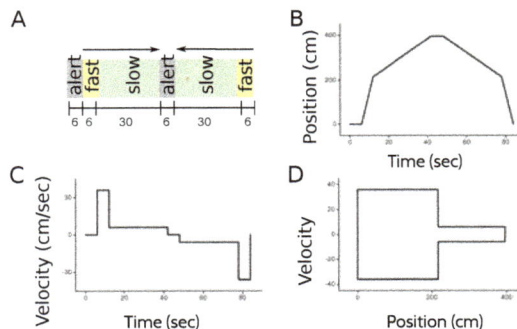

Figure 9. Behavioral task. (A) Each trial consisted of two episodes, the first running to the right, and the second to the left. Each episode began with an alert auditory signal lasting for 6 s, indicating the imminent start of the run. The animal was then compelled to run for 36 s at two different velocities, the order of which depended on the direction of motion; **(B,C)** Position and velocity of the cart as a function of time. The resting interval between the go and return journeys is not represented in the graph, because its duration was variable. During this interval, the animal took the reward, and was turned around in the cart; **(D)** Velocity as a function of position.

4.2. Representation of Kinematic Variables

To estimate the amount of information that different neural codes carry about position, velocity, direction of motion, or combinations of these variables, a suitable representation of the kinematic state of the animal is needed. Each trial was divided in 82 temporal bins (Figure 10A), each lasting for 1 s. In this paper, the specification of the temporal bin along the trial defines the variable *Time* (Figure 10A), and provides the most complete description of the kinematic state of the animal, since the position, velocity and direction of motion are all functions of time.

The variable *Position* specifies the spatial location of the animal (Figure 10B), and is divided into 12 bins. Two of them correspond to the location of the cart at the beginning and the end of the track while the alarm is ringing. Another five locations bins are assigned to the state of fast running, each of them labeling the location of the animal in one of the 5 s in which the cart moves at 36 cm/s. There are five more position bins during slow running, each associated to the location of the cart during 6 consecutive seconds in which the cart moves at 6 cm/s. Position bins during the running state have all equal length, but correspond to time intervals of different durations. Since the running velocity varies throughout the trial, position bins during slow running are associated to time intervals 6 times longer than those of fast running. The marginal probability of the slow-running bins, hence, is 6 times larger than that of the fast-running bins.

We also worked with a kinematic variable defined as *Signed position* (Figure 10C), with positive positions corresponding to the locations in which the animal is running to the right (or preparing to run to the right), and negative positions for left-directed runs. There are 22 bins of signed position. The variable *Direction of motion* has 2 bins: left, right (Figure 10D). When the animal is waiting at the beginning of the track for the run to start, the direction is the one of the imminent run, that is, the one towards the head is pointing to. The feature *Velocity* has 3 bins: fast, slow and still (Figure 10E). *Signed velocity* has 6 bins: still right, fast right, slow right, still left, fast left and slow left. (Figure 10F).

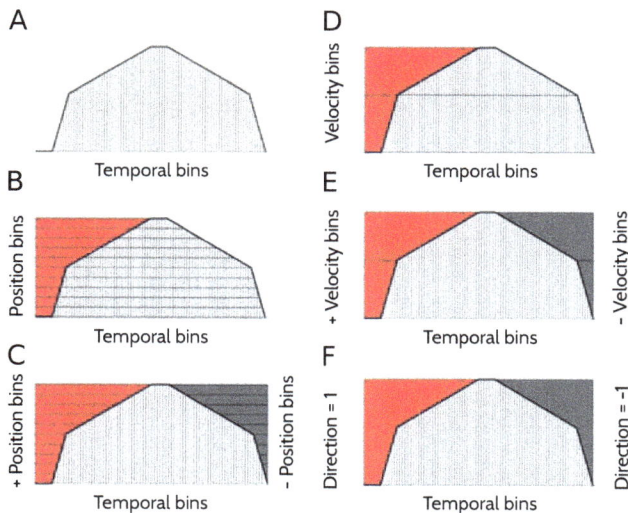

Figure 10. Binning of kinematic variables. (**A**) Temporal bins; (**B**) Position bins on the vertical axis, as a function of time; (**C**) Signed position bins (vertical axes on the left and right) as a function of time; (**D**) Direction of motion bins, on the left and right vertical axes. There is an additional bin for the alert periods where the rat has velocity zero; (**E**) Velocity bins, on the left vertical axis. There is also a zero-velocity bin; (**F**) Signed velocity bins, on the left and right vertical axes. There is also a zero velocity bin.

4.3. Amount of Spectral Power in Different Frequency Bands of the LFP

The LFP was filtered in the theta (6–12 Hz) and delta (1.5–4 Hz) bands. Filtering consisted of (a) applying a fast Fourier transform to the LFP, (b) assigning amplitude 0 to bins with frequency outside the desired band and (c) transforming the result back to the temporal domain. Since we often also needed the phase of the signal, the bandpass filter was sometimes followed by the Hilbert filter (assigning 0 amplitude to negative frequencies in the Fourier domain) to obtain the discrete-time analytic signal as a result. The exact form of the hilbert filter can be found in the MATLAB function hilbert(). The output of the Hilbert filter was used as an argument to the MATLAB function angle(), with which the phase was calculated.

Phase codes are only possible for frequency bands with significant power in the LFP. To identify the frequency bands with non-negligible power, in Figure 11 we show the spectral power density in both the hippocampus and the entorhinal cortex.

In both areas, the spectrum contained a clear maximum in the theta band (Figure 11). An additional and less prominent peak was observed in the delta band of the LFP recorded in the entorhinal cortex, comprising between one third and one half of the total power in the theta band. In the hippocampus, slow oscillations did not give rise to a low-frequency peak, and yet, the power in the delta band was far from negligible, comprising about one third of the power in the theta band. The first harmonic of the theta frequency gave rise to an additional local maximum at about 16 Hz, after which the power spectrum decayed monotonically.

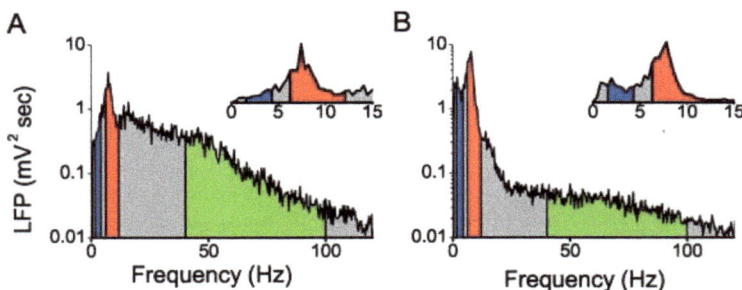

Figure 11. Average spectral power density of the LFPs. (**A**) Hippocampus; (**B**) Entorhinal cortex. Logarithmic scales in the *y*-axis of the main figures, and linear scale in the amplified view of the insets. Delta band: 1.5–4 Hz. Theta band: 6–12 Hz. Gamma band: 60–100 Hz.

4.4. Circular Statistics

The phase of the LFP is a circular variable, here defined between $-\pi$ and π. The mean $\langle\phi\rangle$ and the variance σ_ϕ^2 of a collection of angles α_i with $1 \leq i \leq n$ is

$$\langle\phi\rangle = \text{Arc Tan}(S/C) \tag{1}$$

$$\sigma_\phi^2 = 1 - \sqrt{S^2 + C^2}, \tag{2}$$

where S and C are the average sine and cosine values, respectively [37]

$$S = \frac{1}{n}\sum_{i=1}^{n}\sin\alpha_i,$$

$$C = \frac{1}{n}\sum_{i=1}^{n}\cos\alpha_i.$$

The circular variance is always between 0 and 1. When all angles α_i coincide, the circular variance vanishes. When they are evenly distributed on the circle, the circular variance reaches its maximal value. A small number of samples, however, need not represent faithfully the statistics of the underlying distribution. For example, when a few angles are sampled from a uniform probability distribution, it is highly unlikely to obtain values that are evenly distributed in the interval $(-\pi, \pi)$. So even if the circular variance associated to the uniform distribution is unity, the estimated value is typically lower. The circular variance is in fact negatively biased: On average the value obtained for a finite set of samples is smaller than the value of the underlying distribution. It is, therefore, important to determine whether the circular variance estimated from a few samples provides enough evidence to conclude that the underlying distribution is uneven or, on the contrary, whether that same value could have arisen from a uniform distribution. We therefore worked with a null hypothesis, under which the underlying distribution is uniform. To assess how likely it would be to obtain the measured circular variance under this hypothesis, we compared the measured value with the distribution of values that are expected from the null hypothesis when sampled the same number of times as the real data. If the real circular variance is smaller than 99% of the values obtained from a uniform distribution, the measured value is considered significant, and the underlying distribution, non-uniform.

4.5. Estimation of Encoded Information

The amount of information that a given physiological variable Y (as firing rate, or phase-of-firing with respect to the filtered LFP) and a kinematic variable X (as Time, Position, Direction, etc.) was computed by Shannon's mutual information [38,39]

$$I(X;Y) = \sum_x P(x) \int P(y|x) \, \log_2 \left[\frac{P(y|x)}{P(y)} \right] \, dy,$$

where $P(x)$ is proportional to the amount of time that the animal spent with kinematic variable $X = x$. In other words, if each trial has duration T_{trial}, and if in each trial the animal spends an amount of time T_x with $X = x$, then $P(x) = T_x / T_{\text{trial}}$. In turn, $P(y|x)$ is the probability that the physiological variable Y take the value y conditional on $X = x$.

The number of trials with the electrodes in the same position was around 15. Without additional assumptions, this small number of samples rules out the possibility of making a reliable estimation of the distributions $P(y|x)$. We therefore assumed that in the firing rate code, the distribution $P(\text{rate}|\text{time})$ could be approximated by a Gaussian distribution with the experimentally measured mean and variance. In the two phase codes, $P(\phi|\text{time})$ was modeled as a Von Mises distribution with the mean and concentration values extracted from the data. These assumptions were based on the observation that the conditional probabilities were unimodal, and that the data were typically sufficient to make reasonable estimations of the mean value and the width of the distributions, as shown in the examples of Figure 12.

We checked that in the case of the firing rate code, using a Poisson fit (instead of a Gaussian fit) produced only a minor variation of the obtained information values. The Gaussian model had the advantage to be easily extensible to the bivariate case, when evaluating the amount of information encoded by pairs of simultaneously recorded neurons. In this case, the distribution $P(r_1, r_2|\text{time})$ was modeled as 2-dimensional Gaussian, which only requires the additional estimation of the correlation coefficient between the rates r_1 and r_2 of the two neurons, for each temporal bin. The parametrization of response probability distributions conditional to a given stimulus with specific models has been widely employed before in the framework of Fisher-information-based measures (for example, ref. [40,41] used a multivariate Gaussian model), and sometimes also for the estimation of Shannon-based measures (for example, ref. [4] employed a binary model, whereas [42] used a Gaussian model.)

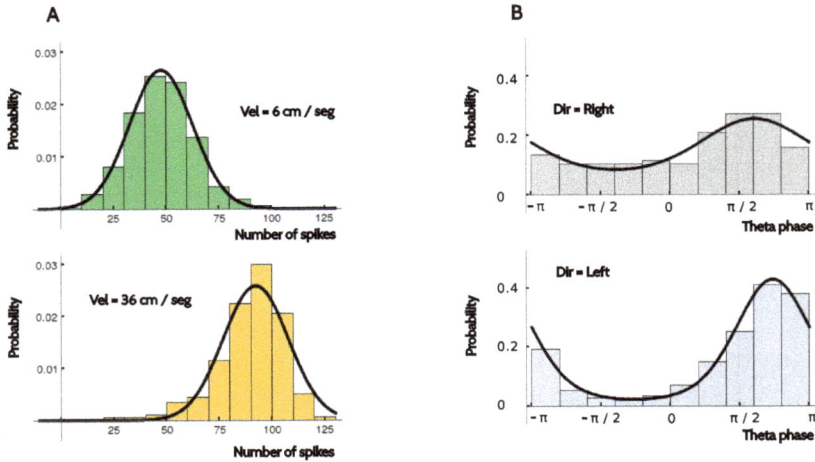

Figure 12. Modeling the conditional response distributions. Comparison between the histogram of the measured response variables and the corresponding fits, for two example neurons and codes. (**A**) Firing rates obtained from an entorhinal neuron for two different running speeds. Note the different in scale of the *x* axes. The two velocities induce markedly different mean firing rates; (**B**) Phase of firing with respect to theta of an entorhinal neuron for the two different running directions. The two directions give rise to distributions of markedly different widths.

The Gaussian fit of the probability $P(r|\text{Time} = t)$ of firing at rate r conditional on the temporal bin Time $= t$ was calculated by estimating the mean spiking rate and the variance at each time t. If the number of spikes in that bin was smaller than 10 (considering all trials) or the number of trials containing spikes in that bin was smaller than 5, the variance was taken as the minimum value of the variances corresponding to Time $\neq t$. This choice ensured conservative information estimates, since variances obtained with only a few samples are unreliable.

To calculate the probability $P(r|S = s)$ of firing r spikes conditional on the kinematic feature S taking the value s, we marginalized over all the Gaussian fits of the distributions $P(r|\text{Time} = t)$ at time bins t compatible with the condition $S = s$. Hence, the distribution $P(r|S = s)$ was typically not Gaussian. With this definition of $P(r|S = s)$, we ensured the validity of the data processing inequality: Since given the value of Time, the value of S is determined, $I(R; \text{Time}) \geq I(R; S)$.

When estimating the information encoded by the firing rates of pairs of neurons recorded simultaneously, the conditional probability $P(r_1, r_2|\text{time})$ was modeled as a bivariate Gaussian distribution characterized by the means $\langle r_1 \rangle$ and $\langle r_2 \rangle$, variances σ_1^2 and σ_2^2 and correlation coefficient ρ. To ensure a reliable estimate of ρ, if the number of spikes in that bin was smaller than 20 (considering all trials) or the number of trials containing spikes in that bin was smaller than 5 (of any of the two neurons), we set $\rho = 0$. By marginalizing $P(r_1, r_2|\text{time})$ in the value of "Time" and keeping r_1 and r_2 fixed, we obtained the distributions $P(r_1, r_2)$ that were used to estimate the information $I(R_1; R_2)$ between the rates.

The von Mises fit of the probability $P(\phi|\text{Time} = t)$ of firing at phase ϕ conditional on the temporal bin Time $= t$ was calculated by estimating the circular mean and the circular variance of all spiking phases obtained at that particular time bin. If fewer than 10 spikes existed, or if less than 5 trials contained spikes, the bin Time $= t$ was discarded from the list of values that was taken by kinematic variable Time in the estimation of information. Just as with the firing rate, to calculate the probability $P(\phi|S = s)$ of firing at phase ϕ conditional on the kinematic feature S taking the value s, we marginalized over all the von Mises fits of the distributions $P(\phi|\text{Time} = t)$ at time bins t compatible with the condition $S = s$. Hence, the distribution $P(\phi|S = s)$ was typically not von Mises. The marginal distributions $p(r)$ and $p(\phi)$ were also obtained by marginalizing the conditional distributions $p(r|t)$ and $p(\phi|t)$, respectively.

4.6. Estimation of the Amount of Synergy

When either the kinematic or the neuronal variable (or both) are defined in terms of two components, one may wonder whether the information encoded by the pair of components is equal, more, or less than the sum of the informations encoded by each component separately. For example, the variable "signed velocity" can be interpreted as a 2-dimensional vector (Direction, Velocity), where the component Direction can take the values $\{1, -1\}$, depending on whether the head of the animal points to the right or to the left, and the component Velocity can be 0, 6 or 36 cm/s, depending on the running speed. If the response variable Y represents the conjunction of direction and velocity, but not each variable separately, $I(\text{Signed vel}; Y)$ is expected to surpass $I(\text{Direction}; Y) + I(\text{Velocity}; Y)$. In this case, direction and velocity are encoded synergistically. If, instead, the two aspects are represented independently, $I(\text{Signed vel}; Y)$ should be approximately equal to $I(\text{Direction}; Y) + I(\text{Velocity}; Y)$. Finally, if the encoding of the running speed could be used to infer the direction, $I(\text{Signed vel}; Y)$ should be smaller than $I(\text{Direction}; Y) + I(\text{Velocity}; Y)$, and the code is redundant.

The amount of synergy between two kinematic variables X_1 and X_2 mediated by the response variable Y is defined as

$$\begin{aligned} S_k &= I(X_1, X_2; Y) - I(X_1; Y) - I(X_2; Y), & (3) \\ &= \langle I(X_1; X_2|Y) \rangle_Y - I(X_1; X_2), & (4) \end{aligned}$$

where the subscript in S_k stands for "synergy in the kinematic variables", and the equality between the two expressions was demonstrated by Brenner et al. [26]. Analogously, when considering the responses of two neurons recorded simultaneously with activities Y_1 and Y_2, the synergy observed in the encoding of kinematic variable X is

$$\begin{aligned} S_r &= I(X; Y_1, Y_2) - I(X; Y_1) - I(X; Y_2), & (5) \\ &= \langle I(Y_1; Y_2|X) \rangle_X - I(Y_1; Y_2), & (6) \end{aligned}$$

where the subscript in S_r stands for "synergy in the response variables".

4.7. Statistical Significance of Information Measures

Estimates of mutual informations obtained with a limited number of samples are positively biased [43–46]. To assess whether each obtained value was significant, we shuffled all spikes in each trial, mimicking the situation in which $P(r|s) = P(r)$ and $P(\phi|s) = P(\phi)$. Information values were calculated for 100 independent realizations of the shuffling. The information value obtained in the real experiment was considered significant if it was larger than all, or at most a single, of the shuffled values. The same criterion was used to assess the significance of the information encoded by pairs of neurons, based on the bivariate distributions $P(r_1, r_2|s)$.

For each recorded area, neural code, and kinematic feature, we determined the fraction of cells encoding significant amounts of information. To assess whether the fractions obtained in two different conditions were significantly different or not, we evaluated the null hypothesis that they were equal, and estimated the probability to obtain two fractions that differed in at least as much as the difference observed in the experimental data, just because the number of cells is limited. Fractions of cells were considered significantly different if the experimental difference was larger than 99% of the trials evaluated under the null hypothesis. In the case of the information between rates $I(R_1; R_2)$, we considered significant the experimental values that surpassed 50 shuffled trials.

To assess whether an estimated synergy is significant, we shuffled the response vectors (y_1, y_2) obtained in different temporal bins 50 times, mimicking the situation in which the conditional probability $P(y_1, y_2|\text{time})$ was drawn from the marginal probability $P(r_1, r_2)$. A synergy value was considered significant if the number obtained with the real data was larger than all but one of the values derived from the 100 shuffled instances.

4.8. Assessing the Degree of Joint Encoding with Different Neural Codes

Each cell did or did not encode a significant amount of information about a specific kinematic feature with the rate, the theta and the delta codes. To verify whether codes tended to co-occur in the same cell or, on the contrary, tended to be segregated in different neurons, for each cell we defined three binary variables C_r, C_θ, C_δ that took the values 1 or 0, depending on whether the cell did or did not transmit a significant amount of information about the variable Time in the rate code, the theta code and the delta code, respectively. We assumed that each area could be characterized by a joint probability distribution $P(c_r, c_\theta, c_\delta)$, and that the collection of cells recorded in each area constituted a sample of the corresponding distribution. Marginalizing over one of the three codes, we obtain bivariate distributions $P(c_r, c_\theta)$, $P(c_\theta, c_\delta)$ and $P(c_\delta, c_r)$. In Figure 13, we illustrate the distributions obtained for an idealized case in which there are two neural codes, each of which recruits half of the neurons in the populations. Panel B represents the case in which the two codes are independent of one another, panel C when they tend to coexist in the same cells, and panel D when they tend to be segregated in different neurons.

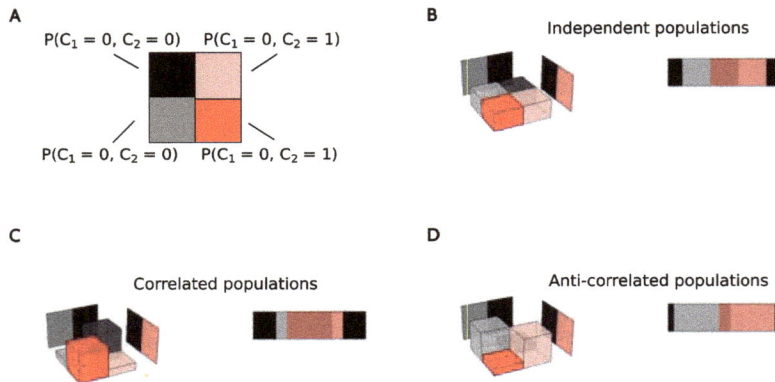

Figure 13. Idealized examples of pairs of codes with varying overlap throughout the population of neurons. (**A**) Color convention used in panels (**B–D**); (**B**) Joint probability distributions (blocks) and the corresponding marginal (planes) of an example case with two codes that are independent. The rectangular stripe shows that the fraction of neurons employing both codes is the one expected from the total number of neurons; (**C**) Codes with a tendency to coexist in the same neurons; (**D**) Codes with a tendency to be segregated into different cells.

If the population of cells employing one neural code is independent from the population using another code, then the knowledge that a given cell has $C_i = 1$ cannot be used to predict the value of C_j, and vice versa. This is shown graphically in Figure 13B, where $P(c_i, c_j) = P(c_i)P(c_j)$. Mathematically, this means that the mutual information $I^b(C_i; C_j)$ between C_i and C_j vanishes, with

$$I^b(C_i; C_j) = \sum_{ab} P_{ij}(ab) \, \log_2 \left[\frac{P_{ij}(ab)}{P_i(a) \, P_j(b)} \right], \tag{7}$$

where the sum runs over all pairs $ab \in \{00, 01, 10, 11\}$, and the supra-script in I^b stands for "binary". A vanishing information implies that the collection of cells that encodes information with one of the tested codes is independent from the other. Positive information values may either indicate that the co-occurrence of two codes in one same cell is more frequent than random (Figure 13C), or alternatively, that if one code is used, then the other is not (Figure 13D). To discriminate between these two situations, we calculate pairwise linear correlations between the variables C_r, C_θ and C_δ. Positive values of the Pearson correlation coefficient implies that codes tend to co-occur in the same cell, whereas negative values are found when cells using each code tend to be segregated. The Pearson correlation between the set of neurons encoding information with codes i and j is defined as

$$
\begin{aligned}
c_{ij}^{b} &= \frac{\langle C_i C_j \rangle - \langle C_i \rangle \, \langle C_j \rangle}{\sqrt{\left(\langle C_i^2 \rangle - \langle C_i \rangle^2\right)\left(\langle C_j^2 \rangle - \langle C_j \rangle^2\right)}} \\
&= \frac{P_{ij}(11) - P_i(1)P_j(1)}{\sqrt{P_i(1)[1-P_i(1)]\,P_j(1)[1-P_j(1)]}} \\
&= \frac{P_{ij}(00) - P_i(0)P_j(0)}{\sqrt{P_i(1)[1-P_i(1)]\,P_j(1)[1-P_j(1)]}}
\end{aligned}
\tag{8}
$$

where the supra-script b stands for "binary".

For each pair of physiological variables (i, j)—that is, for the pair $(i, j) = (r, \theta)$, the pair $(i, j) = (\theta, \delta)$ and the pair $(i, j) = (\delta, r)$—we selected the cells that encoded significant amounts of information about the variable Time in both physiological variables, and with these cells, we calculated the Pearson correlation coefficient between $I(\text{Time}; i)$ and $I(\text{Time}; j)$. The result is displayed in Figure 5C. For each pair of codes i/j, we also performed a linear fit $I(\text{Time}; j) = \alpha\, I(\text{Time}; i) + \beta$. In Figure 5D we report the slopes α of the fits obtained with different physiological variables (i/j) and in the two brain areas.

4.9. Data Selection

Information values were only calculated for sessions where the LFP contained a well-defined peak in the theta band. The peak value of the spectral power density in the theta band was required to be at least twice as larger than the density at any of the inflection points at either side of the peak. This condition was fulfilled by 220 trials (out of 649) in H, and 1186 (out of 1249) in EC.

5. Conclusions

The study of the neural code aims at revealing the dictionary with which external stimuli are translated into neural responses [5,14]. This endeavour implies the identification of the stimulus and response features that are encoded, and the probabilistic mapping between them. One research line explores the dictionary in terms of the adequacy with which the original stimuli may be recovered after a decoding procedure [30,47–50], whereas another line focuses only on the encoding process [27]. In the end, the evolutionary advantage of perceptual systems is to produce adequate behavior, which may or may not require the original stimulus to be decoded. Here we studied the neural code of hippocampal and entorhinal cells of an awake and behaving rat in the framework of encoding. We demonstrated that the use of a broad repertoire of candidate input and response features allowed us to identify novel encoding strategies. In particular, the phase-of-firing with respect to the delta-filtered LFP was shown to encode kinematic information. To a large extent, this information refers to specific time points along the trajectory, that cannot be easily associated with a single position, velocity, or direction of motion, but rather, a conjunction of several features. We interpret these results as a proof of concept, that should be further confirmed with more varied experimental protocols, and a larger number of experimental animals.

Author Contributions: Conceptualization, M.B.M.C., E.K. and I.S.; Data curation, E.K.; Formal analysis, M.B.M.C. and I.S.; Funding acquisition, E.K. and I.S.; Investigation, M.B.M.C., E.K. and I.S.; Methodology, M.B.M.C. and E.K.; Resources, E.K.; Software, M.B.M.C.; Supervision, I.S.; Validation, M.B.M.C.; Visualization, M.B.M.C. and I.S.; Writing—original draft, I.S.; Writing—review & editing, M.B.M.C., E.K. and I.S.

Funding: This work was supported by Consejo Nacional de Investigaciones Científicas y Técnicas of Argentina (06/C444), Universidad Nacional de Cuyo (PIP 0256), and Agencia Nacional de Promoción Científica y Tecnológica (grant PICT Raíces 2016 1004 and PICT 2015 1273).

Acknowledgments: We thank Edvard and May-Britt Moser for agreeing in the use of data that EK collected as a post-doc in their lab.

Conflicts of Interest: The authors declare no conflict of interest. The founding sponsors had no role in the design of the study; in the collection, analyses, or interpretation of data; in the writing of the manuscript, and in the decision to publish the results.

Abbreviations

The following abbreviations are used in this manuscript:

H   Hippocampus
EC   Entorhinal cortex
LFP   Local field potential

Appendix A. Significance Analysis

Figures 3 and 6 display the number of cells encoding significant amounts of information (panel A), and also the histograms of information values (panels B and C). To assess whether two percentages of cells were significant, we ran the test described in Section 4.5. To determine whether the mean information values corresponding to two different histograms were significantly different or not, we performed a Student *t*-test with a significance threshold of $p = 0.01$. For each comparison, we defined a significance variable η that could be either 0 (not significantly different) or 1 (significantly different). There are 6 kinematic variables (Time, Position, Velocity, Direction, Signed Position and Signed Velocity), 3 neural codes (rate, theta and delta) and 2 brain areas (H and EC), so in total, there are 36 percentages of informative cells, 36 information values, and 36 normalized information values. All comparisons define 36×36 η values for each quantity.

Figure A1 displays the obtained η values averaged across kinematic features (left column) or code and area (right column). The diagonal is always black, because all tests result in $\eta = 0$ when the comparison involves two equal conditions. In A1 and B1 we note that the mean information values obtained with the theta and delta codes are not significantly different in the hippocampus (black boxes). Moreover, the mean information of the phase codes in H is rarily significantly different from the phase codes in EC (gray squares). In addition, the mean information values in H is also similar to the values in EC (gray). In A2 we see that direction and velocity (the two features with smallest information) have similar mean information values, all other kinematic features with more distinguishable means. When information values are normalized with their maximal attainable value (the entropy of the stimulus set), however, most kinematic features have non-significantly different means, except for the attribute Time, and (to a lesser extent), Signed Position (panel B2). Panels C1 and C2 show that all pairs of fractions of cells encoding significant amounts of information are always significantly different.

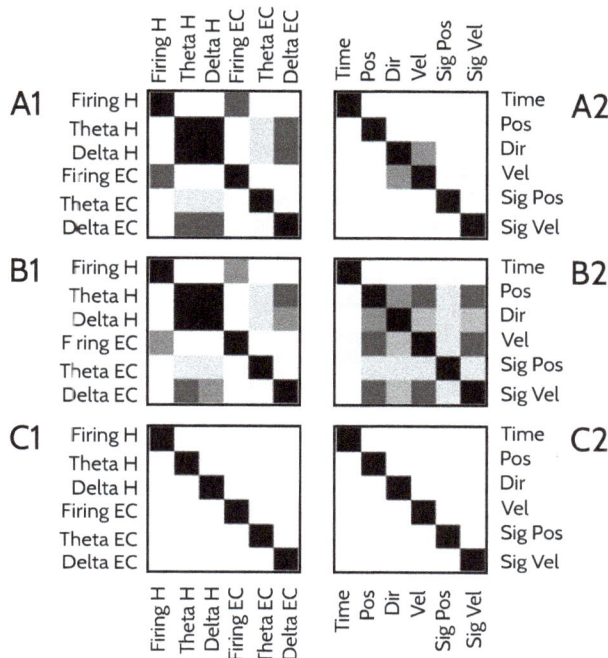

Figure A1. Significance between the different values reported in Figures 3 and 6. Left panels: Grey-scale representation of the significance matrices (see Methods) obtained for each pair of neural code and brain area, averaged across the 6 different kinematic features. Right panels: Gray-scale representation of the significance matrices obtained for each pair of kinematic feature, averaged across the 6 different conditions (Firing H, Theta H, Delta H, Firing EC, Theta EC, Delta EC). (**A**) Significance between the collections of information values expressed in bits, as in Figures 3B and 6B; (**B**) Significance between the collections of information values expressed as a fraction of the maximal information, as in Figure 6C; (**C**) Significance between the fractions of cells that encode significant amounts of information, as in Figures 3A and 6A.

Appendix B. Tables

Table A1. Mean, standard deviation and maximal information values of Figure 3B.

Region	Code	N Cells	Mean (bits)	St. Dev. (bits)	Maximum (bits)
H	firing	107	0.45	0.20	0.96
H	theta	63	0.13	0.066	0.37
H	delta	38	0.098	0.11	0.74
EC	firing	754	0.44	0.22	1.4
EC	theta	297	0.11	0.064	0.43
EC	delta	188	0.089	0.038	0.35

Table A2. Mean, standard deviation and maximal information values of Figure 6B.

Region	Code	Feature	N Cells	Mean (%)	St. Dev. (%)	Maximum (%)
H	firing	Time	107	8.22	3.87	17.31
H	firing	Position	148	4.65	3.61	18.52
H	firing	Direction	72	5.05	6.47	33.42
H	firing	Velocity	116	5.34	3.73	18.41
H	firing	Signed Pos	115	6.36	3.48	16.85
H	firing	Signed Vel	147	5.00	3.81	21.30
H	theta	Time	62	3.96	4.80	36.62
H	theta	Position	47	0.77	1.18	7.58
H	theta	Direction	23	0.19	0.13	0.51
H	theta	Velocity	41	0.48	0.49	1.87
H	theta	Signed Pos	53	1.77	2.83	20.95
H	theta	Signed Vel	40	0.54	0.54	2.28
H	delta	Time	37	2.76	4.21	26.34
H	delta	Position	17	0.33	0.22	0.84
H	delta	Direction	19	0.13	0.09	0.30
H	delta	Velocity	16	0.36	0.23	0.93
H	delta	Signed Pos	20	0.58	0.52	2.15
H	delta	Signed Vel	22	0.31	0.26	1.09
EC	firing	Time	754	7.83	4.22	27.71
EC	firing	Position	870	3.74	2.92	19.37
EC	firing	Direction	601	5.96	7.27	58.63
EC	firing	Velocity	742	4.42	4.82	46.30
EC	firing	Signed Pos	778	5.49	3.17	22.04
EC	firing	Signed Vel	949	5.14	4.92	34.07
EC	theta	Time	295	2.81	2.19	24.51
EC	theta	Position	274	0.41	0.47	3.51
EC	theta	Direction	216	0.36	0.70	6.29
EC	theta	Velocity	295	2.81	2.19	24.51
EC	theta	Signed Pos	261	1.61	1.19	7.58
EC	theta	Signed Vel	307	0.40	0.45	3.86
EC	delta	Time	188	2.26	2.05	25.70
EC	delta	Position	114	0.18	0.13	0.81
EC	delta	Direction	62	0.15	0.16	1.12
EC	delta	Velocity	104	0.18	0.13	0.89
EC	delta	Signed Pos	76	1.19	0.89	6.26
EC	delta	Signed Vel	130	0.18	0.12	0.80

Table A3. Mutual informations between the binary variables defined in Section 4.8 and shown in Figure 5A–C.

Region	Code A	Code B	$I(A; B)$ in bits	$I(A; B)/\text{Min}[H(A), H(B)]$
H	firing	theta	0.032	3.7%
H	theta	delta	0.15	23 %
H	delta	firing	0.29	43 %
EC	firing	theta	0.042	5.2%
EC	theta	delta	0.11	18%
EC	delta	firing	0.25	39 %

References

1. Moser, E.I.; Kropff, E.; Moser, M.B. Place Cells, Grid Cells, and the Brain's Spatial Representation System. *Annu. Rev. Neurosci.* **2008**, *31*, 69–89. [CrossRef] [PubMed]
2. O'Keefe, J.; Dostrovsky, J. The hippocampus as a spatial map. Preliminary evidence from unit activity in the freely-moving rat. *Brain Res.* **1971**, *34*, 171–175. [CrossRef]
3. Fyhn, M.; Molden, S.; Witter, M.P.; Moser, E.; Moser, M.B. Spatial representation in the entorhinal cortex. *Science* **2004**, *305*, 1258–1264. [CrossRef] [PubMed]

4. Skaggs, W.E.; McNaughton, B.L.; Gothard, K.M.; Markus, E.J. An information-theoretic approach to deciphering the hippocampal code. In *Advances in Neural Information Processing Systems*; Hanson, S.J., Cowan, J.D., Giles, C.L., Eds.; Morgan Kaufmann Pub.: Burlington, MA, USA, 1993; Chapter 5, pp. 1030–1037.

5. Dayan, P.; Abbott, L.F. *Theoretical Neuroscience*; The MIT Press: Cambridge, MA, USA, 2001.

6. Battaglia. F.P.; Sutherland, G.R.; McNaughton, B.L. Local Sensory Cues and Place Cell Directionality: Additional Evidence of Prospective Coding in the Hippocampus. *J. Neurosci.* **2004**, *24*, 4541–4550. [CrossRef] [PubMed]

7. Kropff, E.; Carmichael, J.E.; Moser, M.B.; Moser, E.I. Speed cells in the medial entorhinal cortex. *Nature* **2015**, *523*, 419–424. [CrossRef] [PubMed]

8. Solstad, T.; Boccara, C.N.; Kropff, E.; Moser, M.B.; Moser, E.I. Representation of Geometric Borders in the Entorhinal Cortex. *Science* **2008**, *322*, 1865–1868. [CrossRef] [PubMed]

9. Muller, R.U.; Bostock, E.; Taube, J.S.; Kubie, J.L. On the directional firing properties of hippocampal place cells. *Annu. Rev. Neurosci.* **1994**, *14*, 7235–7251. [CrossRef]

10. McNaughton, B.L.; Barnes, C.A.; O'Keefe, J. The contributions of position, direction, and velocity to single unit activity in the hippocampus of freely-moving rats. *Exp. Brain Res.* **1983**, *52*, 41–49. [CrossRef] [PubMed]

11. O'Keefe, J. Hippocampues, theta, and spatial memory. *Curr. Opin. Neurobiol.* **1993**, *6*, 917–924. [CrossRef]

12. Hafting, T.; Fyhn, M.; Bonnevie, T.; Moser, M.B.; Moser, E. Hippocampus-independent phase precession in entorhinal grid cells. *Nature* **2008**, *453*, 1248–1252. [CrossRef] [PubMed]

13. Souza, B.C.; Pavão, R.; Belchior, H.; Tort, A.B.L. On Information Metrics for Spatial Coding. *Neuroscience* **2018**, *375*, 62–73. [CrossRef] [PubMed]

14. Rieke, F.; Warland, D.; de Ruyter van Steveninck, R.; Bialek, W. *Spikes: Exploring the Neural Code*; The MIT Press: Cambridge, MA, USA, 1999.

15. Hubel, D.H. David H. Hubel's Nobel Lecture. Evolution of Ideas on the Primary Visual Cortex, 1955–1978: A Biased Historical Account. In *Nobel Lectures, Physiology or Medicine: 1981–1990*; Frangsmyr, T., Lindsten, J., Eds.; World Scientific Publishing Co.: Singapore, 1993; Chapter 1981, pp. 24–56.

16. Gross, C.G.; Rocha-Miranda, C.E.; Bender, D.B. Visual properties of neurons in inferotemporal cortex of the macaque. *J. Neurophysiol.* **1972**, *35*, 96–111. [CrossRef] [PubMed]

17. O'Keefe, J. John O'Keefe—Nobel Lecture: Spatial Cells in the Hippocampal Formation. 2014. Available online: http://www.nobelprize.org/nobel_prizes/medicine/laureates/2014/okeefe-lecture.html (accessed on 2 May 2018).

18. Chichilnisky, E.J. A simple white noise analysis of neuronal light responses. *Netw. Comput. Neural Syst.* **2006**, *12*, 199–213. [CrossRef]

19. Samengo, I.; Gollisch, T. Spike-triggered covariance revisited: Geometric proof, symmetry properties and extension beyond Gaussian stimuli. *J. Comput. Neurosci.* **2013**, *34*, 137–161. [CrossRef] [PubMed]

20. Rust, N.C.; Schwartz, O.; Movshon, J.A.; Simoncelli, E.P. Spatiotemporal Elements of Macaque V1 Receptive Fields. *Neuron* **2005**, *46*, 945–956. [CrossRef] [PubMed]

21. Estebanez, L.; Bertherat, J.; Shulz, D.E.; Bourdieu, L.; Léger, J.F. A radial map of multi-whisker correlation selectivity in the rat barrel cortex. *Nat. Commun.* **2016**, *17*, 13528. [CrossRef] [PubMed]

22. Strong, S.P.; Koberle, R.; de Ruyter van Steveninck, R.R.; Bialek, W. Entropy and Information in Neural Spike Train. *Phys. Rev. Lett.* **1998**, *80*, 197–200. [CrossRef]

23. Hardcastle, K.; Maheswaranathan, N.; Ganguli, S.; Giocomo, L.M. A Multiplexed, Heterogeneous, and Adaptive Code for Navigation in Medial Entorhinal Cortexs. *Neuron* **2017**, *94*, 375–387. [CrossRef] [PubMed]

24. Diehl, G.W.; Hon, O.J.; Leutgeb, S.; Leutgeb, J.K. Grid and Nongrid Cells in Medial Entorhinal Cortex Represent Spatial Location and Environmental Features with Complementary Coding Schemes. *Neuron* **2017**, *94*, 83–92. [CrossRef] [PubMed]

25. Gawne, T.J.; Richmond, B.J. How Independent Are the Messages Carried by Adjacent Inferior Temporal Cortical Neurons? *J. Neurosci.* **1993**, *13*, 2758–2771. [CrossRef] [PubMed]

26. Brenner, N.; Strong, S.P.; Koberle, R.; Bialek, W.; de Ruyter van Steveninck, R.R. Synergy in a neural code. *Neural Comput.* **2000**, *12*, 1531–1552. [CrossRef] [PubMed]

27. Schneidman, E.; Bialek, W.; Berry, M.J. Synergy, Redundancy, and Independence in Population Codes. *J. Neurosci.* **2003**, *23*, 11539–11553. [CrossRef] [PubMed]

28. Montemurro, M.A.; Rasch, M.J.; Murayama, Y.; Logothetis, N.K.; Panzeri, S. Phase-of-firing coding of natural visual stimuli in primary visual cortex. *Curr. Biol.* **2008**, *18*, 375–380. [CrossRef] [PubMed]

29. Kayser, C.; Montemurro, M.A.; Logothetis, N.; Panzeri, S. Spike-phase coding boosts and stabilizes information carried by spatial and temporal spike patterns. *Neuron* **2009**, *61*, 597–608. [CrossRef] [PubMed]

30. Samengo, I. The information loss in an optimal maximum likelihood decoding. *Neural Comput.* **2002**, *14*, 771–779. [CrossRef] [PubMed]

31. Elijah, D.; Samengo, I.; Montemurro, M.A. Thalamic neurons encode stimulus information by burst-size modulation. *Front. Comput. Neurosci.* **2017**, doi:10.3389/fncom.2015.00113. [CrossRef] [PubMed]

32. Constantinou, M.; Gonzalo Cogno, S.; Elijah, D.A.; Kropff, E.; Gigg, J.; Samengo, I.; Montemurro, M.A. Bursting neurons in the hippocampal formation encode features of LFP rhythms. *Front. Comput. Neurosci.* **2016**, doi:10.3389/fncom.2016.00133. [CrossRef] [PubMed]

33. Reifenstein, E.T.; Kemptner, R.; Schreiber, S.; Stemmler, M.B.; Herz, A.V.M. Grid cells in rat entorhinal cortex encode physical space with independent firing fields and phase precession at the single-trial level. *Proc. Natl. Acad. Sci. USA* **2012**, *19*, 6301–6306. [CrossRef] [PubMed]

34. Buzsáki, G.; Buhl, D.L.; Harris, K.D.; Csicsvari, J.; Czéh, B.; Morozov, A. Hippocampal network patterns of activity in the mouse. *Neuroscience* **2003**, *116*, 201–211, doi:10.1016/S0306-4522(02)00669-3. [CrossRef]

35. Barry, C.; Burgess, N. To be a Grid Cell: Shuffling procedures for determining "Gridness". *biorXiv* **2017**, doi:10.1101/230250. [CrossRef]

36. Ismakov, R.; Barak, O.; Jefferey, K.; Derdikman, D. Grid Cells Encode Local Positional Information. *Curr. Biol.* **2017**, *27*, 2337–2343. [CrossRef] [PubMed]

37. Fisher, N.I. *Statistical Analysis of Circular Data*; Cambridge University Press: Cambridge, UK, 1996.

38. Shannon, C.E. A Mathematical Theory of Communication. *Bell Syst. Tech. J.* **1948**, *27*, 623–656. [CrossRef]

39. Cover, T.M.; Thomas, J.A. *Elements of Information Theory*; Wiley-Interscience: Hoboken, NJ, USA, 2006.

40. Ecker, A.S.; Berens, P.; Tolias, A.S.; Bethge, M. The Effect of Noise Correlations in Populations of Diversely Tuned Neurons. *J. Neurosci.* **2011**, *31*, 14272–14283. [CrossRef] [PubMed]

41. Mathis, A.; Herz, A.V.M.; Stemmler, M.B. Multiscale codes in the nervous system: The problem of noise correlations and the ambiguity of periodic scales. *Phys. Rev. E* **2013**, *88*, 022713. [CrossRef] [PubMed]

42. Butts, D.A.; Goldman, M.S. Tuning Curves, Neuronal Variability and Sensory Coding. *PLoS Biol.* **2006**, *4*, doi:10.1371/journal.pbio.0040092. [CrossRef] [PubMed]

43. Treves, A.; Panzeri, S. The Upward Bias in Measures of Information Derived from Limited Data Samples. *Neural Comput.* **1995**, *7*, 399–407. [CrossRef]

44. Samengo, I. Estimating probabilities from experimental frequencies. *Phys. Rev. E* **2002**, *65*, 046124. [CrossRef] [PubMed]

45. Paninski, L. Estimation of Entropy and Mutual Information. *Neural Comput.* **2003**, *15*, 1191–1253. [CrossRef]

46. Panzeri, S.; Senatore, R.; Montemurro, M.A.; Petersen, R.S. Correcting for the sampling bias problem in spike train information measures. *J. Neurophysiol.* **2007**, *98*, 1064–1072. [CrossRef] [PubMed]

47. Seung, H.S.; Sompolinsky, H. Simple models for reading neuronal population codes. *Proc. Natl. Acad. Sci. USA* **1993**, *90*, 10749–10753. [CrossRef] [PubMed]

48. Eyherabide, H.G.; Samengo, I. When and why noise correlations are important in neural decoding. *J. Neurosci.* **2013**, *33*, 17921–17936, doi:10.1523/JNEUROSCI.0357-13.2013. [CrossRef] [PubMed]

49. Nirenberg, S.; Latham, P.E. Decoding neuronal spike trains: How important are correlations? *Proc. Natl. Acad. Sci. USA* **2003**, *100*, 7348–7353, doi:10.1073/pnas.1131895100. [CrossRef] [PubMed]

50. Quiroga, R.Q.; Panzeri, S. Extracting information from neuronal populations: Information theory and decoding approaches. *Nat. Rev. Neurosci.* **2009**, *10*, 173–185, doi:10.1038/nrn2578. [CrossRef] [PubMed]

MDPI

St. Alban-Anlage 66

4052 Basel

Switzerland

Tel. +41 61 683 77 34

Fax +41 61 302 89 18

www.mdpi.com

Entropy Editorial Office

E-mail: entropy@mdpi.com

www.mdpi.com/journal/entropy